Research Reports in Physics

Research Reports in Physics

Nuclear Structure of the Zirconium Region
Editors: J. Eberth, R. A. Meyer, and K. Sistemich

Ecodynamics Contributions to Theoretical Ecology
Editors: W. Wolff, C.-J. Soeder, and F. R. Drepper

Nonlinear Waves 1 Dynamics and Evolution
Editors: A. V. Gaponov-Grekhov, M. I. Rabinovich, and J. Engelbrecht

Nonlinear Waves 2 Dynamics and Evolution
Editors: A. V. Gaponov-Grekhov, M. I. Rabinovich, and J. Engelbrecht

Nonlinear Waves 3 Physics and Astrophysics
Editors: A. V. Gaponov-Grekhov, M. I. Rabinovich, and J. Engelbrecht

Nuclear Astrophysics   Editors: M. Lozano, M. I. Gallardo, and J. M. Arias

Optimized LCAO Method and the Electronic Structure of Extended Systems
By H. Eschrig

Nonlinear Waves in Active Media   Editor: J. Engelbrecht

Problems of Modern Quantum Field Theory
Editors: A. A. Belavin, A. U. Klimyk, and A. B. Zamolodchikov

Fluctuational Superconductivity of Magnetic Systems
By M. A. Savchenko and A. V. Stefanovich

Nonlinear Evolution Equations and Dynamical Systems
Editors: S. Carillo and O. Ragnisco

Nonlinear Physics   Editors: Gu Chaohao, Li Yishen, and Tu Guizhang

Nonlinear Waves in Waveguides with Stratification   By S. B. Leble

Quark-Gluon Plasma   Editors: B. Sinha, S. Pal, and S. Raha

Symmetries and Singularity Structures
Integrability and Chaos in Nonlinear Dynamical Systems
Editors: M. Lakshmanan and M. Daniel

Modeling Air-Lake Interaction Physical Background
Editor: S. S. Zilitinkevich

Nonlinear Evolution Equations and Dynamical Systems NEEDS '90
Editors: V. G. Makhankov and O. K. Pashaev

Solitons and Chaos   Editors: I. Antoniou and J. F. Lambert

Electron-Electron Correlation Effects in Low-Dimensional Conductors and Superconductors   Editors: A. A. Ovchinnikov and I. I. Ukrainskii

Signal Transduction in Photoreceptor Cells
Editors: P. A. Hargrave, K. P. Hofmann, and U. B. Kaupp

Nuclear Physics at the Borderlines
Editors: J. M. Arias, M. I. Gallardo, M. Lozano

J. M. Arias   M. I. Gallardo
M. Lozano   (Eds.)

Nuclear Physics at the Borderlines

Proceedings
of the Fourth International Summer School,
Sponsored by the Universidad Hispano-Americana,
Santa María de la Rábida, La Rábida, Huelva, Spain,
June 17–29, 1991

With 92 Figures

Springer-Verlag
Berlin Heidelberg New York London Paris
Tokyo Hong Kong Barcelona Budapest

Dr. José Miguel Arias
Dr. María Isabel Gallardo
Professor Dr. Manuel Lozano
Dept. Física Atómica, Molecular y Nuclear, Facultad de Física,
Univérsidad de Sevilla, Apdo. 1065, 41080 Sevilla, Spain

ISBN 3-540-55074-7 Springer-Verlag Berlin Heidelberg New York
ISBN 0-387-55074-7 Springer-Verlag New York Berlin Heidelberg

This work is subject to copyright. All rights are reserved, whether the whole or part of the material is concerned, specifically the rights of translation, reprinting, reuse of illustrations, recitation, broadcasting, reproduction on microfilm or in any other way, and storage in data banks. Duplication of this publication or parts thereof is permitted only under the provisions of the German Copyright Law of September 9, 1965, in its current version, and permission for use must always be obtained from Springer-Verlag. Violations are liable for prosecution under the German Copyright Law.

© Springer-Verlag Berlin Heidelberg 1992
Printed in Germany

The use of general descriptive names, registered names, trademarks, etc. in this publication does not imply, even in the absence of a specific statement, that such names are exempt from the relevant protective laws and regulations and therefore free for general use.

Typesetting: Camera ready by authors
57/3140-543210 – Printed on acid-free paper

Preface

This volume contains the lectures presented by invited speakers at the IV La Rábida International Summer School on Nuclear Physics. This was the IV edition of a summer school organized by our group every three years on topics related to nuclear physics. This Summer School was aimed mainly at young nuclear physicists, both theoreticians and experimentalists, engaged in research work at predoctoral or recent postdoctoral level.

The topics treated in the three previous editions of the School were: "Heavy Ion Collisions", "Theory of Nuclear Structure and Reactions" and "Nuclear Astrophysics". This year's School was entitled "Nuclear Physics at the Borderlines". Special emphasis was placed on those topics along which nuclear physics is expected to develop in the next few years. The aim of the School was to provide the attendants with an opportunity to get into close contact with experienced researchers and listen to their account of the present state-of-the-art in nuclear physics and the main future lines of development.

We would like to express our appreciation to all the lecturers who kindly accepted our invitation to give a talk at the IV La Rábida International Summer School on Nuclear Physics at the Borderlines. They made an important effort and succeeded in presenting their talks in a pedagogical way at the School itself as well as in the written version presented in this volume. Their presence during the School facilitated many informal scientific discussions during the beach and bar "sessions" and contributed to the excellent atmosphere among all the participants.

The organization found a main sponsor in the Universidad Hispano-Americana Santa María de La Rábida, which provided important financial support. The School was held in its buildings and we are most grateful to the entire staff. Considerable financial support was provided also by the Spanish government through the Dirección General de Investigación Científica y Técnica (DGICyT) and by the regional government through the Consejería de Educación de la Junta de Andalucía. We also received financial assistance from the following banks: Banco Español de Crédito and Caja de Ahorros y Monte de Piedad de Huelva y Sevilla. We express our deep gratitude to all of them.

Special thanks are given to the participants from the Departamento de Física Atómica, Molecular y Nuclear at Seville University for their enthusiastic cooperation before, during and after the School. Our gratitude to Elena Arbella and José Díaz for their continuous assistance in the organiztion of the School.

Sevilla, Spain
September, 1991

J.M. Arias
M.I. Gallardo
M. Lozano

Contents

Dispersion Relation Approach to the Nuclear Mean Field
By C. Mahaux and R. Sartor   1

Nuclear Collective Motion
By D.M. Brink   15

Heavy-Ion Interactions at Intermediate Energies
By A. Vitturi, S.M. Lenzi and F. Zardi   31

New Physics Far From Stability
By A. Poves   45

Recent Developments in High Spin Physics
By J.L. Egido   51

Recent Spectroscopy of Exotic Nuclear Shapes at High Spin
By M.A. Bentley   70

Nuclear Phase Transitions at Finite Temperature
By A.L. Goodman   94

Algebraic Approach to Molecular Spectra
By A. Frank   111

Algebraic Model of Hadronic Structure
By F. Iachello   130

Ultrarelativistic Heavy-Ion Collisions: Present and Future
By G. Baym 155

Electron Scattering
By I. Sick 172

The Role of Coloured Quarks and Gluons in Hadrons and Nuclei
By F.E. Close 194

Quarks in Nuclei
By A. Ferrando, P. González and V. Vento 212

Chiral Models of Low Energy QCD
By G. Ripka 233

Random-Matrix Modelling of Stochastic Nuclear Properties
By H.A. Weidenmüller 255

Subject Index 261

List of Participants 263

Introduction

J.M. Arias, M. Gallardo and M. Lozano

Departamento de Física Atómica, Molecular y Nuclear
Facultad de Física, Universidad de Sevilla
Apdo. 1065. 41080 Sevilla, Spain

This volume contains the lectures presented by the invited speakers at the IV La Rábida International Summer School on Nuclear Physics. It was held in La Rábida (Huelva), Spain from June 17th to 29th, 1991.

The title of this edition of the School was Nuclear Physics at the Borderlines. Special emphasis was devoted to topics along which presumably most nuclear physicists will look into in the next few years. The title was obviously too ambitious. It was certainly impossible to cover in one School all branches of nuclear physics and its connections with other fields. We chose some topics that are indeed borderlines, but unfortunately many others (as important as astrophysics or material analysis) had to be left out.

When selecting the topics to be presented at La Rábida School, we decided to consider two basic ways of approaching the borderlines of nuclear physics. One of them is through the new and more accurate data available due to the development of experimental techniques. The other is provided by the application of theoretical models or techniques borrowed from other fields or, reciprocally, the application of the models and techniques developed in nuclear physics in different scientific areas.

In the first group we include the study of high spin, temperature and exotic nuclear phenomena which are being measured due to more sensitive detector arrays. Lectures by Profs. M.A. Bentley, J.L. Egido and A.L. Goodman dealt with this topic.

- The development of large arrays of high resolution low signal-to-noise γ-ray detectors has allowed the observation of weakly populated exotic structures in nuclei, as those related to nuclear superdeformed shapes. Prof. Bentley presented this topic and also included a discussion on future developments in γ-ray detector arrays as EUROGAM and GAMMASPHERE.

The theoretical treatment used to study high spin in nuclear physics was discussed by Profs. Goodman and Egido.

- Prof. Goodman's lectures were devoted to study temperature effects. He uses the language of statistical mechanics applied to small systems and studies the way that transitions in some collective order parameters (shape transition, Δ transition, etc.) are induced by temperature.

- Prof. Egido presents the study of new phenomena related to octupole nuclear shapes and superdeformation. The rotational damping produced at moderate temperatures is also discussed in his lectures. Theories beyond the mean field approach are used to calculate microscopically the strength function of hot rotating nuclei. Also in this first group we include the study of the structure of the traditional "elementary" particles in nuclear physics (nucleons) revealed in experiments exploring nuclear collisions at higher energy. The lectures by Profs. G. Baym, F.E. Close, G. Ripka, I. Sick and V. Vento treats this topic from different points of view.

- Prof. Baym presents a detailed discussion of the program on ultrarelativistic heavy-ion collisions underway at CERN and Brookhaven National Laboratory. He analyzes how new physical phenomena produced by fundamental interactions can be found in those systems. An example is the production of a new state of matter: the quark-gluon plasma.
- Prof. Vento starts giving a short review of some properties of Quantum Chromodynamics (QCD) and then introduces low energy models of hadron structure. Those models are applied to exotic nuclei, quark matter, deep inelastic scattering, etc.
- The complexity of "deriving" nuclear structure from QCD is emphasised in Prof. Close's lectures. He presents a comparison between QED-molecular forces and QCD-nuclear forces, where only colour confinement breaks the analogy. Applications of colour degree of freedom, Pauli's principle and spin-flavour correlations are discussed in his lectures, in particular the magnetic moment of hadrons and proton spin.
- Electron scattering has been shown to be one of the most efficient tools to study hadron structure and there are important programs of development underway such as CEBAF. This field is covered in Prof. Sick's lectures. He presents the topic and then concentrates on two special cases: the study of charge distributions in the lead region and the determination of neutron form factors.
- Prof. Ripka presents two possible descriptions of a nuclear process: using either nucleons or quarks as intermediate fermions. He discusses the possibility of obtaining the same answer with both treatments. Two processes related to the propagation of hadrons in a dense baryonic medium are presented: the polarization of the medium and the change in the hadron's quark structure. The last part of Prof. Ripka's lectures is devoted to the relation between the scaling of effective lagrangians and the changes in the quark constituent masses.

Finally in this first group, the use of radioactive beams and the improvement of detection techniques have allowed to extend the nuclear domain to very neutron rich nuclei far from the line of β stability.

- This is the topic discussed by Prof. A. Poves in his lectures.

In the second group we include the following lectures.
- Prof. D.M. Brink deals with the finite size effects in nuclei and other quantum systems (particle physics, condensed matter physics, astrophysics,...). He presents a detailed discussion of these effects in relation with the quark-gluon plasma (droplets), the pairing phase transitions in nuclei, two nucleon transfer reactions and superdeformation.
- Prof. A. Vitturi uses Glauber-type models to describe nuclear grazing collisions at intermediate energies, $E/A \sim 50 - 1000$ MeV/A. It is shown in his lectures that this model can be used to determine microscopic heavy-ion optical potentials in this energy range. The treatment of both elastic and inelastic scattering is presented.

The power of group theory is well known in physics. In particular, it has been very successful in connection with models of nuclear structure. The application of similar techniques to systems of lower (molecular) and higher (hadronic) energies was presented at the School.
- In his lectures, Prof. A. Frank shows how to apply group theoretical technology to the description of molecular physics. He studies molecular excitations, described in terms of products of $U(2)$ and $U(4)$ groups and presents a unified way of treating vibrational, rotational and electronic excitations in diatomic molecules.
- Prof. F. Iachello introduces the use of group theoretical techniques to the study of hadron structure. The algebraic model of hadron structure exploits the algebra $U(4) \otimes SU_s(2) \otimes SU_f(n) \otimes SU_c(3)$. The different groups correspond to orbital, spin, flavour and colour, respectively. The model can accomodate previous results and describe new possibilities.

Also in this set of topics, we include the description of chaos and its application to nuclear physics problems. This was the content of the lectures presented by Profs. H.A. Weidenmüller and O. Bohigas.
- Prof. Weidenmüller discusses the use of random matrices and its applications to nuclear physics and other fields.
- Prof. Bohigas introduces the treatment of the chaotic dynamics and its relation with quantum systems.

Finally, there are still very interesting problems in "traditional" nuclear physics. One of them, the use of mean field techniques, was treated by Professor C. Mahaux in his lectures.
- Prof. Mahaux studies the mean field theory for systems with both positive and negative energies. An imaginary part describes the coupling between single-particle and more complicated degrees of freedom. Real and imaginary parts are linked by dispersion relations which allows an extrapolation of the mean field towards negative energies.

We feel that, although certainly incomplete, many of the important borderlines of nuclear physics were discussed in the School and that the interested reader will find in this volume a useful overview of the current situation in this field. In addition, the lectures are written in a pedagogical way so as to allow the reader to understand the main ideas behind these exciting research areas.

Dispersion Relation Approach to the Nuclear Mean Field

C. Mahaux and R. Sartor
Institut de Physique B5
Université de Liège, B-4000, Liège (Belgium)

Abstract : The mean field discussed here covers both large negative and large positive energies. It can be identified with the optical-model potential at positive energies and with the shell-model potential at negative energies. At negative as well as at positive energies, it contains an imaginary part which describes, in an average way, the coupling between single-particle and more complicated degrees of freedom. The real and imaginary parts of the mean field are connected by a dispersion relation. The latter predicts that the real part presents a characteristic energy dependence in the transition domain between the optical-model potential and the shell-model potential. It also provides a constraint which enables one to accurately extrapolate the mean field from positive towards negative energies. This is quite useful since much more empirical information is available at positive than at negative energies. The energy dependence of the resulting mean field can be related to properties of single-particle and of quasiparticle excitations. In particular, one can predict the single-particle energies, the spectroscopic factors and the distribution of single-particle strengths in the two nuclei obtained either by adding a nucleon to, or subtracting a nucleon from, a doubly-closed shell nucleus. Examples are given and compared with empirical data.

1. Introduction

The approximate validity of the shell model is one of the most striking features in nuclear physics. It was first considered as a miracle that, in first approximation, one can assume that each nucleon only feels a "shell-model potential" created by the other ("core") nucleons. The optical model is physically akin to the shell model in the sense that one nucleon is assumed to move in a complex "optical-model potential" created by the core nucleons. It is now understood that the success of these models is mainly due to the Pauli principle, which considerably modifies the "effective interaction" that acts between two nucleons in the nucleus as compared to the interaction between two isolated nucleons. Methods for handling strong interactions and the Pauli principle have been developed, chiefly by Brueckner and followers. These techniques have mostly been applied to nuclear matter. They enable one to evaluate semi-quantitatively some properties of the shell-model potential. However, they are not sufficiently reliable for yielding accurate results. A purely empirical determination of the shell-model potential is also quite difficult, since only little experimental information is available at negative energies. This situation has become unsatisfactory. Indeed, the interpretation of new experimental data on (e,e'p) knockout reactions as well as on one-nucleon transfer reactions requires a better knowledge of the shell-model potential and a better understanding of the limitations of the shell model. These problems are discussed in the present lectures. In particular, we shall outline a *"dispersion relation approach"* which enables one to accurately determine the shell-model potential, to make the junction between the shell-model and the optical-model potentials, to evaluate the degree of validity of the shell model and to predict many single-particle properties.

At first sight, one might believe that the optical-model and the shell-model potentials have little in common because of the following differences : (a) The optical-model potential applies to nucleons with a positive (scattering) energy, while the shell-model potential concerns nucleons with a negative (binding) energy. (b) The optical-model potential contains an imaginary part, while the shell-model potential is usually taken real. On second thought, one realizes that the underlying physical assumptions are practically the same in both cases. (a) There exists no reason why the average potential felt by a nucleon should be different if its energy is small and positive rather than small and negative. (b) The imaginary part of the optical-model potential accounts for the fact that the configuration formed by adding a nucleon to the core ground state is not a stationary state; the same feature should be taken into account for the configuration formed by taking out a nucleon from the core ground state. These comments suggest that the shell-model potential and the optical-model potential are two facets of the same mean field, observed at negative energies or at positive energies, respectively. This can be established by using the techniques of many-body theory, which can be used to show that the mean field is intimately related to the "mass operator".

Our presentation will be as follows. Basic concepts and definitions are recalled in sect. 2. Section 3 deals with the dispersion relation which connects the real and imaginary parts of the mass operator. In sect. 4, we describe how this dispersion relation can be used to extrapolate the empirical mean field from positive towards negative energies, i.e., to construct the shell-model potential by extrapolating the optical-model potential. In sect. 5, we illustrate that this "dispersion relation approach" also enables one to evaluate spectroscopic factors at low excitation energies, their average distributions at higher excitation energies, and the occupation probabilities of the shell-model orbits. Finally, sect. 6 mentions some other applications and several open problems. The present lectures are based on a recent review [1], in which one can find details as well as references to earlier works.

2. Concepts and definitions

2.1. Single-particle energies

Let us consider the example of neutrons in ^{208}Pb. In this case, the "core" is the ground state of ^{208}Pb. We refer to it as the "A-nucleon system" and denote its wavefunction by Ψ_0.

When a neutron with wavefunction φ_α is added to the ^{208}Pb core, one obtains a model "configuration" that belongs to ^{209}Pb, in the sense that it involves 82 protons and 127 neutrons. This "configuration" $\Phi_{\alpha(+)} = \{ \varphi_\alpha \otimes \Psi_0 \}$ is not an eigenstate of the Hamiltonian for (A+1) = 209 nucleons. However, if a mean field model is fairly accurate, this configuration is "not very different" from a stationary eigenstate provided that the single-particle wavefunction φ_α is calculated from the mean field. Here, "not very different" refers to the property that, in a time-dependent picture, a rather long time will elapse before the (A+1)-nucleon system leaves the configuration $\Phi_{\alpha(+)}$. Two main cases can be distinguished. (a) If φ_α is a scattering state, the generic index α contains the nucleon energy $E > 0$. Then, the mean field from which φ_α should be calculated is the real part of the optical-model potential, whose imaginary part mainly accounts for the finite "lifetime" of the configuration $\Phi_{\alpha(+)}$. (b) If φ_α is a bound state, the index α stands for the familiar quantum numbers $\{ n, \ell, j \}$. It is only when the single-particle φ_α is nearly unoccupied in the core, i.e., if the single-particle energy E_α is larger than the Fermi energy E_F, that the "configuration" $\Phi_{\alpha(+)}$ can be a good approximation to a bound state $\Psi_{\alpha(+)}$ of ^{209}Pb. The smallest energy that fulfills the inequalities $E_F < E_\alpha < 0$ corresponds to the ground state of ^{209}Pb; it is denoted by $E_{F(+)}$ in Fig. 1. Configurations associated with $E_{F(+)} < E_\alpha < 0$ can be good approximations to bound states of ^{208}Pb with excitation energy $E^* = E_\alpha - E_{F(+)}$. For instance, the single-particle state labelled (4s1/2) in Fig. 1 corresponds to the (1/2)$^+$ level at 2.04 MeV excitation energy in ^{209}Pb, whose ground

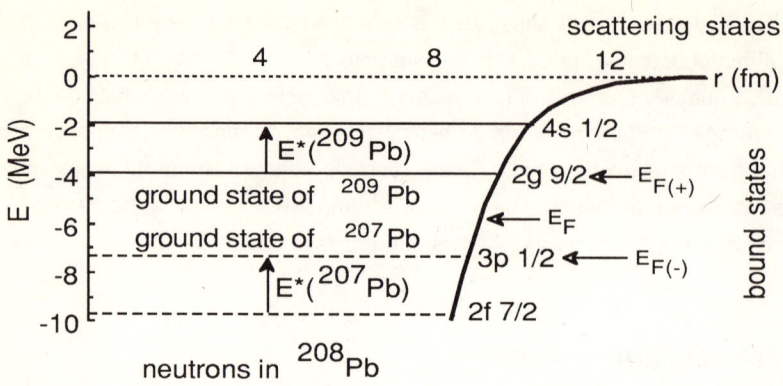

Fig. 1. Adapted from ref. [1]. Single-particle type energies in the example of neutrons in ^{208}Pb. If the neutron energy is positive, $\Phi_{\alpha(+)}$ is a scattering "configuration" that belongs to ^{209}Pb. If E_α is negative and larger than the Fermi energy E_F, the bound "configuration" $\Phi_{\alpha(+)}$ is associated with a bound state of ^{209}Pb. If E_α is negative and smaller than the Fermi energy, the bound "configuration" $\Phi_{\alpha(-)}$ is associated with a bound state of ^{207}Pb. The Fermi energy E_F is defined as the average between the single-particle energy $E_{F(+)}$ associated with the ground state of ^{209}Pb and that of $E_{F(-)}$ associated with the ground state of ^{207}Pb.

state has angular momentum and parity $(9/2)^+$ and is associated with the 2g9/2 single-particle orbit.

A similar reasoning can be carried out for the model "configuration" constructed by taking out a nucleon from the ^{208}Pb core, namely $\Phi_{\alpha(-)} = \{\varphi_\alpha^{-1} \otimes \Psi_0\}$. The corresponding single-particle energies must be smaller than the Fermi energy. If a mean field picture is valid, the "configuration" $\Phi_{\alpha(-)}$ is a good approximation to a bound state $\Psi_{\alpha(-)}$ of ^{207}Pb, with excitation energy $E^* = E_{F(-)} - E_\alpha$. Here, $E_{F(-)}$ denotes the largest value of E_α which is smaller than E_F; it is associated with the ground state of ^{207}Pb. For instance, the single-particle state labelled 3p1/2 in Fig. 1 corresponds to the ground state of ^{207}Pb, while that labelled 2f7/2 corresponds to the $(7/2)^-$ level at 2.34 MeV excitation energy in ^{207}Pb.

One of the main features which emerges from this discussion is that the mean field is related to states of both the (A+1)- and the (A-1)-nucleon systems, in the present example of both ^{209}Pb and ^{207}Pb. Hence, a theoretical approach that would only deal with either the (A+1)- or the (A-1)-nucleon system would not be appropriate.

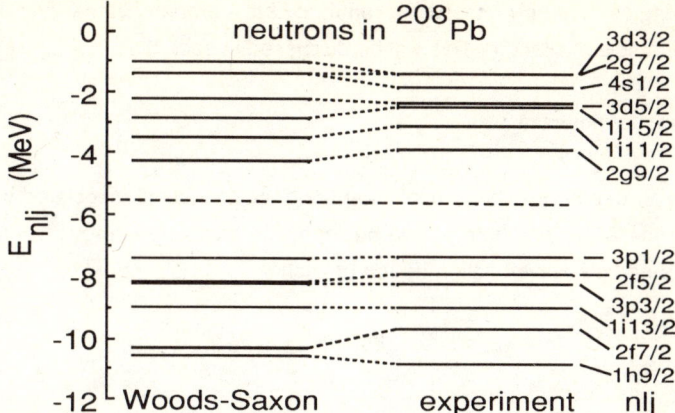

Fig. 2. Adapted from ref. [1]. Comparison between experimental values of the neutron single-particle particles in ^{208}Pb and levels computed from an energy-independent Woods-Saxon potential [2] to which a spin-orbit coupling has been added.

2.2. Shell-model potential

Experimental neutron single-particle energies in ^{208}Pb are gathered in Fig. 2. Their quantum numbers, ordering and average spacings are seen to be in good agreement with the energies computed from a "simple" empirical Woods-Saxon potential, to which a spin-orbit component has been added. Here "simple" refers to the fact that the depth and the radius of the Woods-Saxon potential are the same for all levels. In particular, its depth and radius are "independent of energy".

The theoretical definition of the shell-model potential will be briefly discussed in sect. 5.1. At the present stage, we introduce the "overlap functions" by means of the following projections :

$$\lambda_{\alpha(+)}(r) = <\Psi_0|\Psi_{\alpha(+)}> \quad , \quad \lambda_{\alpha(-)}(r) = <\Psi_{\alpha(-)}|\Psi_0> \quad . \tag{1}$$

By definition, the shell-model potential is required to have these overlap functions as eigenstates (here and below, we omit explicit reference to the kinetic energy and spin-orbit operators as well as to spin, isospin and angular variables). In eq. (1), we used the same index for labelling an eigenstate $\Psi_{\alpha(\pm)}$ of the (A±1)-nucleon Hamiltonian as for the model "configuration" $\Phi_{\alpha(\pm)}$. This implies that one can establish a one-to-one correspondence between a true state and a model configuration. This is possible only for the "single-particle excitations", which lie at low energy. At higher excitation energies, one only observes a remnant of this property in the form of a so-called "quasiparticle excitation". This will be discussed in sect. 5.2.

2.3. Optical-model potential

Let $\Psi_{E(+)}$ denote the many-body scattering wavefunction associated with the scattering by the A-nucleon core of one nucleon with incoming energy $E > 0$. The "elastic single-particle wavefunction" is the projection

$$\lambda_{E(+)}(r) = <\Psi_0 | \Psi_{E(+)}> \ . \tag{2}$$

By definition, the optical-model potential is the single-particle potential whose eigenstate at the energy E is equal to $\lambda_{E(+)}(r)$. Here, we omit the complication that one usually has to introduce an energy average.

This definition does not uniquely define the optical-model potential. We have recently shown that there exists an infinite number of single-particle potentials which all share this property [3]. In order to specify the potential, we shall impose requirements at both negative and positive energies.

2.4. The mean field

The analogy between eqs. (1) and (2) suggests to define the mean field as a potential which has all these overlap functions as eigenstates. It is not yet known whether this requirement uniquely defines the mean field. However, it has been proved that there exists at least one single-particle potential which fulfills this requirement, namely the so-called "mass operator". Henceforth, we shall thus identify the mean field with the mass operator. Accordingly, we shall write φ_α for the normalized value of $\lambda_{\alpha(\pm)}$ and φ_E for the value of $\lambda_{E(+)}$ (properly normalized at large distance).

3. Dispersion relation

The most general form of a one-body wave equation reads ($\hbar = 1$)

$$-(2m)^{-1} \nabla^2 \varphi_E(r) + \int dr' \, M(r,r';E) \, \varphi_E(r') = E \, \varphi_E(r) \ . \tag{3}$$

We recall that we omit reference to the spin-orbit coupling and that we use the same notation for a vector as for its length. The fact that the mean field operator depends upon E means that it is "energy-dependent". The fact that it depends upon two spatial coordinates (namely r and r') means that it is "nonlocal".

The real part (V) and the imaginary part (W) of the mass operator M are related by the following subtracted dispersion relation :

$$V(r,r';E) = V_{HF}(r,r') + \Delta V(r,r';E) \quad , \tag{4a}$$

where

$$V_{HF}(r,r') = V(r,r';E_F) \quad , \tag{4b}$$

$$\Delta V(r,r';E) = \pi^{-1} \int_{-\infty}^{\infty} dE' \, W(r,r';E') \left[\frac{1}{E'-E} - \frac{1}{E'-E_F} \right] . \tag{4c}$$

In these relations, E_F could be replaced by any other fixed reference energy.

The first term on the right-hand side of eq. (4a) is nonlocal and independent of energy. In practice, a potential of this type can be replaced by a "local equivalent" which is local but depends upon energy. This energy dependence is a smooth one if the nonlocality is not complicated. We shall thus replace $V_{HF}(r,r')$ by its local equivalent that we denote by $V_{HF}(r;E)$. We call $V_{HF}(r;E)$ the "Hartree-Fock type" (HF) component of the mean field because its main characteristics are the same as those of phenomenological Hartree-Fock potentials.

The second term on the right-hand side of eq. (4a) vanishes at $E = E_F$. It is expressed in terms of the imaginary part of the mean field. It thus reflects deviations from the independent-particle model. We call it the "dispersive" component of the mean field. Below, we shall use the approximation that $W(r,r';E')$ is local. This appears justified because we shall mainly be interested in the energy dependence of the imaginary part of the phenomenological mean field at low energy. It can be shown that this energy dependence is due to the energy dependence proper of the imaginary part of the mass operator rather than to its nonlocality.

4. Extrapolation of the optical-model potential towards negative energies

4.1. Practical form of the dispersion relation

In keeping with the above, we write eqs. (4a)-(4c) in the following simplified form :

$$V(r;E) = V_{HF}(r;E) + \Delta V(r;E) \quad , \tag{5a}$$

$$\Delta V(r;E) = \pi^{-1} \int_{-\infty}^{\infty} dE' \, W(r;E') \left[\frac{1}{E'-E} - \frac{1}{E'-E_F} \right] . \tag{5b}$$

At positive energy, we identify $W(r;E')$ with the imaginary part of the phenomenological optical-model potential, that we parametrize in such a way that it vanishes at an energy somewhat larger that $E_{F(+)}$. We shall determine $W(r;E')$ for $E' < E_F$ from the assumption that $W(r;E')$ is approximately symmetric with respect to E_F. This assumption is based on the belief that the lifetime of the "hole-configuration" $\Phi_{\alpha(-)}$ should be close to that of the "particle-configuration" $\Phi_{\alpha(+)}$ provided that their excitation energies are approximately the same and are not very large. Within that assumption, the dispersive component is skew-symmetric about the Fermi energy : it is attractive for E somewhat larger than E_F and repulsive for E somewhat smaller than E_F. As a consequence, the particle-hole energy gap as calculated from the full mean field is smaller than that computed from its HF component. This is in keeping with the fact that most HF calculations yield too large particle-hole energy gaps.

Once the dispersive component $\Delta V(r;E)$ has been calculated from the dispersion relation, the sole remaining unknown is the HF component $V_{HF}(r;E)$. We assume that it has a Woods-Saxon radial shape with a typical diffuseness fixed *a priori*, an energy independent radius R_{HF} and a linearly energy dependent depth $U_{HF}(E) = \alpha_{HF} + \beta_{HF} E$. This leaves three unknowns, namely R_{HF}, α_{HF} and β_{HF}. We determine them by using the following two requirements. (a) At positive energy, the mean field should be in keeping with the observed strength and energy dependence of the optical-model potential. (b) At negative energy, it should closely reproduce the experimental value of the Fermi energy. In practice, these two requirements are sufficient to accurately determine $V_{HF}(r;E)$, and thereby the full real part $V(r;E)$.

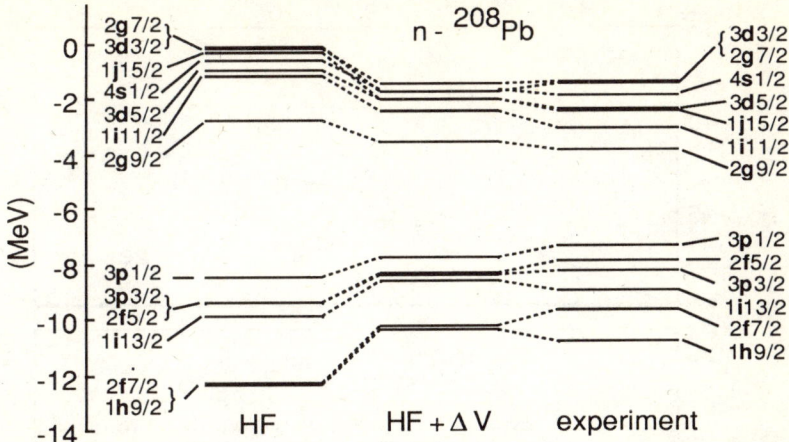

Fig. 3. Single-particle energies for neutrons in ^{208}Pb. The right-hand columns contains experimental values. The middle column gives energies computed from the full mean field, while those calculated for its HF component are shown on the left-hand side.

4.2. The neutron-^{208}Pb potential

Once $V(r;E)$ has been constructed, its reliability can be checked by comparing with experiment the predicted values of the scattering cross sections and of the single-particle energies. In Fig. 3. we show the single-particle energies in the example of the n-^{208}Pb system. As expected, the particle-hole energy gap is smaller when computed from the full mean field rather than from its HF component alone. The agreement between calculated and experimental single-particle energies is as good as that shown in Fig. 2. Note, however, that the constructed field is now energy dependent, in contrast to the assumption which had been *a priori* adopted for the phenomenological shell-model used in Fig. 2.

In order to exhibit the latter feature, it is convenient to approximate the calculated $V(r;E)$ by a Woods-Saxon potential with a typical diffuseness fixed *a priori*. Its depth U_V and radius parameter $r_V = R_V A^{-1/3}$ are shown in Fig. 4. Note the characteristic energy dependence of the radius parameter. At the Fermi energy the calculated values are $U_V = -47.1$ MeV and $r_V = 1.225$ fm. These are quite different from the purely phenomenological values which had been used in Fig. 2, namely $U_V = -40.6$ MeV and $r_V = 1.347$ fm. This illustrates the practical interest of imposing the constraint that the shell-model potential should joint the optical-model potential as the nucleon energy changes sign.

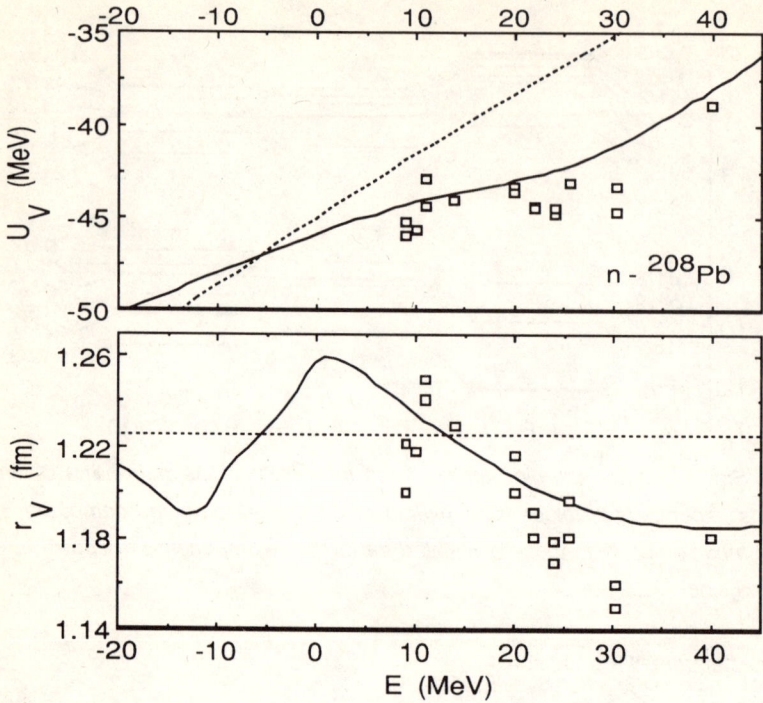

Fig. 4. Adapted from ref. [1]. The squares represent the depths and the radius parameters of phenomenological optical-model potentials for neutron-^{208}Pb scattering. The solid curves have been calculated from the dispersion relation approach. The dotted lines are associated with the Hartree-Fock component of the field.

5. Spectroscopic factors and spectral functions

5.1. Spectroscopic factors

We first better specify the approximate definitions introduced in sect. 2.2. Let $\Psi_{\lambda(\pm)}$ be the normalized bound eigenstates of the Hamiltonian for $(A\pm 1)$-nucleons. The "overlap functions" are defined by

$$\lambda_{(+)}(r) = <\Psi_0 | \Psi_{\lambda(+)}> \quad , \quad \lambda_{(-)}(r) = <\Psi_{\lambda(-)} | \Psi_0> \; . \tag{6}$$

By definition, the "spectroscopic factor" of the state $\Psi_{\lambda(\pm)}$ is equal to the norm of the overlap function :

$$S_{\lambda(\pm)} = <\lambda_{(\pm)} | \lambda_{(\pm)}> . \qquad (7a)$$

Recent (e,e'p) reactions give information on the radial shape of $\lambda_{(-)}(r)$. They support the basic assumption of the mean field model, namely that most overlap functions $\lambda_{(\pm)}(r)$ are proportional to the model single-particle wavefunction $\varphi_{\alpha(\pm)}(r)$ when the difference between the single-particle type energy $E_{\lambda(\pm)}$ and the model energy E_α is not larger than a few MeV. Then, eq. (7a) amounts to

$$|\lambda_{(\pm)}(r)|^2 = S_{\lambda(\pm)} |\varphi_{\alpha(\pm)}(r)|^2 . \qquad (7b)$$

We recall that $\varphi_\alpha(r)$ is normalized to unity.

Until recently, spectroscopic factors were mainly measured by means of stripping or pickup reactions. These processes take place at a large distance from the nuclear centre, say at $r \approx R$. These data thus essentially determine $|\lambda_{(\pm)}(R)|$. It turns out that the ratio $S_{\lambda(\pm)} = |\lambda_{(\pm)}(R) / \varphi_\alpha(R)|^2$ is a very sensitive function of the average potential adopted for calculating $\varphi_{\alpha(\pm)}(r)$, whose choice is somewhat arbitrary. Accordingly, the corresponding empirical spectroscopic factors are not reliable. More specifically, it appears that only the relative rather than the absolute values of the spectroscopic factors can be determined from direct one-nucleon transfer reactions.

This situation has recently improved in the case of proton hole states. Indeed, the (e,e'p) knockout reaction involves the values of the overlap functions in a wider range of radial distances than direct one-nucleon transfer reaction. Accordingly, it provides more information on the radial shape of $\lambda_{(-)}(r)$ or, equivalently, on the single-particle wavefunction $\varphi_{\alpha(-)}(r)$, see eq. (7b). The analysis of the data is delicate, in particular in heavy nuclei in which the Coulomb field sizeably distorts the wavefunctions of the incoming and outgoing electrons. For instance, a preliminary analysis [4] of the data yielded $S_{3s1/2} = (0.49 \pm 0.05)$ for the spectroscopic factor of the ground state of ^{207}Tl , while a recent improved analysis of the same data gives $S_{3s1/2} = 0.65$ [5].

Spectroscopic factors can be evaluated from the dispersion relation approach, because it includes an imaginary part which accounts, on the average, for deviation from the independent-particle model. In the example above, the method yields [1] $S_{3s1/2} = 0.69$, which is in quite good agreement with the experimental value. We do not discuss the spectroscopic factors of the other levels because the improved analysis of the (e,e'p) data has not yet been carried out for the single-particle excitations in ^{207}Tl.

5.2. Spectral functions

For excitation energies larger than several MeV, one observes closely spaced levels with same spin and parity. Each of these levels has a small spectroscopic factor. The prediction of the mean field model is then that the average distribution of the spectroscopic factors presents a "quasiparticle peak" near the single-particle energy E_α calculated from the real part of the mean field. The width of the peak is approximately equal to $(-2 W_\alpha)$, where W_α is the expectation value of $W(r;E_\alpha)$ with respect to φ_α. The average distribution can, for instance, be defined as the sum of the spectroscopic factors of all the levels with the same angular momentum and parity contained in an energy interval of one MeV. It is measured in units (MeV)$^{-1}$ and is called "the spectral function". Figure 5 illustrates the quality of the agreement between calculated and measured spectral functions. Note that the calculated curve is a prediction, i.e., involves no adjusted parameter.

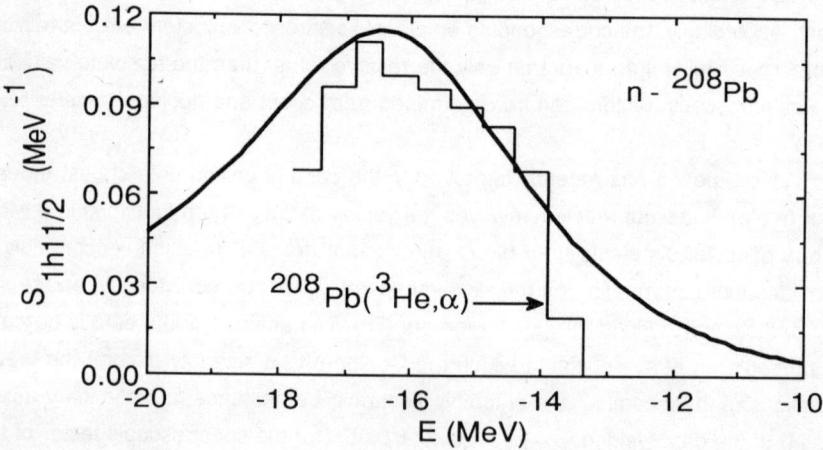

Fig. 5. Adapted for ref. [1]. Spectral function associated with the 1h11/2 neutron quasiparticle peak in ^{207}Pb. The histogram has been measured by means of the ^{208}Pb(^{3}He,α) pickup reaction; these data do not extend below -17.8 MeV, i.e., above 10 MeV excitation energy [6, 7]. The curve has been calculated from the dispersion relation relation approach.

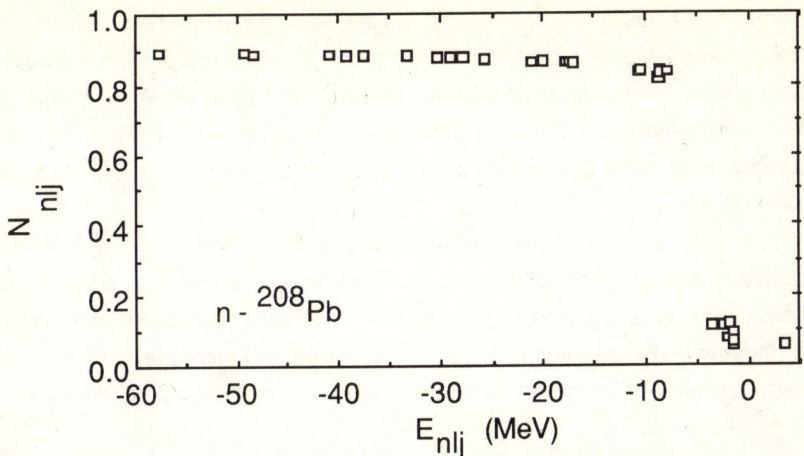

Fig. 6. Adapted from ref. [1]. Occupation probabilities of single-particle neutron orbits in ^{208}Pb, as predicted by the dispersion relation approach.

5.3. Occupation probabilities

The occupation probability of a single-particle orbit φ_α in the core is given by

$$N_\alpha = <\Psi_0|a_\alpha^\dagger a_\alpha|\Psi_0> \ . \tag{8}$$

These quantities can also be evaluated from the dispersion relation approach. In Fig. 6, we show the results in the example of neutrons in ^{207}Pb. In the independent particle model, the occupation probabilities would be equal to unity below the Fermi surface and would vanish above the Fermi surface. Figure 6 supports the validity of the independent particle model but also illustrates its limitation. The quantity N_α is often identified with the integral of the spectral function over all energies $E < E_F$, but this appears questionable. We therefore doubt that N_α can be extracted from the analyses of experimental data.

6. Outlook

In recent years, one has gained considerable theoretical and empirical insight into the mean field model. The experimental progress is mainly due to the recent (e,e'p) measurements and, to a lesser extent, to improved direct one-nucleon transfer experiments. The theoretical progress mainly lies in a better understanding of the definition of the mean field as well as of the limitations of this concept. This has, in part, been made possible by the development of the dispersion relation approach. The latter enables one to accurately construct the mean field at negative energy, i.e. the shell-model potential. It thereby yields single-particle wavefunctions. It also predicts spectroscopic factors, spectral functions and occupation probabilities of single-particle orbits.

A dispersion relation between the real and imaginary parts of the mean field is expected to hold even for complicated systems, because it can be viewed as a result of a causality condition. For instance, dispersion relations have been used in the case of the mean field between two heavy ions, especially in the vicinity of the Coulomb barrier where the imaginary part rapidly depends upon energy, as in the nucleon-nucleus case [8].

The dispersion relation approach has limitations, at least in its present stage of development. In the case of nucleons, for instance, it should be extended to deformed nuclei, and it should be investigated how pairing effects could be taken into account. In the case of heavy ions, the microscopic definition of the mean field is not yet well understood.

References

[1] C. Mahaux and R. Sartor, Advances in Nuclear Physics, vol. 20 (1991) 1
[2] J. Dudek, Z. Szymanski and T. Werner, Phys. Rev. **C23** (1981) 920
[3] C. Mahaux and R. Sartor, Nucl. Phys. (in press)
[4] E. Quint, Ph.D. Thesis, University of Amsterdam (1988)
[5] J.P. McDermott, Phys. Rev. Lett. **65** (1990) 1991
[6] S. Galès, G.M. Crawley, D. Weber and B. Zwieglinski, Phys. Rev. **C18** (1978) 2475
[7] J. Guillot, J. Van de Wiele, H. Langevin-Joliot, E. Gerlic, J.P. Didelez, G. Duhamel, G. Perrin, M. Buenerd and J. Chauvin, Phys. Rev. **C21** (1980) 879
[8] G.R. Satchler, Phys. Reports **199** (1991) 147 and references contained therein.

Nuclear Collective Motion

D.M. Brink

Sub-department of Theoretical Physics
1 Keble Road, Oxford OX1 3NP, UK.

Abstract: Nuclei are small many-body systems and finite size effects play a very important role in controlling their properties. These lectures present a selection of examples. Surface effects and shell effects are typical of finite systems. There is a large kinetic contribution to the surface energy of a nucleus which has its origin in a surface contribution to the single nucleon density of states. The situation is expected to be quite different for quark-gluon plasma droplets because of the different boundary conditions. Superconductivity in metals has similarities with pairing in nuclei but there major differences because nuclei are so small. The pairing phase transition is a singular feature of macroscopic superconductors while in nuclei it is smoothed out because of quantum or thermal fluctuations. On the other hand band crossing, which is associated with the interplay of pairing and nuclear rotations, is a prominent and systematic feature of nuclear structure. There is an analogous effect in the magnetic properties of superconducting grains, but it is very difficult to observe because it is a finite size effect. Two other examples are discussed. One concerns the transfer of pairs of nucleons between nuclei in a heavy ion reaction and the other is about a property of superdeformed bands in nuclei.

1. Introduction

The general topic of this school is Nuclear Physics at the Borderlines. We are concerned with the interaction of nuclear physics with neighbouring fields like particle physics, condensed matter physics, astrophysics or atomic and molecular physics. Traditionally nuclear physics has been concerned with the study of the structure of atomic nuclei and of reactions involving them. Over the past decade nuclear physicists have expanded their interests and at the present time the central territory of the subject includes the study of the structure and interaction of nucleons, relativistic heavy ion physics and the study of the interaction of particles with nuclei.

The subject of these lectures is nuclear collective motion. Nuclei are small quantum systems and quantum size effects play a very important role in the study of their properties. Quantum size effects are also important for the study of atomic clusters, and large molecules. All these areas involve systems which are too small to be treated by the traditional methods of condensed matter physics and too large for the exact methods of the few body problem to be applied. They are systems which are intermediate between microscopic and macroscopic and are often called mesoscopic systems. Special theoretical techniques are needed for the study of mesoscopic systems and some of these are well developed in nuclear physics. In these lectures I discuss some topics in nuclear physics each of which illustrates a principle or technique associated with the mesoscopic character of nuclei.

In nuclei quantum size effects manifest themselves in many ways. First there are surface effects and shell effects. We discuss them with applications to the quark-gluon plasma and to pairing phase transitions in nuclei. We also consider a simple model for the transfer of nucleon pairs between nuclei in a heavy ion reaction. This is rather analogous to Josephson tunnelling between superconductors. Superdeformations are associated with shell effects in deformed nuclei. Here we examine the problem of transitions out of superdeformed bands.

2. Finite size effects in Fermion systems

A large system of non-interacting fermions contained in a region of volume V is characterized by a distribution of single particle states which, to a first approximation, is independent of the shape of the region and of the boundary conditions satisfied at the surface. The lowest single particle states are occupied up to the Fermi level with Fermi momentum k_f. The distribution of states can be characterized by a density of states $\rho(k)$ in k-space. In a smaller system it is modified by surface effects and shell effects.

$$\rho(k) = g\left(\frac{Vk^2}{2\pi^2} + \text{surface corrections} + \text{shell corrections}\right). \tag{2.1}$$

The surface contribution is a smooth correction which depends on the boundary conditions and on the surface area but not on the shape of the region. The shell effects are associated with irregularities in the density of states and depend on the shape of the region. For example the levels in a spherical nucleus are bunched together and there are gaps between major shells, while the levels in a deformed nucleus are much more uniformly distributed.

Both the surface corrections and the shell corrections have been discussed by Balian and Bloch [1,2]. Here I concentrate on the surface corrections. In the case of a scalar field satisfying the equation

$$\nabla^2 \phi + k^2 \phi = 0. \tag{2.2}$$

Bloch and Balian give the surface corrections for two limiting forms of the boundary conditions as

$$\rho(k) = g\left(\frac{vk^2}{2\pi^2} - \frac{\sigma k}{8\pi}\right) \qquad \phi = 0 \quad \text{on} \quad \sigma, \tag{2.3}$$

$$\rho(k) = g\left(\frac{vk^2}{2\pi^2} + \frac{\sigma k}{8\pi}\right) \qquad \partial\phi/\partial n = 0 \quad \text{on} \quad \sigma. \tag{2.4}$$

Here g is a degeneracy factor and σ is the surface area of the region.

In the nuclear case the surface corrections to the density of states contribute to the nuclear surface energy. The average kinetic energy of a nucleon in a Fermi gas with Fermi momentum k_f is

$$\bar{\epsilon} = \bar{E}/N = \frac{3}{10}\frac{\hbar^2 k_f^2}{m}\left[1 + \frac{\pi\sigma}{16Vk_f}\right], \tag{2.5}$$

where the boundary condition (2.3) has been used. There is an increase in the kinetic energy due to the surface correction to the density of states which gives a positive contribution to the nuclear surface energy.

3. Quark-gluon plasma droplets

Quarks are confined inside hadrons in normal nuclear matter because of restrictions on the colour degree of freedom. Calculations based on quantum chromodynamics suggest that the colour should no longer be confined at sufficiently high temperatures or at high enough quark density. Quark deconfinement resulting in the formation of a quark-gluon plasma probably requires temperatures in excess of 200 MeV at low densities, or densities in excess of five times the nuclear density at low temperatures. It is hoped that such high temperatures or densities can be reached in relativistic heavy ion collisions. If such a plasma is formed it will evolve like a relativistic fluid and at some point will break up into hadrons. Bertsch et al [3] suggested that the hadronization of the quark-gluon plasma occurs in two steps: first the fluid becomes unstable and breaks up into droplets, then these decay by evaporating pions or other hadrons. Properties of droplets will be determined mainly by the equation of state of the quark-gluon plasma but surface effects may also be very important.

The MIT bag model gives a simple equation of state for the quark-gluon plasma at a temperature T. In the deconfined phase the quarks and gluons are assumed to form a gas of zero mass fermions and bosons moving in a region of QCD vacuum with volume V. The entropy of the plasma is just the entropy of a non-interacting gas

$$S = c_{qg}VT^3 = \frac{4\pi^2 g_{qg}}{90\hbar^3}VT^3. \tag{3.1}$$

In this equation $g_{qg} = g_g + (7/8)g_q$ is a degeneracy factor depending on the numbers of kinds, g_g and g_q, of quarks and gluons. The factor 7/8 comes from the difference in statistics between the quarks and gluons. The chemical potentials of both the quark and gluons are assumed to be zero. If there are two quark flavours (u and d) and eight gluon colours then $g_q = 24$ and $g_g = 16$ giving $g_{qg} = 37$ [4].

Overall confinement in the quark model is achieved in the bag model by giving the QCD vacuum a constant energy density B which restricts the quarks and gluons to a small region of space. The free energy in the bag model is

$$F = -(1/4)c_{qg}T^4V + BV, \tag{3.2}$$

The entropy and pressure can be obtained from the free energy by using the thermodynamic relations

$$S = -(\partial F/\partial T)_V, \qquad P = -(\partial F/\partial V)_T. \tag{3.3}$$

Using the first of these equations gives eq.(3.1) for the entropy while the second yields an expression for the pressure

$$P = (1/4)c_{qg}T^4. \tag{3.4}$$

The energy is given by

$$E = F + TS = (3/4)c_{qg}T^4 V + BV. \tag{3.5}$$

A droplet in equilibrium has $P = 0$ and a temperature

$$T = T_c = (4B/c_{qg})^{1/4}. \tag{3.6}$$

Such a droplet has an entropy $S = c_{qg}VT_c^3$, an energy $E = ST_c$ and a volume $V = E/4B$. It can decay by evaporating pions. As it evaporates it shrinks in size but its temperature remains constant. Numerical estimates can easily be obtained from these formulae. A quark-gluon plasma with a temperature $T = T_c \approx 200$ MeV has $T/\hbar c \approx 1$ fm^{-3}. In this case the entropy density is $s \approx 16$ mf^{-3} and the energy density is $\varepsilon \approx 3.2$ fm^{-3}. The entropy density of a pion gas at the same temperature is $s \approx 1.3$ fm^{-3} and is much less than for the plasma because the degeneracy factor g is much smaller.

The expressions for F, S, and E for small droplets of quark-gluon plasma are modified by the inclusion of surface terms. For example the expression for the free energy for a finite drop can be expanded as

$$F(T) = F_0(T)V + F_1(T)A + F_2(T)C + ... \tag{3.7}$$

where A is the surface area of the droplet and C is a measure of the average curvature of the surface. These surface corrections could have important consequences for the behaviour of droplets of quark-gluon plasma. For example the temperature of a droplet at zero pressure would depend on its size. Thus the temperature of the droplet would change, rather than remain constant, as it loses energy by evaporating hadrons.

The contributions to F_1 and F_2 could have an intrinsic or a dynamic origin. Farhi and Jaffe [5] argue that the intrinsic contribution to F_1 should be small at $T = 0$. The dynamic contributions to F_1 and F_2 are due to the surface corrections to the density of states of quarks and gluons in a finite droplet. It turns out that the term proportional to A in the density of states of massless quarks and gluons is zero and that the curvature term is the leading correction. The density of states of non-interacting gluons with boundary conditions corresponding to a perfect chromomagnetic conductor is the same as for the density of states of the electromagnetic field in a cavity with conducting walls.

Balian and Bloch [2] give a formula for the electromagnetic case. The expression for the density of states with surface corrections for massless quarks is quoted in a convenient form by Farhi and Jaffe [5]. The density of states in wave number k is

$$\rho(k) = g\left(\frac{Vk^2}{2\pi^2} + C_s kA + \frac{C_R}{8\pi}\int d^2s\left(\frac{1}{R_1} + \frac{1}{R_2}\right) + ..\right). \tag{3.8}$$

Here V, A, R_1 and R_2 are the volume, surface area and radii of curvature and g is the degeneracy factor. For zero mass non-interacting quarks $C_s = 0$, $C_R = -\pi/6$, while for massless gluons $C_s = 0$, $C_R = 2/3\pi$. The leading correction to the density of states for massive quarks is proportional to A and a formula is given by Berger and Jaffe [6]. Thus if quarks have a non-zero mass there will be a dynamic contribution to the surface tension of a quark matter droplet.

4. Finite size effects in superconductors

Electrons near the Fermi surface in a superconductor interact to form correlated pairs Cooper [7]. The pairs are constructed from time reversed single particle states. The two electrons in a Cooper pair have zero total spin and equal and opposite linear momentum $\mathbf{k}$ and $-\mathbf{k}$. Each pair has a binding energy 2Δ which is much smaller than the Fermi energy ε_F and the main components of the wave function come from electron states with energies ε within Δ of the Fermi energy,

$$\varepsilon_F - \Delta < \varepsilon < \varepsilon_F + \Delta. \tag{4.1}$$

The energy spread $\delta\varepsilon \approx 2\Delta$ corresponds to a momentum range $\delta p \approx 2\Delta/v_F$ where v_F is the Fermi velocity. The uncertainty relation $\delta x \approx \hbar/\delta p \approx \hbar v_F/2\Delta$ gives an estimate of the size of a Cooper pair. In superconductivity the quantity

$$\xi_o = \frac{\hbar v_F}{\pi \Delta} \tag{4.2}$$

is called the coherence length or correlation length of the superconductor. It is of the same order of magnitude as δx and is a measure of the size of a Cooper pair. The coherence length ξ_o is much larger than the crystal lattice spacing ($\sim 5\text{Å}$), in Type I superconductors. The Fermi velocity of electrons in these materials is normally large ($v_F \approx 10^6$ ms^{-1}) and the energy gap is small leading to a large coherence length. For example $\xi_o \approx 2300\text{Å}$ for Sn and $\xi_o \approx 830\text{Å}$ for Pb.

Superconductivity in solids is analogous to pairing in nuclei. In superconductors the coherence length is large compared with the interatomic spacing in the material but small compared with the typical size of a piece of superconducting material. The situation is very different in a nucleus. Using the appropriate Fermi wave number ($k_F \approx 1.36$ fm^{-1}) we get $\hbar v_F = 52$ MeV fm. Then eq. (4.2) gives a coherence length

$$\xi_0 \approx \frac{16.7}{\Delta}\text{fm} \tag{4.3}$$

For a medium mass nucleus with $A = 140$, $\Delta \approx 1$ MeV and $\xi_0 \approx 17$ fm. The nuclear radius is $R = 1.2\, A^{1/3}$fm ≈ 6.3fm. Thus the coherence length is larger than the nuclear radius. The same result holds for all nuclei in the periodic table. Thus, in a nucleus the size of a Cooper pair is given by the nuclear size rather than by the coherence length.

Finite size effects in nuclei smooth out some of the striking effects associated with phase transitions in superconductors, but at the same time there are new phenomena associated with the finite size which are unknown in superconductors. Shell effects are a consequence of the finite size of nuclei. The spacing $\hbar\omega_0$ between major shells in a nucleus can be estimated from the formula

$$\hbar\omega_0 \approx 41 A^{-1/3} MeV \approx \frac{49}{R} MeV \qquad (4.4)$$

where we have used $R = 1.2\, A^{1/3}$fm. Eqs (4.3) and (4.4) give a relation

$$\frac{R}{\xi_0} \approx 2.9 \frac{\Delta}{\hbar\omega_0} \qquad (4.5)$$

Thus the condition that the nuclear radius is small compared with the coherence length is related to a condition that the pairing strength 2Δ is less than the shell spacing $\hbar\omega_0$. In nuclei a phase transition from normal into superfluid states can take place at T=0 as a function of particle number. In closed shell nuclei $\Delta \ll \delta = \hbar\omega_0$, while in open shell nuclei $\Delta > \delta = \hbar\omega_o/10$. Thus pairing is important in open shell nuclei but its effects are very weak in closed shell nuclei. The interplay between pairing and shell effects is responsible for the band crossing or "backbending" phenomena observed in rotating deformed nuclei.

Properties of a sample of a superconductor depend on its dimension. A two-dimensional film or a one-dimensional wire behave differently from a three-dimensional sample. The meaning of a thin film is that the thickness is small compared with the coherence length. Similarly, a wire is effectively one-dimensional if its radius is small compared with the coherence length. Using the same criteria a nucleus should be regarded as a zero-dimensional superconductor.

If the dimensions of a superconducting particle become much smaller than the coherence length other effects come into play. Anderson [9] suggested that there is a lower limit in size for a particle still to be superconducting (cf Perenboom et al [10]). A relevant parameter for this regime is the ratio of the mean spacing of single particle states δ with the same spin to the transition temperature T_c

$$\bar{\delta} = \frac{\delta}{kT_c} = \frac{2}{\rho(\epsilon_F) T_c}, \qquad (4.6)$$

where $\rho(\epsilon_F)$ is the density of states at the Fermi level. Mühlschlegel et al [8] have calculated the effects of thermal fluctuations on the superconducting phase transitions using

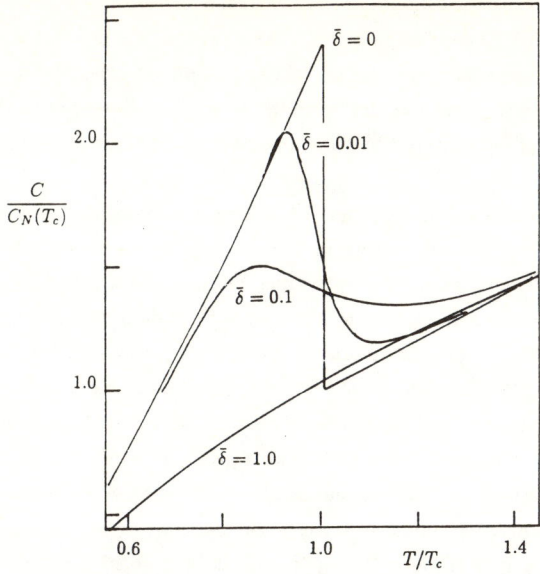

Fig. 1. Theoretical temperature dependence of the heat capacity C of a small superconducting particle compared with the heat capacity $C_N(T_c)$ of the normal state at the transition temperature. The quantity $\bar{\delta}$ is a measure of the particle size and is defined in eq.(4.6). This sketch is taken from a figure in [8].

Ginzburg-Landau theory. They show that the fluctuations smooth out the discontinuity in the thermal capacity at the transition temperature (cf. Fig. 1).

The smoothing is complete when $\bar{\delta} = 1$, but is already significant if $\bar{\delta} \approx 0.01$. This smoothing has been observed experimentally by Tsuboi and Suzuki [11]. They measured the electronic specific heat of small particles of Sn with an average diameter ranging from 25 up to 220 nm and find an effect consistent with the theoretical results shown in fig. 1.

A typical medium mass nucleus with $A = 160$ has a level spacing $\delta \approx 400$ keV. The gap parameter $\Delta \approx 0.95$ MeV which corresponds to a transition temperature $T_c \approx 0.54$ MeV. Thus $\bar{\delta} \approx 0.75$ and fig.1 shows that the phase transition is almost completely smoothed out. Quantum size effects are significant in nuclei and no sharp pairing phase transition is expected. Pairing correlations should definitely become weaker as the excitation energy is increased but there will be no sudden transition.

Mottelson and Valatin [12] argued that there is a close formal correspondence between equations of motion in a constant magnetic field and in a rotating reference system. They suggested that critical magnetic field phenomena in superconductors should have their counterpart in the rotational spectra of nuclei. The Coriolis forces in a rotating nucleus tend to decouple pairs of particles in time reversal states. When the angular velocity is sufficiently large then pairing correlations should be destroyed completely. Mottelson and Valatin estimated a critical angular velocity ω_c above which there would no longer be any pairing correlation. This is analogous to the critical magnetic field H_c for a superconductor.

The correspondence between the effect of a magnetic field on a superconductor and the influence of rotations on pairing in a nucleus is not complete. The London penetration depth plays an important role in superconductors but it has no analogue in the nuclear case. A nucleus is like a superconductor with an infinite penetration depth.

Superconductivity in a spherical metal particle, with a size small compared with the London penetration depth, is destroyed by an applied magnetic field in two stages. In the absence of a magnetic field all the electrons are paired in the superconducting ground state. Excited states are formed by breaking pairs. The two quasiparticle states have an excitation energy about 2Δ, four quasiparticle states have twice that excitation energy and so on. The magnetic field produces a Zeeman splitting of the excited states and reduces the energy gap. The splitting is largest in a quasi- particle state with maximum angular momentum. This is $p_f R$, where R is the radius of the particle and p_f is the Fermi momentum. When the field has a strength B_1 given by

$$\frac{e}{2m}(2p_f R) B_1 = 2\Delta \tag{4.7}$$

the lowest two-quasiparticle state becomes degenerate with the fully paired ground state. In these circumstances the field is strong enough to reduce the energy gap to zero but not strong enough to destroy the superconductivity. It is an example of gapless superconductivity (Perenboom et al [10]).

The-two-quasi particle state with highest angular momentum has a magnet moment $(e/2m)2p_f R$ while the largest magnetic moment of a four-quasiparticle state is almost twice that value. Thus when the field increases slightly above B_1 the four-quasiparticle state becomes degenerate with the fully paired state. As the field increases further more and more pairs are broken. The resultant blocking reduces the effective strength of the pairing interaction and eventually the pairing disappears. Calculations reviewed in ref. [10], based on the BCS theory with a Fermi gas density of states and including no shell effects give the critical field as

$$B_c = 2.57 B_1. \tag{4.8}$$

The first of these size effects exists in rotating nuclei. The largest two-quasiparticle angular momentum is $j_1 + (j_1 - 1) = 2j_1 - 1$ where j_1 is the maximum single particle angular momentum available near the Fermi level. Normally it corresponds to the intruder state with $j_1 = l_1 + 1/2$ which is pushed down from the next shell by the

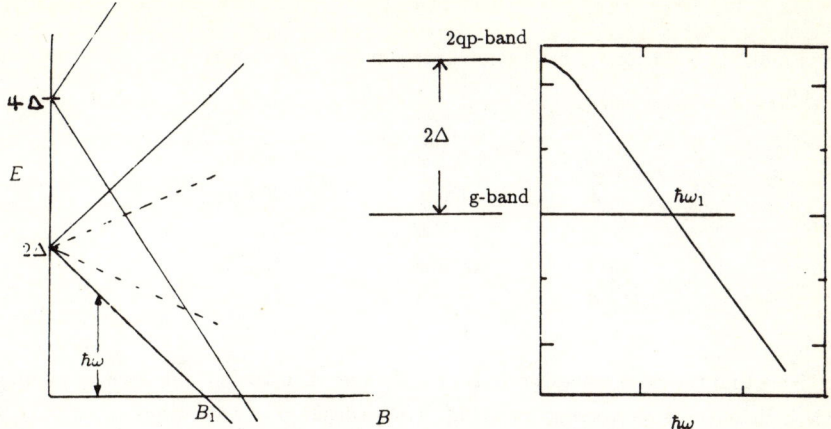

Fig. 2. The left hand part of the figure is a sketch of the Zeeman splitting of the two- and four-quasiparticle levels in a small superconducting particle. The lowest two-quasiparticle level becomes degenerate with the fully paired state when the applied field $B = B_1$ given by eq.(4.7). The right hand part shows the effect of the Coriolis plus centrifugal forces on a highly alignable high-j two-quasiparticle state in the deformed nucleus ^{168}Hf. The band-crossing occurs for an angular velocity $\hbar\omega_1 \approx 0.27$ MeV (after fig.5, Garrett et al [13]).

spin-orbit interaction. This two-quasiparticle state is split by the rotation and becomes degenerate with the fully paired state when

$$\hbar\omega_1 = 2\Delta/(2j_1 - 1). \tag{4.9}$$

Physically this size effect is associated with the band crossing or 'backbends' observed in rotating nuclei and ω_1 should be identified with the band crossing frequency. The two-quasiparticles align their angular momentum with the rotational axis of the nucleus. Systematics obtained from experimental data are fitted well by an empirical relation

$$\hbar\omega_1 \approx 1.67\Delta/j_1. \tag{4.10}$$

Empirical estimates for band crossings for some high-j shells are given in Table I together with the limiting theoretical values from eq.(4.9). They are taken from the review of Garrett et al [13].

Table I

Empirical estimates of $\hbar\omega_1$ from experiment compared with limiting values from eq.(4.9).

j	Empirical	Limiting
9/2	.336 Δ	.250 Δ
11/2	.279 Δ	.200 Δ
13/2	.239 Δ	.167 Δ
15/2	.209 Δ	.143 Δ

Note that the empirical estimate is always larger than the limiting theoretical value. This is because band crossing occurs in deformed nuclei and the limiting value (4.9) assumes a spherical system.

Backbending is a striking effect which is observed in the rotational spectra of many deformed nuclei. The corresponding effect is much more difficult to detect in superconductors because the critical field B_1 (4.7) depends on the radius of the sample and it is difficult to obtain grains of a uniform size. As well as producing backbending, rotations tend to quench the pair correlations in a nucleus. There should be a phase transition to an unpaired state at a critical rotational frequency ω_c. Analogy with the superconducting case eq.(4.8) suggests that $\omega_c \approx 2.6\omega_1$. There is no trace of this phase transition in rotating nuclei. Theoretical investigations show that the sharp phase transition predicted by the BCS theory is smoothed out by the effects of quantum fluctuations. This is analogous to the smoothing effect of the thermal fluctuations on the phase transition at the critical temperature T_c. There are two theoretical approaches. One is based on number projection from BCS wave functions (Egido and Ring [14]) and the other on pairing fluctuations (Shimizu et al [15,16]). Pairing correlations should definitely be reduced as the angular velocity increases but they do not vanish suddenly at $\omega = \omega_c$. The phase transition which is such a prominent effect in super conductors is difficult to observe in nuclei.

5. Pair transfer

This topic is not strictly related to finite size effects but it is a borderline subject and is connected to the theory of Josephson junctions [17]. It concerns the transfer of pairs of nucleons between two nuclei in a heavy ion collision. We follow the analysis of Josephson junctions given by de Gennes [18]. The problem of pair transfer between heavy ions hs been studied by Dietrich and Hara [19], Broglia and Winther [20] and other authors.

Consider two even nuclei A and B in their ground states in close proximity with fixed positions. The internal state of the system is denoted by $|0\rangle$. The ground states of neighbouring even systems can be obtained by moving pairs from A to B or vice-versa. The internal state with N pairs transferred is denoted by $|N\rangle$ where N is positive for transfer from A to B and negative for transfer from B to A. A more general state is a linear superposition of these states

$$\Psi = \sum_N a_N |N\rangle. \tag{5.1}$$

de Gennes uses a pair transfer hamiltonian in the sub-space of states of the form in (5.1). It is defined by

$$H_t |N\rangle = -\frac{1}{2} J_0 (|N+1\rangle + |N-1\rangle), \tag{5.2}$$

where J_0 is a pair transfer amplitude, assumed to be independent of N. de Gennes expresses it in terms of single particle transfer amplitudes. The Schrödinger equation can be solved by putting $a_N = \exp(iN\phi)$. The corresponding wave functions and energy eigenvalues are

$$\Psi = \sum_N e^{iN\phi} |N\rangle, \qquad E = -J_0 \cos\phi. \tag{5.3}$$

Here we have neglected any limitation on the value of N. The wave function in (5.3) is not an eigenstate of N but describes a distribution of pairs over the two nuclei. This solution is written in terms of a gauge angle ϕ and corresponds to a transformation from the number representation to the gauge representation. The state Ψ is specified by the amplitudes a_N and in the gauge representation by a wave function

$$f(\phi) = \frac{1}{\sqrt{2\pi}} \sum a_N e^{iN\phi}, \quad \text{where} \quad a_N = \frac{1}{\sqrt{2\pi}} \int_0^{2\pi} e^{-iN\phi} f(\phi). \tag{5.4}$$

The gauge angle ϕ is a conjugate variable to the number operator N in the same way as the rotation angle about an axis is conjugate to the angular momentum component along that axis. In the gauge representation the number operator N is represented by

$$N = -i \frac{\partial}{\partial \phi}. \tag{5.5}$$

This is analogous to the relation between angle and angular momentum and corresponds to the commutation relation $[\phi, N] = i$. There is the usual kind of uncertainty relation between ϕ and N. If ϕ is specified exactly then N is completely uncertain and vice-versa.

A more general hamiltonian has a term depending on the number of pairs moved as well as the transfer term (5.2)

$$H = H_t + \epsilon(N). \tag{5.6}$$

The term $\epsilon(N)$ is the total ground state energy of the two nuclei when N pairs have been transferred. The Schrödinger equation with the hamiltonian (5.6) can not be solved analytically for a general ϵN but the Heisenberg equations of motion for the operators ϕ and N have an interesting form. They are

$$i\hbar \frac{dN}{dt} = [N, H] = [N, H_t] = iJ_0 \frac{d\cos\phi}{d\phi}$$

$$i\hbar \frac{d\phi}{dt} = [\phi, H] = [\phi, \epsilon(N)] = i\frac{d\epsilon(N)}{N}$$

and simplify to give

$$\frac{dN}{dt} = -(J_0/\hbar)\sin\phi, \qquad \frac{d\phi}{dt} = \epsilon'(N)/\hbar. \qquad (5.7)$$

These are Josephson's two relations. The first connects the Josephson tunnelling current with the gauge angle and the second gives the rate of change of ϕ in terms of the potential difference between A and B.

The transfer ampiltude J_0 is time dependent for pair transfer between heavy ions. It is small when the nuclei are far appart and a maximum at the point of closest approach. We can write a time dependent Schrödinger equation for the wave function $f(\phi)$ in the gauge representation as

$$i\hbar \frac{\partial f}{\partial t} = H_t f = -J_0(t)\cos(\phi)f, \qquad (5.8)$$

where we have used the form of H_t in the gauge repesentation give in equation (5.3). It should be solved with the initial condition $N = 0$, $(f(\phi) = 1/\sqrt{2\pi}$ as $t \to -\infty$. The solution as $t \to +\infty$ is

$$f(\phi) = \frac{1}{\sqrt{2\pi}}e^{ix\cos\phi}, \quad \text{with} \quad x = \frac{1}{\hbar}\int_{-\infty}^{\infty} J_0(t)dt. \qquad (5.9)$$

The amplitudes a_N are the Fourier coefficients in the expansion of $f(\phi)$ and are proportional to Bessel functions

$$a_N = i^N J_N(x).$$

Thus the probability that N pairs have been transferred after the collision is

$$P_N = |a_N|^2 = |J_N(x)|^2.$$

Unfortunately in the heavy ion case x is normally small so that the probability that $|N| > 1$ is very small.

6. Decay of superdeformed nuclei

Collective rotational bands associated with superdeformed states have been observed in nuclei in several regions of the periodic table. The first band was found in ^{152}Dy by Twin et al [21]. Subsequently SD states were discovered in neighbouring nuclei and in also in nuclei near ^{194}Hg. States in the SD band in ^{152}Dy with spin near

$J = 60$ are populated in a heavy in reaction. The cascade of collective E2 transitions is observed down to about $J = 24$. Then it stops due to decays out of the band. A similar pattern exists in bands in other nuclei. In all cases the SD bands decay into normal states over 2-4 transitions within the band. Some authors have argued that the sudden transition out of the band is associated with the onset of pairing. Here we examine a mechanism for the decay which treats it as a two step process [22-26].

The decay out of the band is assumed to be due to the mixing of SD states with normal states. In the absence of mixing a SD state decays to the next state in the band by a collective E2 transition with a width Γ_Q while a normal state decays with a width Γ_c. After mixing the strength of a SD state is spread. We let p_i denote the proportion of the SD state in the mixed state i. With these assumptions the total probability to decay out of the band is given by Vigezzi et al [25] as

$$P_{out} = \sum_i \frac{p_i(1-p_i)\Gamma_c}{p_i\Gamma_Q + (1-p_i)\Gamma_c}. \tag{6.1}$$

In the limit of very weak coupling most of the SD strength is in one state and we have $p_0 \approx 1$ and $(1 - p_0) \ll 1$. First there is a probability p_i that the SD state mixes with an ordinary state I. In this limit writing $P_{in} = 1 - P_{out}$ we have

$$\frac{P_{out}}{P_{in}} \approx (1-p_0)\frac{\Gamma_Q + \Gamma_c}{\Gamma_Q}. \tag{6.2}$$

Thus there is a strong transition out of a SD band even in the weak coupling limit provided Γ_c/Γ_Q is large enough.

The transition from weak to intermediate or strong coupling can be examined in a simple numerical model [26]. A SD state was coupled to a set of 20 normal states with a constant matrix element V. The normal states were taken to have equal spacing D and the unperturbed SD state was separated by $D/4$ and $3D/4$ from the central two normal states. The hamiltonian was diagonalized and the transition probability out of the band calculated from eq.(6.1). The results are shown as a contour plot of P_{out}/P_{in} in Fig. 3. The results depend on the ratios V/D and Γ_c/Γ_Q.

Next we estimate the parameter values for the case of ^{152}Dy. The last observed transition is thought to be due to the decay of the $J = 24$ state and the excitation energy above yrast is estimated as $E^* \approx 4$ MeV. The E2 widths have been measured. For the $J = 24$ decay the gamma ray energy is 602 keV and $\Gamma_Q = 8$ meV. The width is proportional to the transition energy to the fifth power and changes markedly from one state to the next in the band. The decays out of the band could be due to E1 or E2 gamma transitions, but probaby the E1 transitions are stronger. The E1 widths at neutron threshold $E^* \approx 8$ MeV have been measured in the A=152 region. They are all about 100 meV without much fluctuation. Values at lower energies can be estimated using the methods described by Bjornholm and Lynn [27]. One of their simple

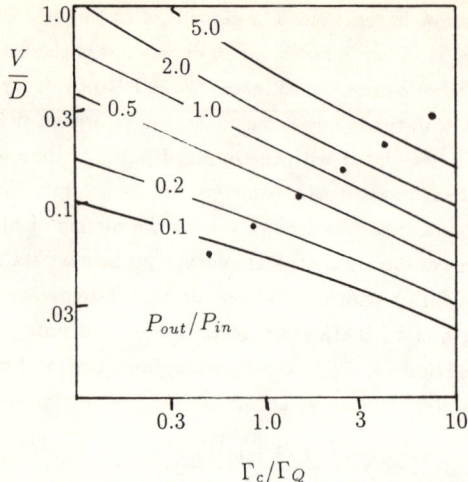

Fig. 3. Contour plot of the branching ratio P_{out}/P_{in} depopulating the SD band as a function of V/D and Γ_c/Γ_Q. The heavy dots show a sequence of transitions appropriate for the decay out of the SD band in ^{152}Dy.

prescriptions is $\Gamma_c \approx E^{*2}$ and gives $\Gamma_c \approx 25$ meV for $E^* = 4$MeV. Using this estimate we find $\Gamma_c/\Gamma_Q \approx 3$. The result is quite sensitive to assumptions about the density of states and could be reduced by a factor of 2 or 3 if pairing effects are important.

The absolute value of the level density D is very uncertain and could lie anywhere in the range 10eV to 2 keV. This large range is due to uncertainties in spin dependence and the status of pairing. Even though there are considerable uncertainties in Γ_c and D the changes in these quantities with E^* and J are quite well determined if pairing and spin dependence are assumed to change slowly with J. Because of the larger moment of inertia of the SD band we expect E^* to increase by about 300 keV when J changes from 24 to 22. Then Γ_c and D change by factors of about 1.16 and 0.54 respectively from one transition to the next. If the intrinsic quadrupole moment of the SD band is constant over the range of interest, then Γ_Q will be reduced by a factor of 0.68 for the next transition in the band.

We also need an estimate for the coupling strength V. To get this we make the hypothesis that the damping width from SD to normal states, $\Gamma^\downarrow = 2\pi V^2/D$, is independent of J over the range of interest. Then $V \sim \sqrt{D}$. Except for statistical fluctuations V/D increases by a factor of 1.4 and Γ_c/Γ_Q by a factor of 1.7 with each tansition.

Fig. 4. shows some results for the decay out of the SD band in ^{152}Dy together with experimental results. Experimentally the decay out of the SD band is much more rapid than the theoretical prediction. There are several comments which can be made. The calculation neglects statistical fluctuations in the coupling matrix elements and in the position of the SD state relative to the background normal states. If, by chance, the SD state is very close to a normal state the mixing will be very large. In other bands the decay out of the band is in fact more gradual. There could be a collective decay mechanism which has been neglected here. It might also be that some new physics is required which gives a rapid change in $\Gamma^\downarrow$. Herskind et al [23] and Schiffer et al [24] argued that this could be due to the pairing phase transition, but the arguments presented in section 4 of these lectures suggest the the pairing strength would change very smoothly with angular momentum.

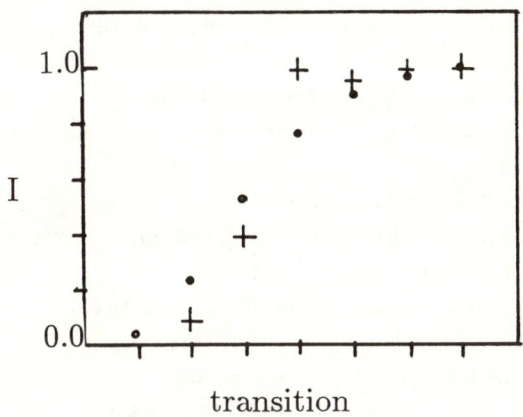

Fig. 4. In-band intensities of SD transitions. The dots are calculated as explained in the text, while the crosses are experimental values.

REFERENCES

[1] Balian, R. and Bloch, C., (1970) Ann. Phys. **60**, 401.

[2] Balian, R. and Bloch, C., (1971) Ann. Phys. **64**, 271.

[3] Bertsch, G., Gong, M., McLerran, L., Ruuskanen, V. and Sarkkinen, E. (1988) Phys.Rev. **D37**, 1202.

[4] Cleymans, J., Gavai, R.V. and Suhonen, E. (1986) Phys.Rep. **130**, 217.

[5] Farhi, E. and Jaffe, R.L., (1984) Phys. Rev. **D30**, 2379.

[6] Berger, M.S. and Jaffe, R.L., (1987) Phys. Rev. **C35**, 213.

[7] Cooper, L. N., (1950), Phys Rev **104**, 1189

[8] Mühlschlegel, B., Scalapino, D. J., and Denton, R. (1972), Phys Rev **B6**, 1767.
[9] Anderson, P.W.,(1959) J.Phys.Chem Solids **11**, 26.
[10] Perenboom, J. A. A. J., Wyder, P., and Meier, F., (1981), Physics Reports **78**, 173.
[11] Tsuboi, T. and Suzuki, T., (1977) J.Phys.Soc.Jap. **42**, 437 and 444.
[12] Mottelson, B.R. and Valatin, J.G.,(1960) Phys.Rev.Lett. **5**, 511.
[13] Garrett, J.D., Nyberg, J., Yu, C.H., Espino, J.M. and Godfrey, M.J., (1988) Int. Conf. on Contemporary Topics in Nuclear Structure Phys., Ed. Cassen, Frank, Moshinsky and Pittel, World Scientific, pp 699-729.
[14] Egido, J.L. and Ring, P. (1982) Nucl.Phys. **A383**, 189; **A388**, 228.
[15] Shimizu, Y.R., Garrett, J.D., Broglia, R.A., Gallardo, M. and Vigezzi, E., (1989) Rev.Mod.Phys. **61**, 131.
[16] Schmizu, Y.R., Vigezzi, E. and Broglia, R.A., (1990) Nucl.Phys. **A509**, 80.
[17] Josephson,B.D., (1962) Phys.Lett. **7**, 251.
[18] de Gennes,P.G., (1966) Superconductivity of Metals and Alloys (Benjamin 1966) Ch.4.
[19] Dietrich, K. and Hara, K. (1973) Nucl.Phys. **A211**, 349.
[20] Broglia, R.A. and Winther, Aage, (1991) "Heavy ion reactions", Addison-Wesley. p430.
[21] Twin, P. et al (1986) Phys.Rev.Lett. **57**, 811.
[22] Ragnarsson, and Aaberg, S., (1986) Phys.Lett. **B180**, 191.
[23] Herskind, B. et al, (1987), Phys.Rev.Lett. **59**, 2416.
[24] Schiffer, K., Herskind, B. and Gason, J. (1989) Z.Phys. **A332**, 17.
[25] Vigezzi, E., Broglia, R.A. and Dossing, T. (1990) **B249**, 163.
[26] Bertsch, G. and Brink, D.M. (1991) Lab. Report, MSU.
[27] Bjornholm, S. and Lynn, J.E. (1980) Rev. Mod. Phys. **52**, 851.

HEAVY-ION INTERACTIONS AT INTERMEDIATE ENERGIES

A. Vitturi [1], S.M. Lenzi [2] and F. Zardi [2]
[1] Dipartimento di Fisica and INFN, Trento, Italy
[2] Dipartimento di Fisica and INFN, Padova, Italy

Abstract: We discuss heavy-ion grazing collisions at intermediate energies (E/A ≈ 50 – 1000 MeV) within a model à la Glauber based on microscopic nucleon-nucleon collisions in the eikonal limit. Applications to elastic and inelastic scattering processes are presented. The model can also be used to determine microscopic heavy-ion optical potentials in this energy range. In addition to the "bare" elastic potential, one also obtains the dynamical contribution arising from the coupling to the excited channels. At variance with the standard behaviour characterizing the energy range around the Coulomb barrier, at high bombarding energies all terms of the optical potentials are complex functions, and their energy dependence as well as the balance between real and imaginary parts are essentially governed by the behaviour of the microscopic nucleon-nucleon scattering at the corresponding energy. Examples of predicted optical potentials at different energies are presented.

1. Introduction

The study of the properties of ion-ion potentials and couplings has become in the recent years one of the central items in heavy-ion physics. Due to the many-body nature of the two colliding objects and the large variety of processes occurring during the collision (e.g. inelastic processes, transfer reactions, fusion, multifragmentation processes, particle production, etc.), potentials and couplings display strong non-locality and energy dependence. They will furthermore depend on the specific nuclear properties of the colliding nuclei, as for example existence of strong collective modes and shell effects. It is therefore not completely surprising that, in spite of fundamental works (cf. e.g. ref. [1] and [2]), the current philosophy for decades has seen the heavy-ion optical potential as a mere object to be phenomenologically determined for each colliding system and energy. Only recently [3–6] large efforts have been devoted, from one side, to the study of the formal properties of the optical potentials and form factors and to the understanding of the origin of their energy dependence and, on the other side, to the development of microscopic methods for their actual evaluation. Note that a correct description of the elastic scattering is a necessary prerequisite for extracting any structure information from the analysis of heavy-ion scattering experiment. In fact, most of heavy-ion collisions can either be described as perturbative processes on a background of elastic diffusion, as the grazing reactions, or rely heavily on the transmission coefficients, as is the case of more central collisions.

The interest so far has been mainly focussed on the energy range around the Coulomb barrier, where one usually assumes a "bare" real potential of the folding type. This displays a smooth energy dependence originating from the non-locality of the elementary nucleon-nucleon interaction. As in the case of the nucleon-nucleus potential (see e.g. the lectures by Mahaux in these proceedings [7]), the phenomenological ion-ion potential shows a much stronger energy dependence originating from the coupling of the entrance elastic channel to the different reaction channels. In fact, when one simulates the coupled-channel situation by a one-channel (elastic) process, the coupling to reaction channels gives origin to both a correction to the real part and the appearance of an absorptive imaginary contribution ("polarization" potential). The magnitude and the balance of these two terms depends on the strengths of the coupling terms, the Q-values of the intermediate reaction channels and the bombarding energy. As a limiting case, for example, at low bombarding energies where reaction channels are closed and the whole flux remains in the elastic channel, the polarization potential is purely real. On the other extreme, i.e. small Q-values or rather large bombarding energies, the correction to the "bare" potential becomes purely imaginary. One can actually show [3] that the two terms are connected by a dispersion-type relation. As a consequence, for example, the real part assumes its largest value in correspondence of the rapid variation of the imaginary part associated with the opening of reaction channels at the barrier threshold (the so-called "threshold anomaly"). A visualization of this effect is given in Fig. 1 for the case of ^{16}O + ^{208}Pb.

Although many aspects and properties of the optical potential have been clarified, its actual determination still requires a rather involved procedure. The situation becomes more favourable at high bombarding energies, where the conditions for the use of the eikonal propagation apply. In this case the inverse problem can be easily solved via the Abel transform, and the equivalent optical potential can therefore be determined from the elastic phase shifts obtained within models à la Glauber based on microscopic nucleon-nucleon collisions. I will discuss in these lectures the application of such models to the description of heavy-ion grazing collisions at intermediate energies and to the determination of the corresponding heavy-ion interactions.

2. One-channel elastic scattering and the "bare" optical potential

I will first consider the case of one-channel elastic scattering disregarding any explicit coupling to other channels. A good description of the elastic processes has been obtained at intermediate energies within the optical limit [8−10] of the nucleus-nucleus Glauber model. This model is the extension to the case of the collision of two com-

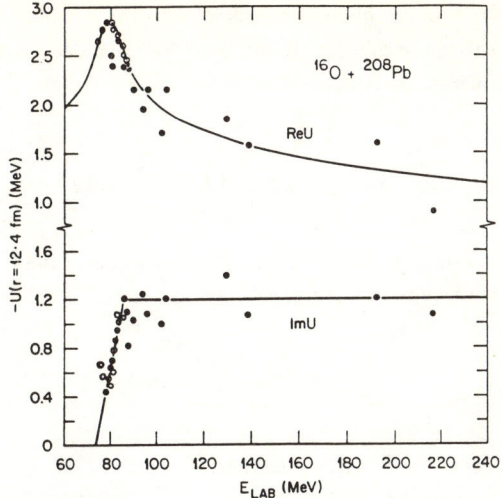

Fig. 1 - *Energy dependence of the real and imaginary parts of the optical potential for the reaction $^{16}O+^{208}Pb$ (adapted from ref. [3]).*

posite systems of the original Glauber model devised for nucleon-nucleus scattering. Basic goal of the approach is to describe the dynamics of the ion-ion process in terms of the contributions arising from the microscopic nucleon-nucleon collisions. All possible sequences of elementary collisions have to be taken into account. Taking profit of the eikonal approximation (or in classical terms of the straight-line propagation) and under the assumption that the system remains in its ground state et each elementary collision ("frozen"-nucleus approximation), one can sum analytically the multiple-scattering series to all orders. In the final expression for the elastic scattering amplitude

$$f(\Delta) = (ik/2\pi) \int d^2b \, e^{i\Delta \cdot \mathbf{b}}(1 - e^{2i\delta(b)}) \qquad (1)$$

the elastic phase shift associated with the impact parameter b within the model takes the simple expression

$$\delta(b) = (i/2) \int \rho_0^P(\mathbf{r}_P)\rho_0^T(\mathbf{r}_T)\gamma_{NN}(\mathbf{b} - \mathbf{s}_P + \mathbf{s}_T)d\mathbf{r}_P d\mathbf{r}_T$$
$$= (1/4\pi k_{NN}) \int \rho_0^P(q)\rho_0^T(q)f_{NN}(q)e^{-i\mathbf{q}\cdot\mathbf{b}}d^2q \,, \qquad (2)$$

where ρ_0^P and ρ_0^T are the projectile and target densities, k and k_{NN} are the ion-ion and the nucleon-nucleon relative momenta in the corresponding center-of-mass system, Δ is the transferred momentum ($\Delta = 2k\sin(\theta/2)$), while $\mathbf{s}_P$ and $\mathbf{s}_T$ are the projections of the coordinate vectors on the plane perpendicular to the incident momentum. The microscopic nucleon-nucleon profile γ_{NN} is connected to the nucleon-nucleon scattering amplitude by the bidimensional Fourier transform

$$\gamma_{NN}(b) = (1/2\pi i k_{NN}) \int d^2q \, e^{-i\mathbf{q}\cdot\mathbf{b}} f_{NN}(q) \, . \tag{3a}$$

An alternative representation of $\gamma_{NN}(b)$ can be given in terms of the two-body interaction V_{NN} and the relative nucleon-nucleon velocity v_{NN} in the form

$$\gamma_{NN}(b) = 1 - e^{-(i/\hbar v_{NN}) \int_{-\infty}^{\infty} V_{NN}(b,z) \, dz} \, . \tag{3b}$$

Note that I have displayed our formalism considering only the central contributions of the elementary scattering amplitude f_{NN} or the corresponding two-body potential V_{NN}, thus leading to an isotropic ion-ion scattering amplitude and consequently to the central part of the ion-ion potential. As a matter of fact one could include in eq. (2) additional terms, as spin-orbit etc., which would lead to more general expressions for the ion-ion scattering amplitude and for the associated potential.

This simple model for elastic scattering has been applied to a large variety of cases, covering wide regions in bombarding energy and mass number of colliding nuclei, obtaining in all cases a good reproduction of the experimental data. This result is particularly satisfactory, since the model have no free parameters, densities and nucleon-nucleon scattering amplitudes at the corresponding energies being taken from literature without any further adjustment. A few typical examples are given in Fig. 2. Note that surprisingly the model seems to give a rather good account of the experimental elastic process also at relatively low bombarding energies, of the order of 30 MeV per nucleon.

Our description yields directly the scattering amplitude without requiring the introduction of an optical potential, as in the standard description. One can however determine the equivalent optical potential, i.e. the potential that yields the same elastic phase shifts within a one-channel description. Within the eikonal approximation the inversion procedure is in fact univoque, the potential being obtained from the phase shifts through the Abel transform [12]

$$V(r) = \frac{2\hbar v}{\pi r} \frac{d}{dr} \int_r^\infty \frac{\delta(b)}{(b^2 - r^2)^{1/2}} b \, db \, . \tag{4}$$

One can use directly this expression in connection with (2) to perform the calculation of the potential. The expression can, however, be further manipulated formally to evidence

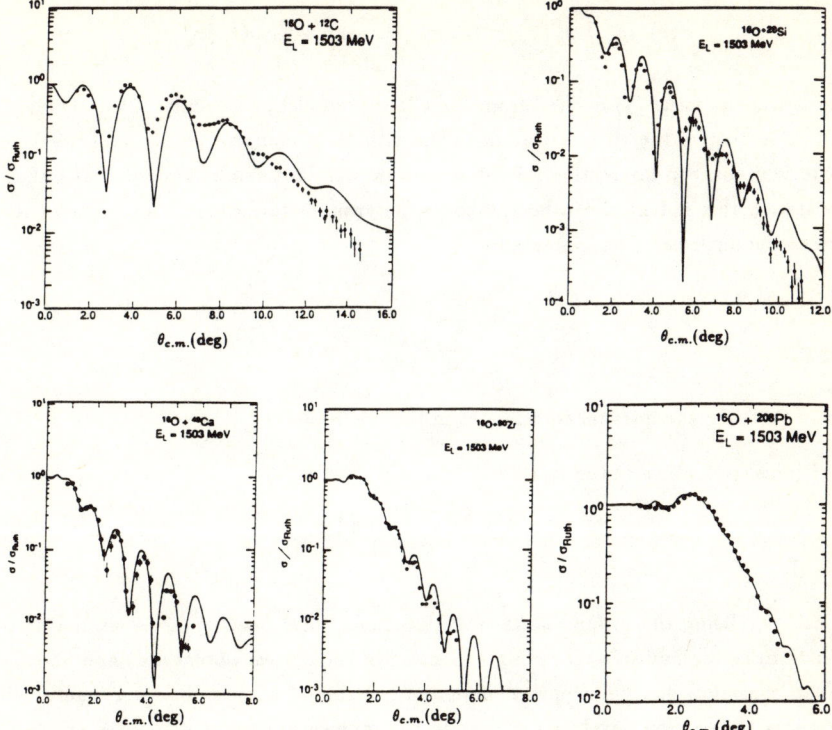

Fig. 2 - *Elastic angular distributions obtained for different systems and bombarding energies, compared with the experimental data. For details on the nuclear densities and N-N scattering amplitudes cf. ref. [11].*

the microscopic structure of the optical potential. With this purpose let me introduce the quantity

$$t_{NN}(r) = \frac{2\hbar v_{NN}}{\pi r} \frac{d}{dr} \int_r^\infty \frac{\gamma_{NN}(b)}{(b^2 - r^2)^{1/2}} b\, db , \qquad (5)$$

which by definition satisfies the inverse relation

$$\int_{-\infty}^\infty t_{NN}(r)\, dz = -2\hbar v_{NN} \gamma_{NN}(b) . \qquad (6)$$

The quantity t_{NN} is therefore just the t-operator which in the eikonal approximation gives the nucleon-nucleon scattering amplitude $f_{NN}(q)$. Note that this t-operator does not coincide with the tridimensional Fourier transform of the $f_{NN}(q)$. Taking into account eqs. (2) and (5), eq.(4) can be given the microscopic form

$$V(r) = \int \rho_0^P(\mathbf{r}_P)\rho_0^T(\mathbf{r}_T)t_{NN}(|\mathbf{r}-\mathbf{r}_P+\mathbf{r}_T|)d\mathbf{r}_P d\mathbf{r}_T \ . \tag{7}$$

The optical potential (4) derived from the Glauber model phase shifts (2) is therefore just the folding of the two nuclear densities with the t-operator, which replaces here the nucleon-nucleon interaction potential used in standard folding procedures. Otherwise stated, this potential can be viewed as the ground state expectation value of the microscopic nucleus-nucleus interaction

$$V(\mathbf{r};\zeta_P,\zeta_T) = \sum_{ij} t_{NN}(|\mathbf{r}-\mathbf{r}_P^i+\mathbf{r}_T^j|) \ , \tag{8}$$

ζ_P and ζ_T being the internal coordinates in the projectile and target.

Optical potentials of the form

$$V = <\Phi_0^P \Phi_0^T | \sum_{ij} t(ij) | \Phi_0^P \Phi_0^T> \ , \tag{9}$$

Φ_0^P and Φ_0^T being the ground state wave functions, have been obtained as the first-order term in the multiple scattering expansion of the optical potential [2] and already used to construct heavy-ion optical potentials. I should remark, however, that such an approach has been developed mainly in the momentum space [2,13] (see e.g. refs. [14] and [15] for more recent developments). Simplified expressions, obtained from eq.(9) by assuming forward-peaked t-matrix, are well known and widely used in the literature, in the form $V(r) = -(4\pi/2m)f^0 \rho^T(r)$ for nucleon-nucleus collisions and in the form $V(r) = -(4\pi/2m)f^0 \int \rho^P(\mathbf{r})\rho^T(\mathbf{r}-\mathbf{r}')d\mathbf{r}'$ for nucleus-nucleus scattering [16]. An explicit realization of eq.(9) into an expression involving the r-representation of t has been suggested, e.g. by Feshbach [17] and by Satchler [18]. Note also that, although the simple linear relation between the optical potential and the elementary scattering amplitude seems to involve only single scattering processes, our potential is generated by the inversion of the phase shift (2) which is obtained by summing the full multiple scattering series.

Expression (7) for the optical potential displays some of the features of the potentials obtained in phenomenological approaches. Since f_{NN} and consequently t_{NN} are complex quantities, the potential includes, in fact, the effect associated with the bulk absorption. Its energy dependence in both real and imaginary parts simply reflects the corresponding functional dependence of the elementary scattering amplitudes. Both the complex nature of the optical potential and its energy dependence should not be confused with the analogous features arising from dynamical couplings. The additional effects due to these couplings within this approach will be introduced and discussed below.

Examples of predicted optical potentials obtained with the inversion procedure in the intermediate-energy range for the system ^{16}O+^{208}Pb are shown in Fig. 3 at different bombarding energies. The figures are representative of similar behaviour obtained for other systems. The parameters defining the nuclear densities are taken from ref. [19]. The standard parametrization of Love and Franey has been used for the nucleon-nucleon scattering amplitude [20]. In the surface region, which is the most relevant to the elastic scattering, the shape of the potential is essentially determined by the nuclear densities and is practically unaffected by the energy variation. Concerning the absolute magnitude, as can be seen from the figure, the strength of both real and imaginary parts is reduced by almost a factor two as the energy moves from 50 to 200 MeV per nucleon. As the energy is still increased, the real part is further decreased and eventually becomes repulsive. The absorption, on the contrary, increases again.

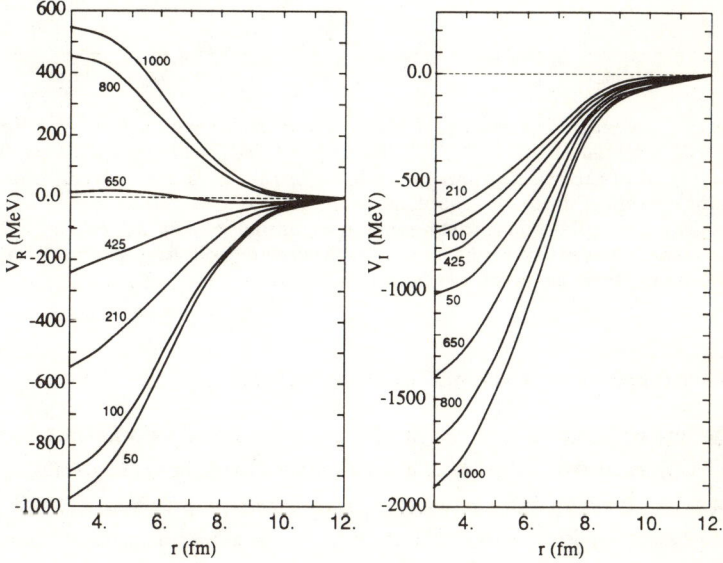

Fig. 3 - *Calculated real (left) and imaginary part (right) of the optical potential for the $^{16}O + ^{208}Pb$ system at different bombarding energies. A Fermi-type form has been used for the ^{16}O and ^{208}Pb nuclear densities, with parameters taken from [19]. For the elementary scattering amplitude one has used the parametrization of Love and Franey from ref. [20].*

The values of the real and imaginary parts at $r=10$ fm are shown in Fig. 4a for the same system, as a function of the bombarding energy. In so doing I can use the fact that outside the strong absorption region all calculated potentials have an exponential behaviour with the same slope, so that one can simply characterize them by a scaling factor. As is apparent from the figure, the trend of the dominant absorptive part reflects directly the behaviour of the nucleon-nucleon cross section, a feature already discussed by DeVries and Peng [21] in connection with the behaviour of the total nucleus-nucleus cross sections.

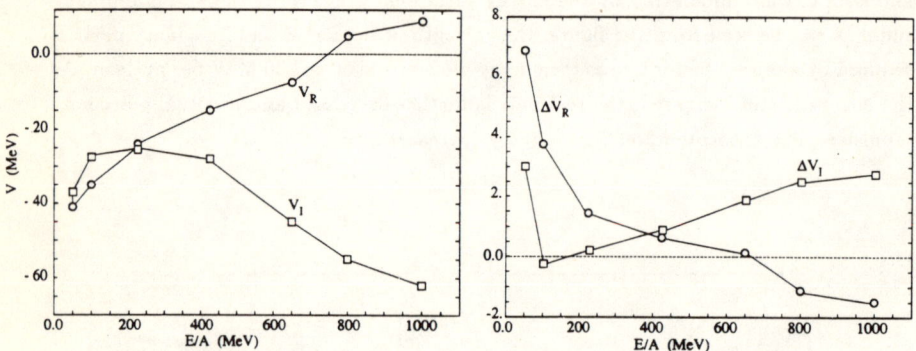

Fig. 4 - *Energy dependence of the real and imaginary parts of the "bare" optical potential for the $^{16}O + ^{208}Pb$ system at fixed ion-ion distance $r=10$ fm (left), and of the real and imaginary parts of the "polarization" contribution (right) resulting from the coupling to the low-lying 3^- state in ^{208}Pb. For details on densities and nucleon-nucleon amplitude cf. caption to Fig. 3. For the transition density we have used a Tassie-type collective parametrization with octupole deformation parameter $\beta_3=0.10$.*

3. Inelastic processes and coupling form factors

The formalism briefly outlined in the previous section for the elastic scattering can be easily extended to the description of other classes of grazing collisions. As a paradigmatic example I will consider the case of inelastic processes. To this purpose let me consider explicitly an excited state in the target with angular momentum L and third component M, connected to the ground state through the transition density $\delta\rho^T_{LM}(\mathbf{r}) = \delta\rho_L(r)Y_{LM}(\hat{r})$. The basic role in the formalism is now played by the non-diagonal matrix element $\mu_{LM}(\mathbf{b})$ of the profile γ_{NN}

$$\mu_{LM}(\mathbf{b}) = \int \rho^P_0(\mathbf{r}_P)\, \delta\rho^T_{LM}(\mathbf{r}_T)\, \gamma(\mathbf{b} - \mathbf{s}_T + \mathbf{s}_P)\, d\mathbf{r}_T d\mathbf{r}_P \;, \qquad (10)$$

which can be written in the form

$$\mu_{LM}(\mathbf{b}) = e^{iM\phi_b}\mu_{LM}(b) = \frac{e^{iM\phi_b}}{ik_{NN}} B_{LM} \int_0^\infty dq\, q\, \hat{\rho}_0^P(q)\, f_{NN}(q)\, \hat{\rho}_L^T(q)\, J_M(qb)\,, \quad (11)$$

in terms of the corresponding functions in the momentum representation [10,22]. The quantities B_{LM} are geometrical factors. One can consider again all possible sequences of elementary collisions and, in lowest-order perturbation, assume that the transition to the excited state is originated at any scattering order by a single elementary inelastic collision, all the others being of elastic character. In this way one can again sum the multiple scattering series to all orders, obtaining for the inelastic scattering amplitude to the excited LM state the expression [22]

$$f_{0 \to LM}(\Delta) = (ik/2\pi) \int d^2b\, e^{i\Delta \cdot \mathbf{b}} e^{2i\delta(b)} \mu_{LM}(\mathbf{b})\,, \quad (12)$$

where the elastic phase shift $\delta(b)$ is given by eq. (2). All necessary information to calculate the inelastic amplitude is therefore again given by densities and elementary N-N amplitude, with the addition of the relevant transition density.

The formalism has been applied to a variety of cases in refs. [22] and [11]. Some typical examples are shown in Fig. 5. In all cases a rather good agreement with experiment has been obtained, using densities and microscopic scattering amplitudes taken from literature and transition densities as in the collective model.

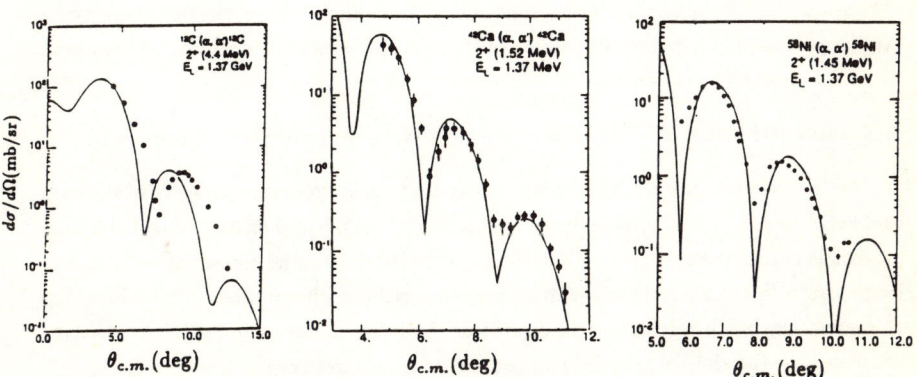

Fig. 5 - *Inelastic angular distributions obtained for different systems, compared with the experimental data. For details on the nuclear densities, transition densities and N-N scattering amplitudes cf. ref. [11].*

As in the case of the optical potential, one can extract the coupling form factor $F_{LM}(\mathbf{r})$ that, within the Distorted Wave Born Approximation, leads to the same inelastic scattering amplitude. Starting from a non-spherical form factor and using the eikonal approximation for the distorted waves, one in fact obtains for the eikonalized DWBA the corresponding expression

$$f_{0\to LM}(\Delta) = (ik/2\pi) \int d^2b \, e^{i\Delta \cdot \mathbf{b}} e^{2i\delta(b)} \int_{-\infty}^{\infty} dz F_L(\mathbf{r}) . \tag{13}$$

Expressing now γ_{NN} in eq. (10) through its representation (6), the quantity $\mu_{LM}(\mathbf{b})$ can be recast in the form

$$\mu_{LM}(\mathbf{b}) = \int_{-\infty}^{\infty} dz \int \rho_0^P(\mathbf{r}_P) \, \delta\rho_{LM}^T(\mathbf{r}_T) \, t_{NN}(|\mathbf{r} - \mathbf{r}_T + \mathbf{r}_P|) \, d\mathbf{r}_T d\mathbf{r}_P , \tag{14}$$

so that we can make the identification

$$F_{LM}(\mathbf{r}) = \int \rho_0^P(\mathbf{r}_P) \, \delta\rho_{LM}^T(\mathbf{r}_T) \, t_{NN}(|\mathbf{r} - \mathbf{r}_T + \mathbf{r}_P|) \, d\mathbf{r}_T d\mathbf{r}_P \tag{15}$$

to be compared with eq.(7). On its turn eq. (12) can be interpreted as the eikonalized DWBA inelastic amplitude relative to the microscopic form factor (15). As for the corresponding optical potential given by eq. (7), the microscopic form factor is linear in the nucleon-nucleon scattering amplitude. Since in our approach the N-N amplitude is the basic source of the energy dependence and of the complex phase, the form factors are expected to display a behaviour similar to the one of the optical potential, so justifying the usual collective assumption.

4. Coupled-channel elastic scattering and the "polarization" potential

Let us consider now the more rich case of a scattering process in presence of strongly coupled channels. This problem in the standard approach based on optical and coupling potentials requires the solution of a system of coupled differential equations. Within our formalism á la Glauber the eikonal propagation reduces the problem to the handling of matrix algebra involving coupling matrix elements of the form (10) [10]. Let me consider here the simplest case of two coupled channels, referring the reader to ref. [10] for other coupling situations which, although involving an arbitrary number of excited states, lead to simple analytic results. Because of the coupling, the elastic scattering amplitude acquires the modified expression [10]

$$f(\Delta) = (ik/2\pi) \int d^2b \, e^{i\Delta \cdot \mathbf{b}}(1 - e^{2i\delta(b)} \cosh[\Lambda(b)^{1/2}]) \,, \qquad (16)$$

where the quantity $\Lambda(b)$ is given by

$$\Lambda(b) = \sum_M [\mu_{LM}(b)]^2 \qquad (17)$$

The expression (16) for the elastic amplitude, to be compared with the uncoupled expression (1), can be obtained by summing in the multiple scattering series all the contributions arising from sequences of excitation-deexcitation processes associated with the transition matrix element μ_{LM} given in (10).

We can now recast the coupled channel problem in terms of an effective elastic optical potential, which simulates the effect of the coupling to the reaction channels. This can be done by expressing the scattering amplitude (16) in the form

$$f(\Delta) = (ik/2\pi) \int d^2b \, e^{i\Delta \cdot \mathbf{b}}(1 - e^{2i\hat{\delta}(b)}) \,, \qquad (18)$$

by means of the effective phase shift

$$\hat{\delta}(b) = \delta(b) - \frac{i}{2} \ln \cosh[\Lambda(b)^{1/2}] \qquad (19)$$

In the case of weak coupling ($\Lambda \ll 1$) we can retain the lowest order term in the expansion of $\ln \cosh[\Lambda(b)^{1/2}]$, leading to the phase shift

$$\hat{\delta}(b) = \delta(b) - (i/4)\Lambda(b) \,. \qquad (20)$$

This corresponds to include in the elastic scattering the two-step contribution arising from the excitation and subsequent deexcitation of the intermediate state. The effective optical potential can be finally obtained by inserting in eq. (4) the above expression (19) for the effective phase shift $\hat{\delta}(b)$. Given the linear dependence in the Abel transform between phase-shifts and potential, one can separate the "bare" contribution associated with $\delta(b)$ from the "polarization" term. As an example the calculated "polarization" corrections to the "bare" optical potential are shown in Fig. 6, and in Fig. 4b at fixed r, for the previously considered cases. The coupling to the 3^- state in ^{208}Pb has been considered (cf. caption to the figure for details). To evidence the relative importance of these "polarization" contributions, the curves have to be compared with the corresponding "bare" values shown in Fig. 1. These polarization corrections are relatively small, but one should keep in mind that, within our scheme, coupling to different channels gives additive effects [10]. Note that since the "polarization" terms involve the transition density via the coupling terms (10), their radial dependence differs from the "bare" one. More precisely, in the case of surface-peaked transition densities as in the case of low-lying collective modes, also the polarization terms turn out to be more effective on the nuclear surface.

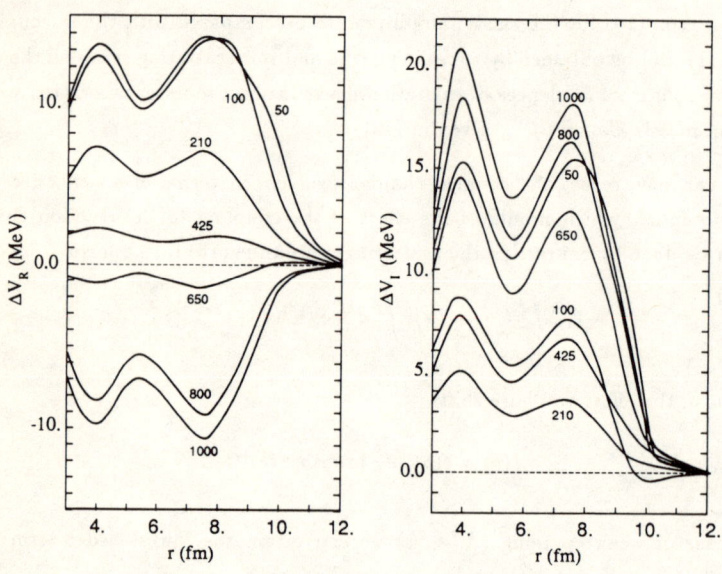

Fig. 6 - *Real (left) and imaginary part (right) of the polarization potential for the $^{16}O + ^{208}Pb$ system at different bombarding energies, resulting from the coupling to the low-lying 3^- state in ^{208}Pb. For densities, transition density and microscopic scattering amplitude cf. captions to figs. 3 and 4.*

As apparent from the figure, also the energy dependence of the "polarization" potential is different from the one of the "bare" potential. This feature can be easily understood in the case of validity of the lower order expression (20) and of isotropic nucleon-nucleon scattering amplitude f_{NN}, parametrized in the usual form $f_{NN} = [k_{NN}\sigma_{NN}(i + \alpha_{NN})]/4\pi$. While the bare optical potential is directly proportional to f_{NN}, the polarization term is in fact proportional to $i(f_{NN})^2$. More explicitly its real part is proportional $\sigma_{NN}^2 \alpha_{NN}$ and the imaginary part to $\sigma_{NN}^2(\alpha_{NN}^2 - 1)$. Since both σ_{NN} and α_{NN} vary with the energy, this leads in general to a different energy dependence of the two contributions.

As mentioned above, the effect on the ion-ion optical potential of the coupling to reaction channels has been extensively discussed in ref. [4], where explicit formulae for the polarization term are given in terms of the coupling form factors, Q-values and bombarding energies. Within that approach all the initial "bare" potentials are real. With this basic difference, in the limit of high bombarding energy the second-order perturbative approach of ref. [4] gives formally the same expression (20) given in this note. At variance with our case, the reality of the potentials in their case leads to a pure "absorptive" contribution, without any real "polarization" term. As a further consequence of the energy dependence and complex character of the nucleon-nucleon t-matrix elements the real and imaginary parts of our polarization potential do not satisfy the usual dispersion relation.

5. Concluding remarks

As a summary of these lectures, we have seen how the use of the eikonal approximation at intermediate energies strongly simplifies the reaction mechanism associated with heavy-ion grazing collisions. A rather simple formalism can be developed to provide, from one side, the bulk "bare" potential in terms of the elementary nucleon-nucleon scattering amplitudes, and, from the other side, the "polarization" contribution arising from the coupling to the internal degrees of freedom of the colliding systems. The success of such an approach in describing the experimental elastic and inelastic data over a wide range of masses and energies opens the real possibility of extracting clear structure information from more complicated heavy-ion reactions, avoiding the usual uncertainties related to the choice of the optical potentials. As a step in this direction, it is for example possible to combine this eikonal description of the reaction mechanism with structure models based on group theoretical approaches. This leads to closed analytical expressions for the population of all coupled channels (see e.g. ref. [23] for the excitation of a rotational band in deformed nuclei and ref. [24] for the case of multipair transfer reactions in superfluid nuclei).

References

[1] H. Feshbach, Ann. of Phys. **5** (1958) 357; **19** (1962) 287.

[2] M.L. Goldberger and K.M. Watson, *Collision Theory* (New York, N.Y., 1964).

[3] M. A. Nagarajan, C. Mahaux and G. R. Satchler, Phys. Rev. Lett. **54** (1985) 1136.
C. Mahaux, H. Ngô and R. G. Satchler, Nucl. Phys. **A449** (1986) 354; **A456** (1986) 134.

[4] R. A. Broglia, G. Pollarolo and A. Winther, Nucl. Phys. **A361** (1981) 307.

G. Pollarolo, R. A. Broglia and A. Winther, Nucl. Phys. **A406** (1983) 369.

C. H. Dasso, G. Pollarolo and S. Landowne, Nucl. Phys. **A443** (1985) 365.

C. H. Dasso, S. Landowne, G. Pollarolo and A. Winther, Nucl. Phys. **A459** (1986) 134.

[5] M. V. Andrés, F. Catara, Ph. Chomaz and E. G. Lanza, Phys. Rev. C **39** (1989) 99.

[6] N. Vinh Mau, Nucl. Phys. **A457** (1986) 413; **A470** (1987) 406.

[7] C. Mahaux, these proceedings.

[8] W. Czyż and L.C. Maximon, Ann. Phys. **52** (1969) 59.

[9] J. Formánek, Nucl. Phys. **B12** (1969) 441.

[10] S. M. Lenzi, F. Zardi and A. Vitturi, Phys. Rev. C **42** (1990) 2079.

[11] S.M. Lenzi, A. Vitturi and F. Zardi, Phys. Rev. C **40** (1989) 2114.

[12] R.J. Glauber, *High-energy collisions theory, Lectures in Theoretical Physics*, Interscience , N.Y. (1959) 315.

[13] A. L. Fetter and K. N. Watson, Advances in Theoretical Physics **1** (1965) 115.

[14] H. F. Arellano, F. A. Brieva and W. G. Love, Phys. Rev. C **41** (1990) 2188.

[15] K. Kaki, Nucl. Phys. **A516** (1990) 603.

[16] J.Y. Hostachy et al., Nucl. Phys. **A490** (1988) 441.

[17] H. Feshbach, Ann. Rev. Nucl. Sci. **8** (1958) 49 .

[18] G.R. Satchler, *Direct Nuclear Reactions*, Oxford, Clarendon Press, 1983.

[19] C.W. de Jager, H. de Vries and C. de Vries, At. Data and NDT **14** (1974) 479

[20] M. A. Franey and W. G. Love, Phys. Rev. C **31** (1985) 488.

[21] R.M. DeVries and J.C. Peng, Phys. Rev. C **22** (1980) 1055.
J.C. Peng, R.M. DeVries and N.J. DiGiacomo, Phys. Lett. **98B** (1981) 244.
N.J. DiGiacomo, R.M. DeVries and J.C. Peng, Phys. Rev. Lett. **45** (1980) 527.

[22] S.M. Lenzi, A. Vitturi and F. Zardi, Phys. Rev. C **38** (1988) 2086.

[23] G. Fäldt and R. Glauber, Phys. Rev. C **42** (1990) 395.

[24] S. M. Lenzi, F. Zardi and A. Vitturi, to be published.

New Physics Far From Stability

A. Poves
Departamento de Física Teórica, UAM
Cantoblanco, 28049 Madrid, Spain

> *Abstract* : Some recent results in nuclei far from the valley of β stability are reviewed. Shape transitions, new spatial distributions of nucleons in nuclei, as well as evidences pointing to the breakdown of the conventional mean field description of nuclei are among the many new phenomena that arise by going far off stability. Furthermore, the study of decays and reactions involving exotic nuclei is a powerful tool to study the interplay between nuclear and particle physics and a major source of information for nuclear astrophysics studies.

INTRODUCTION

The study of nuclei far from stability is nowadays on the forefront of research in nuclear physics. After a long period of systematic work, many new phenomena have been discovered as a consequence of the continuing experimental effort to upgrade the existing facilities and to build - or adapt - others. Besides the classical - and leading - centers, as for instance Isolde at Cern, a number of heavy ion accelerators have disembarked into the far from stability domain, this is the case for GANIL, Daresbury, GSI and others.

The study of nuclei far off stability is a natural borderline of nuclear physics. Very precisely the drip lines represent the limits of actual existence of nuclei of a given mass. In the newly explored regions there have been findings that establish unexpected links with other borderlines such as the study of nuclei at high spin and the interphase between nuclear and particle physics.

The realm of far from stability studies includes as a first stage the production techniques - expallation, fission, heavy ion reactions, projectile fragmentation, etc - and the isotope separation techniques. After this process, the exotic nuclides may be used as secondary radioactive beams or just studied on/off line with sophisticated spectroscopic tools, adapted to the different particles emitted ($\alpha, \beta, \gamma, n, p, e, \ldots$). Of utmost importance for the growth of the field has been the advent of the new generation of hyperfine structure spectroscopic facilities as well as the development of ion traps.

The combined use of these techniques provides data of superior quality on basic properties of nuclei such as:

Ground state masses, radii, spins, magnetic moments and quadrupole moments.

β decay strength functions, β delayed emissions, heavy fragment emissions

Reaction cross section that give information about the matter distributions

Cross sections for reactions of astrophysical interest ...

In ref[1] an excellent review of the present experimental situation can be found.

The consequences of these experimental results on our understanding of the structures of the atomic nucleus will be now illustrated with a few examples.

N=20 FAR FROM STABILITY AND THE STRUCTURE OF THE NUCLEAR MEAN FIELD

The mass measurement on the isotopes of Sodium made by Thibaut et al [2] showed that at N=20, instead of the behaviour of the S_{2n} expected at a shell closure, the results suggested the onset of a deformed regime. The present situation is summarized in figure 1 where more recent data from references [3-5] are collected. In this figure the anomaly in the S_{2n} values is clear for Ne, Na and Mg. An even more convincing evidence comes from the spectroscopic studies. In figure 2 we show the excitation energy of the first 2^+ state along the Mg and Si isotopic chains. The normal trend at a shell closure is exemplified by ^{34}Si whose 2^+ excitation energy increases by 1.5 MeV with respect to ^{32}Si, while in ^{32}Mg an extremely low value for the excitation energy of the 2^+ is found, hence supporting the picture of a small region of large deformation around ^{31}Na.

These experimental facts have been explained in a shell model framework in refs. [6-9]. The local onset of a deformed regime results from the combined effect of two mechanisms. In first place, it happens that the energy gap between different major shells gets largely reduced far from stability. As a consequence, intruder configurations are less disfavoured than closer to the stability. Secondly, the intruder configurations have built in collectivity due to the neutron proton quadrupole interaction, provided the nucleus has a few valence protons. Hence their correlation energy is enough to beat the normal configurations ant to become dominant in the ground state, provoking structural and shape changes in some nuclei.

Only two mass units apart, ^{34}Si is, on the contrary, a new example of doubly magic nucleus [10] (N=20,Z=14), a situation fully compatible with the explanations given above in terms of the interplay of intruder and normal configurations.

figure 1

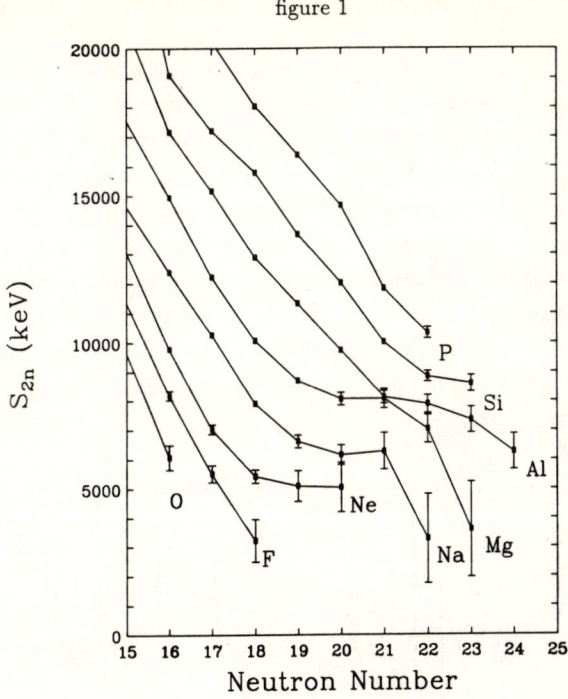

figure 2

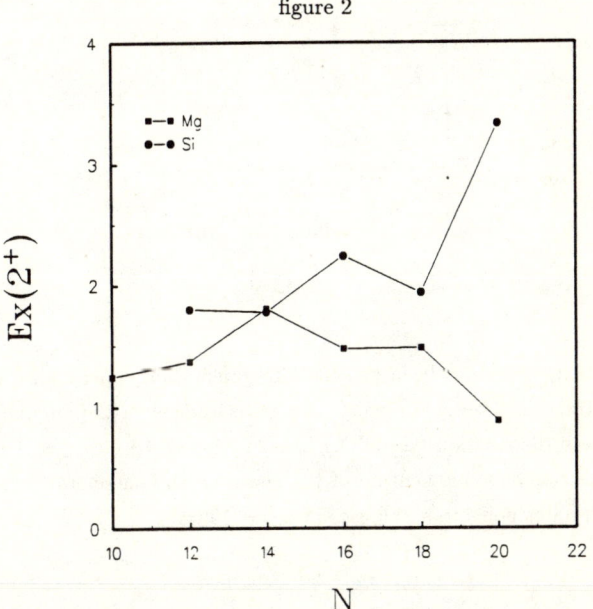

^{11}Li AND THE DISTRIBUTION OF MATTER IN DRIP LINE NUCLEI

^{11}Li enjoy at present a prima donna status in nuclear physics. It is a nucleus whose number of neutrons is magic and reach the drip line simultaneously. ^{10}Li is unbound (or extremely loosely bound). No stable Li isotopes beyond mass 11 have been found. This short lived nuclide has a very large Q_β value and exhibit almost the complete catalogue of β-delayed emissions. However these are not the reasons for its celebrity. It rather comes from the very large reaction cross section - and consequently matter radius - found by Tanihata and coworkers [11] using a ^{11}Li beam, followed by the description proposed by Hansen and Jonson [12] in which ^{11}Li appears as a dineutron bound to ^9Li . In this configuration ^{11}Li is said to exhibit a neutron halo. In figure 3 the mass radii extracted from the experimental interaction cross sections are plotted. Notice the large radius of ^{11}Li and ^{14}Be.

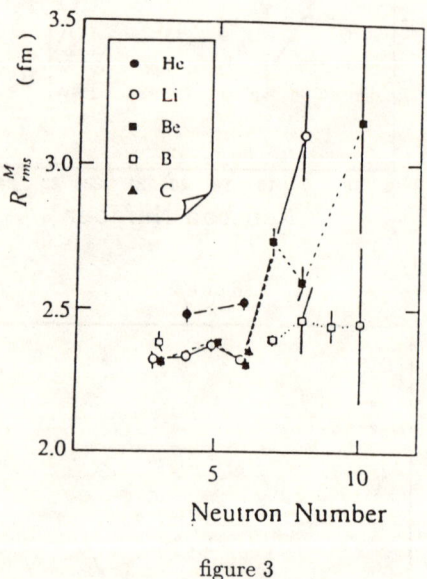

figure 3

Very recently there has been projectile fragmentation experiments at GANIL [13] aimed to extract the size of the halo. The measurement of the angular spread of the two neutrons in the reaction ^{11}Li (X,X) ^9Li+2n, $\Theta = (2.9 \pm 0.4)°$, leads to a very small value for the transverse momentum of the neutrons and therefore to large sizes. The very first estimates point to a halo size of about 10 fm.

We have examined [14] to what extent such a phenomenon may be accounted for by Hartree-Fock calculations using density-dependent effective forces that give good results at the stability valley. We have studied two cases, ^{11}Li and ^{14}Be, which are the nuclei where the halo effects seem to be most prominent. Our first results are the following:

> In ^{14}Be the Ska Skyrme-type force, through slight modifications of the parameters governing the symmetry energy, gives a solution with two neutrons in an extremely large $2s_{1/2}$ orbit, $<r^2>^{1/2} \sim 8$ fm. This result is consistent with the most recent fragmentation experiments [13] that point to ~ 10 fm halos.
>
> In ^{11}Li, on the contrary, as far as the two outer neutrons site is the $1p_{1/2}$ shell, it is impossible to get such a large halo radius. The $1p_{1/2}$ orbit can at most reach $<r^2>^{1/2} \sim 5$ fm. This number is compatible with early estimates [15] from the results of Tanihata et al [11] and also with the result of a recent analysis of the double charge exchange reaction on ^{11}B [16].

More experimental information is clearly needed, but if the new experiments confirm the far out halo result of [13], it would be necessary to accept that in ^{11}Li the ground state configuration is a 2p-2h intruder, with two neutrons in the $2s_{1/2}$ orbit, i.e the same kind of behaviour found at N=20 far from stability [6]. Work is in progress aiming at a more complete description of these nuclei, treating HF selfconsistency and $0\hbar\omega$ correlations on the same footing.

Finally one could wonder whether the anomalies attached to ^{11}Li size could be due to deformation. A high quality measure of the quadrupole hyperfine effect in ^9Li and ^{11}Li made at Isolde [17] has ruled out this possibility by showing that the ratio of the quadrupole moments of ^{11}Li to ^9Li has the value 1.10(16).

CONCLUSION

These are just two examples among the many leading to new physics, that come from far from stability studies. There are others that make links with apparently distinct domains, as for instance the existence of superdeformation in the ground state in very proton rich nuclei (^{80}Zr) [18] or very neutron rich (^{104}Zr) and even the possible presence of twin bands ($^{98}Sr - ^{100}Sr$) [19]. At the interphase of nuclear and particle physics, the occurrence of enhanced $0^- \to 0^+$ β decays as in the $^{50}K \to ^{50}Ca$ beta decay, provides a unique ground to study the renormalization of the axial current in nuclei [20]. To conclude, the detailed study of the $^{37}Ca \to ^{37}K$ β^+ decay has been recently used to calibrate the ^{37}Cl solar neutrino detector and has also reopened the discussion on the quenching of the Gamow-Teller strength in nuclei [21].

REFERENCES

[1] C. Detraz and D. J. Vieira, *Ann. Rev. Nucl. Part. Sci.* 39:407-465(1989).

[2] C. Thibault et al. *Phys. Rev.* C12 (1975) 644.

[3] A. Gillibert et al. *Phys. Lett.* B192 (1987) 39.

[4] D. J. Vieira et al. *Phys. Rev. Lett.* 57 (1986) 3253.

[5] N. A. Orr et al. *Phys. Lett.* B258 (1991) 29

[6] A. Poves and J. Retamosa, *Phys. Lett.* B184 (1987) 311.

[7] A. Poves and J. Retamosa, *Proc. of the workshop: "Nuclear Structure of Light Nuclei far from Stability: Experiment and Theory"*, Ed. G. Klotz, C.R.N., Obernai (1990) p. 31.

[8] A. Poves and J. Retamosa, preprint FTUAM 21/91

[9] E. K. Warburton, J. A. Becker and B. A. Brown, *Phys. Rev.* C41 (1990) 1147.

[10] P. Baumman et al. *Phys Lett B* 228 (1989) 458.

[11] I.Tanihata et al., *Phys. Rev. Lett.* 55 (1985) 2676

[12] P.G.Hansen and B.Jonson , *Europhys. Lett.* 4 (1987) 409

[13] R.Anne et al., *Phys. Lett.* B250 (1990) 19

[14] J. M. G. Gomez, A. Poves and C. Prieto, in preparation.

[15] G.F.Bertch, B.A.Brown and H.Sagawa, *Phys. Rev.* C39 (1989) 1154

[16] W.R. Gibbs and A.C. Hayes , *Phys. Rev. Lett.* 67 (1991) 1395

[17] E. Arnold et al. to be published

[18] C.J. Lister et al, *Phys. Rev. Lett.* 59 (1987) 1270

[19] G. Llersoneau et al *Z. Phys.* A337 (1990) 143

[20] E. Warburton, to be published

[21] A. Garcia et al, to be published

Recent Developments in High Spin Physics [*]

J.L. Egido

Departamento de Física Teórica C-XI
Universidad Autónoma de Madrid
E-28049 Madrid, Spain.

> *Abstract*: Some topics concerning the new developments in Nuclear Structure, mainly at high-spin, are discussed. Based on the mean field approximation -and theories beyond it- we shortly analyze: Octupole and superdeformed shapes at low excitation energies, the rotational damping width and, finally, strength functions of hot rotating nuclei.

1. Introduction

In the last years there has been an explosion of interesting new experimental data on nuclear structure mainly at high angular momenta, a fact that converted this field in one of the most active ones within the community of nuclear physics.

Heavy Ion (HI) fusion reaction has turned out to be a very powerful tool to study the atomic nuclei in terms of the excitation energy (E^*) and the angular momentum (I): At small excitation energies, near to the Yrast line, one has discrete levels forming rotational bands of different character. In this region many new features have been found as, for example, terminating bands, shape changes, octupole shapes and very recently superdeformed bands. On the other hand, the analysis of the very energetic γ-ray of the Giant Dipole Resonance (GDR) has allowed -among other features- the study of the nuclear shapes as a function of E^* and I. The region in between, the quasicontinuum, is characterized by a steeply-rising level density as a function of excitation energy. From the analysis of the gamma rays emitted in the rotational cascades evidence was found for the damping of the rotational motion and discussions are going on concerning the motional narrowing. This region is probably the one where most of the new discoveries are going to be found when the new ball detectors (Euroball and Gammasphere) start to work.

Among all these new developments, for my lectures, I have chosen some characteristics of each of the energy regions mentioned above that are close to my own working field. In the first lecture I shall talk on nuclear shapes (octupole and superdeformed ones). I shall devote most of the time to the octupoles shapes because the superdeformed ones are going to be covered by J. Bentley. In the second one, as a typical quasicontinuum effect, I shall discuss the damping of the rotational motion. In the third lecture, finally, the strength functions of hot rotating nuclei are considered.

The theoretical basis used to discuss these effects will be the mean field approach (mainly Hartree-Fock-Bogoliubov, HFB) -and theories beyond it- at zero and finite temperature. I shall not discuss in detail the theory of HFB because this will be covered by Alan Goodman in his lectures. The theories beyond HFB will be shortly discussed in the corresponding sections.

---

[*]Work supported in part by DGIC y T, Spain, under project PB88 - 0177

2. Nuclear Shapes

The existence of rotational nuclear spectra was the first indication of the deviation from the spherical shape of the nuclear density. The use of a deformed potential has led to the interpretation [1] of an enormous amount of data. The new developments [2] in the experimental techniques have made possible the study of nuclei at high angular momentum where the coriolis force largely affects the shape of the nucleus. Thus, besides the large variaty of oblate and prolate nuclei, many new phenomena, such as terminating rotational bands, high-spin isomeric states, octupole shapes, superdeformed shapes and possibly triaxial shapes have been observed.

To make a detailed description of all these new phenomena is a difficult task, so I shall concentrate on two of them: Octupole shapes and superdeformation. Mainly, there have been two approaches: The so called macroscopic-microscopic appoximation, (Nilsson-Strutinsky and Woods-Saxon-Strutinsky) and the Hartree-Fock-Bogoliubov (HFB) one. I shall refer to the latter one. For a recent review on nuclear shapes in mean field approach see ref. [3] and references therein. The Proceedings of the Oak-Ridge conference (1990) [4] and the Strasbourg meeting (1991) [5] contain contributions with the late developments in Nuclear shapes.

2.1 Octupole Shapes

In the last years, the existence of very low-lying negative parity states 1^- in the light actinides as well as the experimental evidence [6] of the lack of two-phonon octupole states 0^+ in those nuclei has motivated a renewed interest in the study [7-16] of the octupole degree of freedom. Based on the analogy with the quadrupole degree of freedom where the quadrupole correlations, in some nuclei, give rise to a permanent deformation and hence, to a breakdown of the rotational invariance in the intrinsic system, one would expect in the case of strong octupole correlations a permanent octupole shape together with a breakdown of the parity invariance. [For the consequences of a permanent octupole shape on the nuclear spectrum (see ref.[1], p. 19)] The first step in the theoretical description of the octupole degree of freedom we are going to discuss, consists of a HF+BCS mean field calculation using the density dependent force of Gogny [17]. The calculation was performed imposing axial symmetry to the wave function. In addition to the particle number, two further constraints were imposed. To study nuclear properties as a function of the octupole operator ($\hat{Q}_{30} = \sqrt{(4\pi)/7} r^3 Y_{30}$), its mean value $q_3 = <\phi|\hat{Q}_{30}|\phi>$ was constrained. The constraint on $\hat{Q}_{30}$ forces the breaking of the parity invariance of the wave functions and gives rise to the possibility of admixtures of spurious states - associated to the center of mass motion - with the physical states. To prevent this mixing it is necessary to constrain the position of the center of mass ($<\hat{z}>$ in the axially symmetric case) at the origin of coordinates.

Mean field approximations are good to describe bulk properties and expectation values. As an example of our HFBCS calculations we now discuss some results. Other results concerning pairing energies, multipole moments, etc, can be found in refs. [15,16]. In Fig. 1 we see the potential energy as a function of the constrained octupole parameter q_3 for the Radium, Thorium and Barium isotopes. Generally, we observe that the collective potential has a minimum for q_3 different from zero. However, the depth of that minimum is not the same for all the nuclei and it is only significantly different from zero (greater

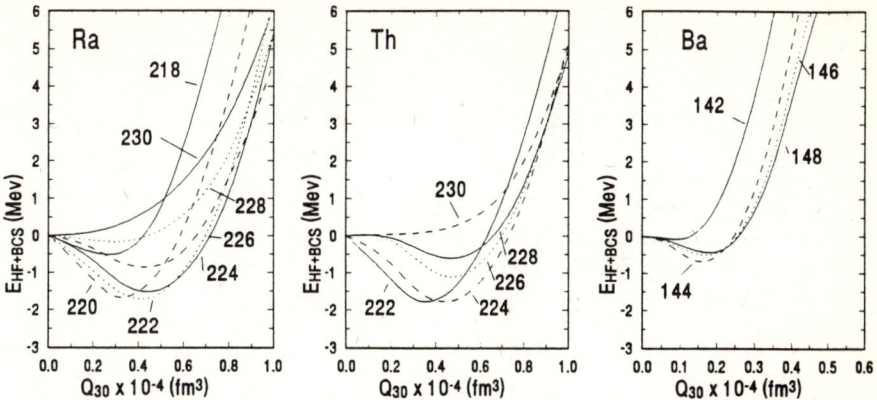

Figure 1: Potential energy curves for the Radium, Thorium and Barium Isotopes

than 0.5 MeV) in the $^{220-226}Ra$, $^{222-226}Th$ and ^{144}Ba nuclei. The depth of the potential displays the degree of "octupolness"

The information given by mean field theories is restricted to the energy and shape of the -generally- deformed ground state. To restore the parity symmetry broken in the mean field approximation and in order to describe collective excitations one has to go beyond the mean field. To achieve this goal, we have used the q_3 collective degree of freedom to build up a collective hamiltonian in the spirit of the Generator Coordinate Method (GCM). Starting from the general equation of the GCM for one collective coordinate, it is possible [18] to reduce, under some assumptions, the GCM equation to a Schrödinger equation

$$\hat{\mathcal{H}}_{coll}\phi_\alpha(q_3) = \epsilon_\alpha \phi_\alpha(q_3). \tag{1}$$

Here the hamiltonian operator $\hat{\mathcal{H}}_{coll}$ is given by

$$\hat{\mathcal{H}}_{coll} = -\frac{1}{\sqrt{G(q_3)}}\frac{\partial}{\partial q_3}\sqrt{G(q_3)}\frac{1}{2B(q_3)}\frac{\partial}{\partial q_3} + V(q_3) - \epsilon_0(q_3). \tag{2}$$

where $G(q_3)$ is the metric, $B(q_3)$ is the mass parameter associated with the collective motion along q_3, $V(q_3)$ is the collective potential and $\epsilon_0(q_3)$ is the Zero Point Energy (ZPE) correction. All these quantities can be computed microscopically from the overlaps and energy kernels. A similar expression to Eq. (2) is also obtained in the framework of the Adiabatic Time Dependent Hartree-Fock (ATDHF) theory, but the values of the collective parameters differ from the ones computed within the GCM framework. One can combine the advantages of both methods by using phenomenological perscriptions used by the Bruyères-le-Châtel group [22] (see also [16]) This method (hereafter referred as ATDHF+ZPE) uses the ATDHF mass, the GCM metric and a ZPE correction given by:

$$\epsilon_0(q_3) = \frac{1}{2}G_{GCM}(q_3)B_{ATDHF}^{-1}(q_3) \tag{3}$$

In the calculation of the collective parameters the "Cranking approximation" [22] has been used. Expectation values and transition probabilities are easily calculated in the above formalism [24]. Now we present the results of collective calculations carried out in the framework of the ATDHF+ZPE for several nuclei in the $Ra(Z=88)$, $Th(Z=90)$ and

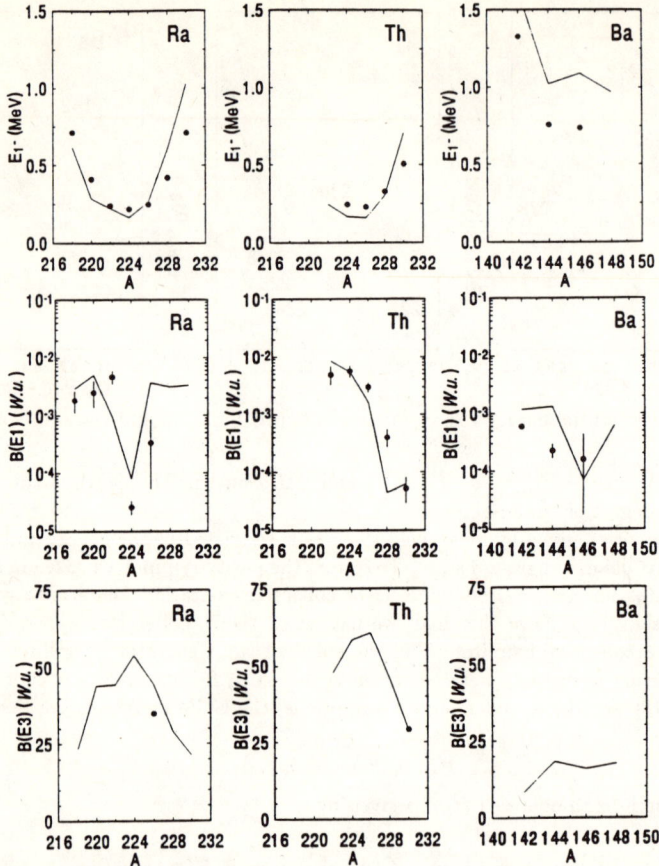

Figure 2: *Upper part*: Theoretical (full line) and experimental (full circles) results for the $0^+ - 1^-$ energy splitting. *Middle part*: Theoretical (full line) and experimental (full circles) results for the $B(E1, 1^- \to 0^+)$ electromagnetic transitions. *Lower part*: Theoretical (full line) and experimental (full circles) results for the $B(E3, 3^- \to 0^+)$ electromagnetic transitions.

$Ba(Z = 56)$ chains of isotopes. Previous results on calculations using different collective parameters as well as the results of two dimensional $(q_2 - q_3)$ collective calculations in the ^{144}Ba and ^{222}Ra nuclei can be found elsewhere [15,16,21,23]. The previous results do not differ qualitatively from the ones presented here.

We are interested in the lowest states of positive and negative parity, which correspond to the 0^+ ground state and the first excited states $1^-, 3^-$ (whose band head is the 0^- state computed in our calculation) observed experimentally.

In the upper part of Fig. (2) the results for the $0^+ - 1^-$ splittings (full line) are compared to the experimental 1^- energies (full circles). As expected, the splittings are smaller for those nuclei with a deep potential well, being the experimental trend well reproduced in all the cases. In the middle part results for the $B(E1, 1^- \to 0^+)$ electromagnetic transitions (full lines) are compared to the experimental results (full circles). The experimental results have been obtained from the recent review of Butler and Nazarewicz [25] and in many cases have been extrapolated from high spin results. The characteristic feature of the three isotope chains is the dramatic decrease of the B(E1) for some nuclei. This behaviour seems to be quite general for isotope chains where octupole correlations are important and it can be understood in terms of the dipole moments, see refs. [15-16]. In the lower part results for the $B(E3, 3^- \to 0^+)$ electromagnetic transitions (full lines) are compared to the experimental results [26] (full circles). In this case, the behaviour of the mean field octupole moment for protons is roughly the same for all the nuclei and a correlation between the magnitude of $B(E3)$ and the "octupolness" of the nucleus is observed. It is worth mentioning that the scarce experimental results agree very well with the theoretical predictions.

2.2 Superdeformed shapes

The experimental detection of the discrete line superdeformed rotational band in ^{152}Dy by Twin [27] and collaborators has been one of the most spectacular realizations in Nuclear Structure. The moment of inertia of the band was found to be close to that of a rigid rotor with a 2:1 axis ratio. The existence of superdeformed shapes was already known from the observation of fission isomers in the actinide region and there were several predictions for Rare Earth nuclei based on potential energy surfaces at high spins. For the new discovered nuclei, the basic mechanism, i.e. the bunching of single-particle states around the 2:1 deformation, is the same as for the fission isomers. The great advantage of the non-fissile nuclei (at least in the energy-angular momentum we are interested in) is that the rotation stabilizes the very deformed configurations.

Two regions of superdeformation have been studied systematically: The ^{152}Dy region [28] and the ^{192}Hg region [29,30,31,32]; further studies have been done in other regions (see S. Aberg in ref. 4). Most of these investigations have been done with the Nilsson (or Woods-Saxon) -Strutinsky approximation. Calculations with realistic forces have been done with the Gogny force and with the Skyrme force [35].

In Fig. 3 we see the potential energy surfaces (PES) for ^{152}Dy (from ref. [33]) and for ^{192}Hg (from ref. [34]), in the (β, γ) plane, calculated with constrained HFB and Gogny forces. In these PES the zero point energy associated with the fluctuations of the collective variable is also included. In the Dy nucleus we see a superdeformed minimum at $\beta = 0.7$ and about 4.5 MeV high, in ^{192}Hg there is a minimum at $\beta = 0.6$ MeV and 3.0 MeV high. The difference in the excitation energy of the superdeformed minimum explains why the ^{152}Dy gets Yrast at $I > 50\hbar$ and in ^{192}Hg around $10\hbar$.

In Fig. 4 (left hand side) we display the protons single particle levels of ^{152}Dy as a function of the deformation parameter $Q_0 = \sqrt{(4\pi)/7} <r^2 Y_{20}>$ for the selfconsistence

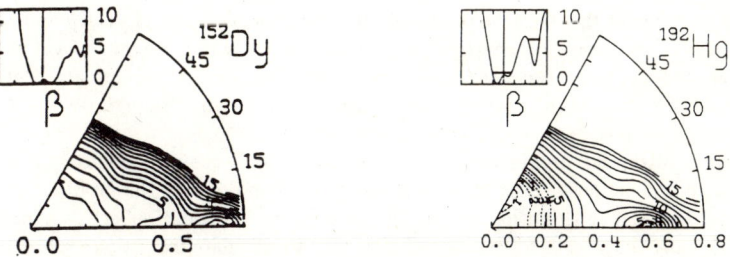

Figure 3: Potential energy surfaces for ^{152}Dy and ^{192}Hg nuclei.

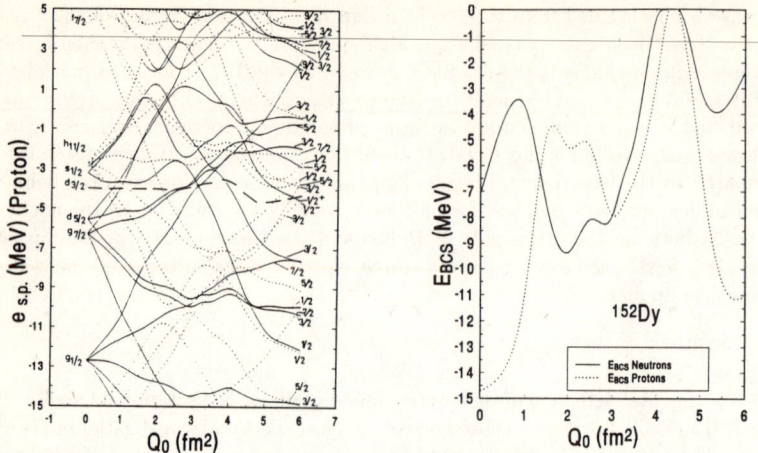

Figure 4: *Left hand side* : Single particle levels for ^{152}Dy. *Rigth hand side*: Pairing energies along the constrained path for ^{152}Dy and Gogny forces.

Gogny field. The large energy gaps at the Fermi surfaces around the equilibrium minimum $Q_0 = 4500 fm^2$ explain the existence of superdeformation for this nucleus. These large gaps in the nucleus also justify the existence of several superdeformed bands in the neighbouring nuclei associated with the low-lying one-particle or one-hole states, for instance, in ^{153}Dy three superdeformed bands have been observed in ^{153}Dy.

In Fig. 4 (rigth hand side) we show the pairing energies calculated in the mean field approximation for the nucleus ^{152}Dy as a function of the quadrupole moment. The oscillatory character of the proton and neutron energies has to do with the level density around the Fermi level as it can be seen in the left hand side.

An interesting point is the fact that the proton and the neutron pairing energy becomes zero in the superdeformed minimum and that at both sides of the minimum again we get large pairing energies, i.e. the pairing fluctuations could be important. This has been shown to be the case in the calculations [37] of the dynamical moment of inertia $\mathcal{J}^{(2)}$, where they get a worse agreement with the experiment if a constant pairing gap was included in the calculation. The general shape of the various $\mathcal{J}^{(2)}$ behaviour of the superdeformed bands in several nuclei is well understood in terms of the different occupations of the high-N orbits. A good agreement with the experimental results only is obtained after the pairing correlations have been taken into account.

A very interesting problem is the one risen by the observation [38] that the excited superdeformed bands found in $^{(151)}Tb$ and $^{(150)}Gd$ had almost identical gamma-ray energies to those in the Yrast bands in the neighbouring Z+1 nuclei, ^{152}Dy and $^{(151)}Tb$, respectively. The transition energies are identical to within an average of about 1 Kev, about one order of magnitud smaller than expected. This similarity is not accidental since other examples have been reported. This intriguing problem, although several solutions [39,5] have been proposed, is not well understood yet.

Another intriguing problem in superdeformed nuclei is the decay out of the superdeformed band. The most realistic calculations have been done by the group of Bruyeres-Le-Chatel (see the contribution to ref. [5]) who, using the Bohr collective hamiltonian, obtained many-body wave functions and were able to calculate transition probabilities for the nucleus ^{192}Hg. They found that the calculated superdeformed intensity disappears in four transitions between $10^+, 8^+, 6^+$ and 4^+, while the last state experimentally observed is 8^+. There have been other calculations [5], but for the moment this problem is not well understood yet.

There has been the prediction [40] on the octupole inestability of ^{152}Dy in the superdeformed minimum and other nuclei. Experimental evidence has been found (see Sharpey-Shafer in [5]) in the superdeformed nucleus ^{193}Hg.

3. Damping of Rotational Motion

In the last few years new detection systems and larger accelerators have allowed to analize the unresolved part of the γ-ray nuclear spectra following heavy-ions fusion rection. At high angular momentum most of these gamma rays seem to be of rotational character (they correspond to stretched E2 transitions and their energies are roughly correlated with the spin). The study of the correlations of these gamma rays seems to indicate [41] [42] [43] that the nuclear rotational motion is damped at modest temperatures. The rotational damping causes each rotational state to emit, rather than a single gamma ray energy, one from a distribution of energies, with a full width at half maximum (FWHM), Γ_{rot}. The existence of damping implies that a mixing of states is taking place. At relatively low temperatures ($T \sim 0.5$ MeV) the level density - the number of rotational bands per unit energy - is so high that even small residual interactions can cause that mixing (see fig. 5)

The first theoretical attempt to describe rotational damping was undertaken by Leander [44]. The most recent attempt [45] has been done by the Copenhaguen group; in their work - which we shall describe in more detail below - they reduce the calculation of the damping width to the evaluation of the strength function of the compound states. Based in the usual strength function formalism they derived a set of coupled equations which they solved in two extreme limits.

We now briefly describe the method of ref. [45]. Let us denote by $|\mu(I)\rangle$ the unperturbed single particle basis states at angular momentum I. These states, for example, could be the many particle-hole excitations obtained by diagonalization of the cranked Nilsson hamiltonian. Each state $|\mu(I)\rangle$ decays [46] by a collective rotational E2 transition to only one state at angular momentum $I - 2$, which we denote by $|\mu(I-2)\rangle$. If we now diagonalize the full hamiltonian, i.e. including residual interactions, we obtain the eigenstates of the compound nucleus, given by

$$|\alpha(I)\rangle = \sum_\mu X_\mu^\alpha(I)|\mu(I)\rangle, \qquad (4)$$

the expansion coefficients, in general, depend on I because so do the basis states. Only in the hypothetical case that all unperturbed bands, at a given I, do have the same rotational frequency and the residual interaction is spin independent, the coefficients for $|\alpha(I-2)\rangle$ in eq. (4) will be the same as for $|\alpha(I)\rangle$. In this hypothetical situation, let us denote these states by $|\alpha(I-2)\rangle_{rot}$. Since the linear combinations are the same, each state $|\alpha(I)\rangle$ decays mainly, by a collective E2 to its rotational partner state $|\alpha(I-2)\rangle_{rot}$. In general, this will not be the general situation, and the residual interaction $\Delta H(I-2)$ will depend on the angular momentum. This will cause a spreading of the states $|\alpha(I-2)\rangle_{rot}$, i.e. in this model the calculation of the rotational damping width is reduced to the calculation of the strength function of the rotational partner state $|\alpha(I-2)\rangle_{rot}$. For that purpose the standard theory [47] for evaluating strength functions is applied, with the results

$$P_\alpha(E) = \frac{1}{2\pi} \frac{\Gamma_\alpha + \Delta}{(E_\alpha + \delta E_\alpha - E)^2 + \frac{1}{4}(\Delta + \Gamma_\alpha)^2}$$

$$\Gamma_\alpha = \Delta \sum_\beta \frac{(\Delta H_{\alpha\beta})^2}{(E - E_\beta)^2 + (\frac{1}{2}\Delta)^2}$$

$$\delta E_\alpha = \sum_\beta \frac{(\Delta H_{\alpha\beta})^2 (E - E_\beta)}{(E - E_\beta)^2 + (\frac{1}{2}\Delta)^2} \qquad (5)$$

where Δ is a smoothing parameter [47] and the states $|\beta\rangle$ are obtained by diagonalization of the total hamiltonian at $I - 2$ among all rotational partner states with the exception of the state $|\alpha(I-2)\rangle_{rot}$. The residual interaction ΔH between the states $|\alpha\rangle_{rot}$ has two contributions, one stemming from the dispersion of rotational frequencies among the $|\mu\rangle$ states, the other one from the change with angular momentum of the residual interaction, the latter one is very small and is usually neglected [45].

To calculate the coefficients X_μ^α, the authors assume (without further justification) they are provided by the GOE. After some approximations, which we shall not discuss (see ref. [45] for details), one obtains for the system Eq. (5)

$$P_0(E) = \frac{1}{2\pi} \frac{\Gamma_0}{(\delta E_0 - E)^2 + (\frac{1}{2}\Gamma_0)^2} \qquad (6)$$

$$\Gamma_0 = 2(2\Delta\omega_0)^2 \int_{-\infty}^\infty dx \frac{\Gamma_\mu}{(E - x)^2 + \Gamma_\mu^2} P_0(x) \qquad (7)$$

$$\delta E_0 = (2\Delta\omega_0)^2 \int_{-\infty}^\infty dx \frac{(E - x)}{(E - x)^2 + \Gamma_\mu^2} P_0(x) \qquad (8)$$

with $P_0(E - E_\alpha) = P_\alpha(E)$, Γ_μ the spreading width of the states $|\mu\rangle$ in the compound states $|\alpha\rangle$, and $\Delta\omega_0$ the average deviation of the frequency ω_μ, of the states $|\mu\rangle$, from the average value ω_0. The quantities Γ_μ and $\Delta\omega_0$ can be calculated in different approximations as we will see below. Notice that Γ_0 and δE_0 depend on E.

Eqs. (3-5), which determine P_0, are non-linear and they have to be solved in a self-consistent way. In ref. [45], they gave the solutions for two limiting cases:

a) For $\Gamma_\mu \gg 2\Delta\omega_0$, one can set

$$\Gamma_0 \approx \frac{2(2\Delta\omega_0)^2}{\Gamma_\mu}, \quad \delta E_0 \approx 0 \qquad (9)$$

which provides for $P_0(E)$ a Breit-Wigner shape with a width

$$\Gamma_{rot} \equiv \Gamma_0 \approx \frac{2(2\Delta\omega_0)^2}{\Gamma_\mu}. \qquad (10)$$

b) For $\Gamma_\mu \ll 2\Delta\omega_0$ one can set

$$\Gamma_0 \approx 2(2\Delta\omega_0)^2 \pi P_0(E), \quad \delta E_0 \approx \frac{E}{2} \qquad (11)$$

and one obtains [45] for $P_0(E)$ a Wigner semicircle

$$P_0(E) = \frac{2}{\pi W^2} \sqrt{W^2 - (E - E_\alpha)^2} \qquad (12)$$

with $W = 2(2\Delta\omega_0)$. The second moment of the distribution of Eq. (12) is given by $(\frac{1}{2}W)^2$, in ref. [45] they relate this value to the rotational damping width by

$$\Gamma_{rot} = W = 2(2\Delta\omega_0) \qquad (13)$$

We would like to point out that in the two cases, (a) and (b), the definition of Γ_{rot} was different. In case (a) Γ_{rot} is identified with the FWHM and in case (b), with half of the second moment of the distribution. These magnitudes, in general, do not coincide for any distribution. This has also to be taken into account when these magnitudes are compared with experimental data where Γ_{rot} is always defined as the FWHM.

Besides these limiting solutions, one can find the exact solution of the non-linear equation system (6-8) by iteration [48]. In this case, for $P_0(E)$ one finds [48] a smooth transition from Wigner semicircles, through gaussians to Breit-Wigner shapes, at variance with the approximate solutions. The approximative predictions coinciding with the exact ones only in the extreme cases.

The approximate as well as the exact solutions depend on the quantities $\Delta\omega_0$ and Γ_μ. The frequency dispersion $\Delta\omega_0$ has been evaluated in several approximations. In the simplest one [45] [49] the dispersion in alignments was calculated within the cranked harmonic oscillator, and the fluctuation in the shape was included by using the rotating liquid drop model. It has the advantage or providing an analytical expression in terms of the excitation energy U, U being measured relative to the Yrast line. For the typical case of a nucleus with a deformation where the relation between the larger and the smaller axis is 1.3 : 1, and for $I = 40\hbar$, one obtains [49]

$$\Delta\omega_0 \sim 0.036\, U^{\frac{1}{4}}\; MeV \qquad (14)$$

A more realistic calculation of the dispersion in alignments was done in [50] with the cranked Nilsson potential approximation. Lastly, a selfconsistent cranked Hartree-Fock-Bogoliubov approximation at finite temperature was done in ref. [51]. Notice that this calculation includes pairing effects. The shape fluctuations were taken into account using the Landau theory of shape fluctuations. The damping width Γ_μ, unfortunately, is much more difficult to calculate and also experimentally very poorly known. It has been evaluated within the Fermi gas model and one obtains [49]

$$\Gamma_\mu \sim \frac{1}{45}\, U^{\frac{3}{2}}\; MeV \qquad (15)$$

The energy dependence seems right for high energies, but for the modest temperatures in which we are interested in probably [52] a constant value of Γ_μ would be more appropiated.

In spite of the quantitative difference between expression Eq. (14) and the results of the realistic calculations [51] for $\Delta\omega_0$, we shall use the first one for simplicity reasons, besides the fact that the main conclusions of this model remain qualitativaly the same in both cases. In the calculations we are going to see we have used the parametrization of Eq. (14) and Eq. (15) for $\Delta\omega_0$ and Γ_μ, respectively. Within this parametrization one can easily estimate the crossing point of the limiting cases $\Gamma_\mu \gg 2\Delta\omega_0$ and $\Gamma_\mu \ll 2\Delta\omega_0$. It corresponds to $\Gamma_\mu = 2\omega_0$, from Eq. (14) and Eq. (15) we obtain $U = 2.6$ MeV.

In Fig. 5 (rhs) we show the rotational damping widths as a function of the excitation energy. The continuous line corresponds to the approximate solution as given in eqs. (7) and (10). The exact rotational damping widths are defined as the FWHM of the exact strength functions $P_0(E)$. One can see that the high energy limit is well reproduced by the approximate solution. This is not the case, however, for the low energy limit ($\Gamma_\mu \ll 2\Delta\omega_0$) where we find larger values, at 1 MeV around 50 %. One could argue that in the approximate solution (and for $\Gamma_\mu < 2\Delta\omega_0$) one should define Γ_{rot} as the FWHM of the semicircle instead of half the second moment as it is done in ref. [45]. In this case at very small energies everything is fine, but at larger energy one gets into trouble, for instance at $U = 2$ MeV one would get for Γ_{rot} around 300 KeV instead of the 200 keV predicted by the exact solution, i.e. again 50 % off.

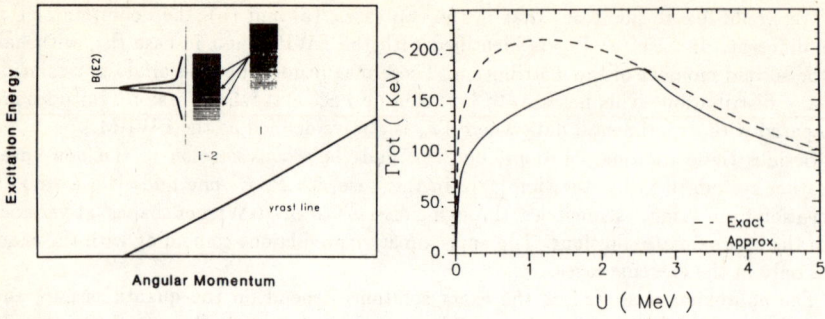

Figure 5: *Left hand side*: The decay of a damped rotational state (darkened) is illustrated, together with a strength function for a damped decay, from [55]. *Right hand side*: The rotational damping width Γ_{rot} as a function of the excitation energy for the exact and the approximate solutions.

Although there is some discrepancy between the approximate and the exact solution, the main feature of Γ_{rot}, i.e. its decrease with increasing energy, remains in both cases. This phenomenom has been related to the motional narrowing [53]. Experimentally, there is evidence [41] [42] [43] for the damping of the rotational motion; however, there are growing doubts [54] [55] about the existence of motional narrowing. From the two quantities entering into the model above, i.e. $\Delta\omega_0$ and Γ_μ, the first one has been calculated in several approximations [45] [51] [50] and, although they differ in magnitude, to use one or another approximation maintaining the same U dependence in Γ_μ, at most would shift the decreasing behaviour of Γ_{rot} in 1 MeV towards higher energies. Clearly the key quantity is Γ_μ, for which there is no reliable calculation and scarse experimental information. These results have been calculated using Γ_μ as given from the Fermi gas model, i.e. Eq. (15). Recently, it has been proposed [52] [48] that one should use constant values of Γ_μ, instead of the Fermi gas energy dependence, since the former is in better agreement with the experiment.

In Fig. 6 we show Γ_{rot} as calculated with a constant value of Γ_μ, and $\Delta\omega_0$ as given in Eq. (14). The value of Γ_μ used in fig. 6a has been obtained by setting $U = 1.5$ MeV in the expression (15). In the same way the Γ_μ values used in figs. (b-c-d) correspond to the Γ_μ values predicted by the Fermi gas model (see eq. (15)) at energies $U = 2.0, 2.5$ and 3.0 MeV, respectively. The dashed lines represent the exact solution, the continuous lines the approximate one, and the dotted ones the "unnarrowed" values as given by $2(2\Delta\omega_0)$ for all U values. The common feature to all four graphs consists in Γ_{rot} being a growing function of U. The fact that we do not find a decreasing value of Γ_{rot} with U does not mean, however, that there is no motional narrowing. As a matter of fact, there is motional narrowing when Γ_{rot} becomes smaller than the "unnarrowed" value $2(2\Delta\omega_0)$. In fig. 6, for cases (a) and (b), we find no motional narrowing. In fig. 6c there is narrowing for $U \leq 2.6$ MeV, and for damping widths $\Gamma_\mu = .12$ MeV (see fig. 6d) and larger, there is motional narrowing for all excitation energies. Widths as depicted in Fig. 6 probably would fit with the experimental data discussed in refs.[54] [55]. It would be also very interesting to make some further theoretical study without invoking the GOE to test this unjustified assumption.

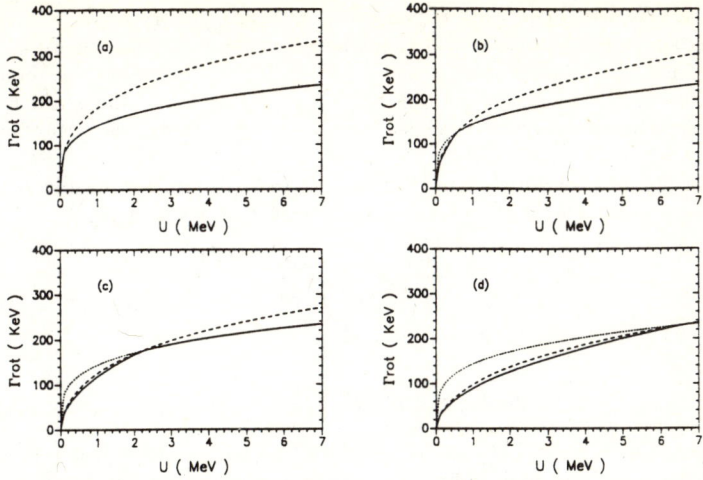

Figure 6: The rotational damping width Γ_{rot} as a function of the excitation energy for constant values of Γ_μ, see text for details.

4. Strength functions of hot rotating nuclei

Experimental investigations of the quasicontinuum γ-rays emited by a compound nucleus formed in a (HI, xn) reaction have opened the door to study high-spin states far above the yrast line. One of the open problems in the interpretation of the data relates to the path in the (E, I) plane on which a nucleus with excitation energy E and spin $\hbar I$ reaches the groud state via γ-emission. More precisely, we would like to know at each point in the (E, I) plane, the transition strengths for the various statistical $E1$, $M1$, and $E2$ decays that compete with the collective $E2$ decay within an excited rotational band roughly parallel to the yrast line.

Calculations based on simple models [56,57,58] unfortunately carry considerable uncertainties. In the statistical model, for instance, the energy dependence of the level density is crucial but is only poorly known . The same applies to the ratio of average $E1$ and $E2$ matrix elements [58]. On the other hand, an exact calculation, taking into account the decay properties of all states with spin $\hbar I$ in the energy interval between E and $E + dE$, is clearly out of the question in the continuum region.

In view of these difficulties, we have employed the following microscopic method to calculate the transition rates: We have combined the random-phase approximation (RPA) with the finite-temperature (FT) cranked Hartree-Fock-Bogoliubov (CHFB) approach in the formalism of the grand canonical ensemble. This is done first by solving the FTCHFB equations for given values of cranking frequency and temperature. Small oscillations around the minima so determined are subsequently taken into account in terms of the FTRPA.

The HFB approximation has been used in nearly all microscopic investigations of states near the yrast line [59,60], where it works very well. The extension to finite temperatures has likewise been very successful [61,62,63]. In principle, this approximation provides us with the information necessary to calculate transition strengths as functions of (E, I). We have added the RPA for two reasons. First, the mean-field approximation is not very suitable for describing collective states, while the RPA shifts the collective strengths of the low-lying vibrations and the giant resonances to their proper places. Second, in the RPA the Goldstone modes stemming from the symmetry broken in the mean-field approximation, separate exactly [64] from the normal modes. The Goldstone modes can be artificially shifted far away in energy; the shift removes the problem of spurious contributions to the transition strength [66]. Finally, the use of the grand canonical ensemble is justified because the level density is very high, already 1 or 2 MeV above the yrast fine.

The decay rate from initial state i with spin I_i to final state f with spin I_f through emission of a photon with multipolarity L and energy E_γ is given by [1]

$$T_{fi}(E_\gamma) = \frac{8\pi(L+1)}{\hbar L[(2L+1)!!]} \left(\frac{E_\gamma}{\hbar c}\right)^{2L+1} B_{fi}(L) \qquad (16)$$

where $B_{fi}(L)$ is the reduce transition probability

$$B_{fi}(L) = \frac{1}{2I_i+1} |\langle fI_f \| \mathcal{M}_L \| iI_i \rangle|^2 \qquad (17)$$

The decay rate for a nucleus *at finite temperature* T for emission of a photon of multipolarity L and energy E_γ is obtained from Eq. (16) by averaging over the initial states i with fixed spin I_i with the thermal occupation probabilities $p_\mu(I_i)$ and allowing for transitions into all states f that have fixed spin I_f and that are accesible by energy conservation

$$T(E_\gamma, I_i \to I_f) = \sum_{\mu\nu} p_\mu T_{\mu\nu}(E_\gamma) \delta(E_\gamma - [E_\nu(I_f) - E_\mu(I_i)])$$

$$= \frac{8\pi(L+1)}{\hbar L[(2L+1)!!]} \left(\frac{E_\gamma}{\hbar c}\right)^{2L+1} S(E_\gamma, I_i \to I_f) \qquad (18)$$

with the strength function given by

$$S(E_\gamma, I_i \to I_f) = \sum_{\mu\nu} p_\mu \frac{1}{2I_i+1} |\langle \nu(I_f) \| \mathcal{M}_L \| \mu(I_i) \rangle|^2 \delta(E_\gamma - (E_\nu(I_f) - E_\mu(I_i))) \qquad (19)$$

this formula holds in the laboratory system, i.e. with eigenstates of the angular momentum. Writing it in the rotating system requires two changes [66,67]. (i) The axis of quantization must be chosen along the rotation axis. Then, the negative of the m-quantum number of the tensor operator $\mathcal{M}$ coincides with the difference $(I_f - I_i)$, between final and inital spins. (ii) The energy argument E_γ in the strength function must be shifted by $(I_f - I_i)\omega$, where ω is the cranking frequency. The result is given by

$$S(E_\gamma, I_i \to I_f) = S_{int}(E_\gamma - \omega(I_f - I_i), I_i \to I_f) \qquad (20)$$

with

$$S_{int}(E, I_i \to I_f) = \sum_{\mu\nu} p_\mu |\langle \nu(I_f) \| Q_{L,I_f-I_i} \| \mu(I_i) \rangle|^2 \delta(E - [E_\nu(I_f) - E_\mu(I_i)]) \qquad (21)$$

and

$$Q_{L,I_f-I_i} = \sum_m d^L_{I_f-I_i,m}(\pi/2) \mathcal{M}^{int}_{Lm} \qquad (22)$$

A crucial simplification is obtained using the response function formalism and separable forces. In the intrinsic system the finite temperature response funtion is given by

$$R_{\sigma\sigma'\tau\tau'} = \sum_{\mu\nu} p_\mu \left(\frac{\langle \mu|a^\dagger_{\sigma'}a_\sigma|\nu\rangle\langle\nu|a^\dagger_\tau a_{\tau'}|\mu\rangle}{E - (E_\nu - E_\mu) + i\eta} - \frac{\langle\mu|a^\dagger_\tau a_{\tau'}|\nu\rangle\langle\nu|a^\dagger_{\sigma'}a_\sigma|\mu\rangle}{E + (E_\nu - E_\mu) + i\eta} \right) \qquad (23)$$

Here, $a_\sigma^\dagger$ and a_σ are quasiparticle operators. Writing Q as a matrix in quasiparticle space, and defining

$$R(Q_{L,I_f-I_i}; E; I_i \to I_f) = Tr(Q_{L,I_f-I_i}^\dagger \mathcal{R} Q_{L,I_f-I_i}) \qquad (24)$$

one finaly obtains

$$S_{int}(E, I_i \to I_f) = -\frac{1}{\pi}\frac{1}{1-e^{-E/T}}\Im m\, R(Q_{L,I_f-I_i}; E; I_i \to I_f) \qquad (25)$$

The separable interaction allows to solve the Lippmann-Schwinger equation

$$\mathcal{R} = \mathcal{R}^0 + \mathcal{R}^0\, \mathcal{W}\, \mathcal{R} \qquad (26)$$

and $\mathcal{R}$ can be expressed in terms of $\mathcal{R}^0$, the free response function. The matrix $\mathcal{W}$ takes into account the residual interaction not contained in the mean-field approximation. The free response function $\mathcal{R}^0$ is given explicitly in terms of the quasiparticle energies E_μ and the temperature-dependent occupation probabilities f_μ by

$$\mathcal{R}^0 = \frac{2(f_\nu - f_\mu)}{E - E_\mu + E_\nu + i\eta}\delta_{\mu\mu'}\delta_{\nu\nu'} \qquad (27)$$

We note in passing that by putting $\mathcal{W} = 0$, and by thereby replacing $\mathcal{R}$ in Eq. (25) by $\mathcal{R}^0$, we obtain the strength functions (and transition probabilities) in the HFB approximation.

In the framework of the RPA, the parameter η in Eq. (27) ought to be put equal to zero. This would produce poles in R^0 and R at the positions of the HFB and of the RPA transition energies, respectively, and a corresponding spike structure of delta-function type in the strength function Eq. (25). For the numerical calculations (where we proceed in finite energy steps ΔE) this choice of η is obviously unacceptable. Choosing η finite is tantamount to smearing each state over a finite energy interval of width $\Gamma = 2\eta$. Such a choice is indicated by theories beyond the RPA which identify Γ with a spreading width, and is therefore physically reasonable. Choosing $\Gamma \gg \Delta E$ we are sure to pick up all the strength numerically. In the calculations, the condition $\Gamma \gg \Delta E$ had to be put in balance with two further requirements: (i) CPU time obviously grows strongly with decreasing ΔE; this puts a lower bound in ΔE. (ii) Values of Γ in excess of several tens of keV are physically unrealistic; they would smear the strength (especially that from the numerous dense-lying states at higher excitations energy) over too wide a region. We recall that R^0 in Eq. (27) has lorentzian shape and a very soft cut-off in the tails.

The RPA scheme just described allows for the calculation of all transition strengths but one: The collective intraband transition. Indeed, within the RPA the latter has the form of a transition from ground state to ground state. But the response function formalism describes only transitions connecting the ground state with an excited state. Fortunately, this difficulty is easily removed. The collectivity of the transition in question is basically due to deformation. Therefore, it can reliably be calculated in the HFB approximation. This is what we have done throughout.

We have applied this scheme to the calculation of $E1, M1$, and $E2$ transitions in the deformed heavy nucleus ^{164}Er [65] and the transitional ^{156}Er [66]. Except for one modification mentioned below, we have used the same hamiltonian and the same configuration space as Baranger and Kumar [68]. This hamiltonian is known to reproduce both ground-state deformations and gap parameters in the rare earth region very well. It has also been used to investigate properties of nuclei in this region at finite temperature [60,69]. In the Kumar-Baranger hamiltonian, the residual interaction $\mathcal{W}$ is separable.

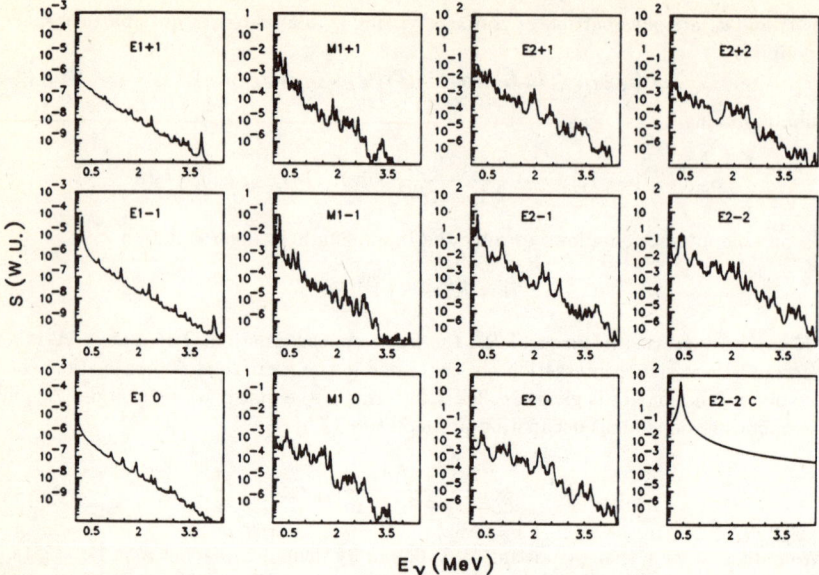

Figure 7: Strength functions (in Weisskopf units) for gamma emission versus gamma energy (in MeV) as labeled in the inserts in the RPA approximation for $\hbar\omega = 0.2 MeV$ and $T = 0.4 MeV$. The selfconsistent parameters are $\beta = 0.202, \gamma = -3.28^0, \Delta_p = 1.30 MeV, \Delta_n = 0.02 MeV, I = 14.42, E^* = 4.51 MeV$

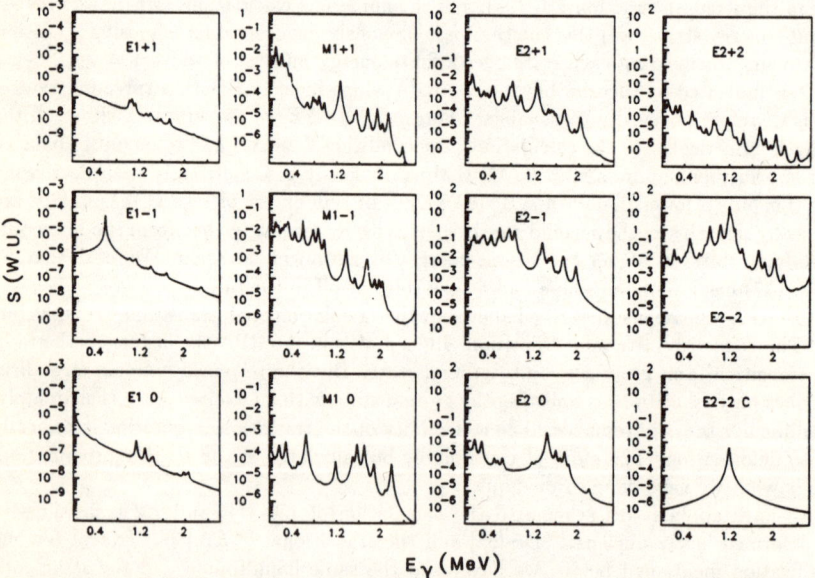

Figure 8: Strength functions (in Weisskopf units) for gamma emission versus gamma energy (in MeV) as labeled in the inserts in the RPA approximation for $\hbar\omega = 0.6 MeV$ and $T = 0.4 MeV$. The selfconsistent parameters are $\beta = 0.123, \gamma = -37.09^0, \Delta_p = 0.00 MeV, \Delta_n = 0.00 MeV, I = 50.0, E^* = 2.60 MeV$

This facilitates solving the Lippmann-Schwinger Eq. (26). We have modified the Kumar-Baranger hamiltonian by adding a dipole-dipole term. The strength of this term was chosen so that the giant dipole resonance in spherical nuclei is close to the experimental value [67]. Adding such a term does not change the self-consistent solutions of the original hamiltonian since the mean-field expectation value of the dipole operator vanishes.

For a given point in the (E, I) plane, our calculations yield 11 strength functions ($2L + 1$ functions for each operator of multipolarity L), each one being a function of the energy E_γ of the emitted γ-ray. A detailed disccussion of the results can be found in ref. [65] for ^{164}Er and in ref. [66] for ^{156}Er. Concerning the comparison of the RPA and HFB calculations one finds that the RPA results for the $E1$ strengths are typically one order of magnitude smaller than those for the HFB. This is because the RPA shifts $E1$ strength to the giant dipole resonance. No such difference exists for the $M1$ transition because not much magnetic vibrational collectivity is expected in those nuclei. For a good rotor, the collective $E2$ strength is much bigger than the statistical one. However, the statistical $E2$ RPA is much increased over its HFB value.

As an illustrative example of the numerical results we show the emission strength functions for ^{156}Er at $T = 0.4$ MeV, $\omega = 0.2$ MeV (Fig. 7) and at $T = 0.4$ MeV, $\omega = 0.6$ MeV (Fig. 8) for the 11 statistical transitions plus the stretched collective one (E2-2 C) located at $E = 2\hbar\omega$ in the RPA approximation. To facilitate the comparison between this $E2$ strength and the other strength functions, we have smeared it out with a Breit-Wigner function, using the same value for the width, $\Gamma = 30$ KeV, as in the rest of the calculation where we also used $\Delta E = 10$ KeV. The inset in each box indicates the type of transition, for instance the box with the inset E1+1 gives the emision strength function E1 from $I \rightarrow I - 2$ as a function of the energy of the emitted photon. The maximum γ-ray energy is always determined by the excitation energy E^* of the emitting nucleus (the energy above the yrast line). The strength functions are shown on a logarithmic scale in W.U.

A feature clearly visible in some of the figures, especially those for the E1 strength, is an increase of the high-energy parts of the strength functions (relative to the low-energy) parts as ω (or I) increase. We ascribe this to increased mixing of levels due to the Coriolis force. As we can see in the figure captions the nucleus ^{156}Er is prolate at $\omega = 0.2$ MeV and oblate at $\omega = 0.6$ MeV, if we compare the corresponding strength functions in both figures we see that in general they are quite different. The most striking differences are in the $E2$'s. In the prolate nucleus we find large collective transition probability (E2-2 C) and small values for the statistical one (E2-2). In the oblate one we find almost no collective strength (E2-2 C) and rather large values for the statistical one (E2-2).

To compare with the statistical model, we display in fig. 9 ratios of RPA strength functions versus E for $T = 0.7 MeV$ and for the given rotational frequencies for the nucleus ^{164}Er. To smooth out the strong fluctuations of the RPA solutions with E we have used a larger value of Γ (100 KeV) than before. We note that the dependence on the level density largely cancels, so that the ratios are also indicative of the relative size of the transition matrix elements and, in the case of the $E1/M1$ ratio, of the ratio of the transition probabilities (the energy dependence of dipole emission is independent of the parity of the radiation) except for an overall factor of about 100. Taking into account this factor, we see that for the dipole modes at $\hbar\omega = 0.075$ MeV, magnetic emission dominates for $E < 3 MeV$ while it is the more important for almost all energies at $\hbar\omega = 0.575$ MeV. Both the $E1/M1$ and $E1/E2$ ratios show a strong secular increase with E_γ over several orders of magnitude. We ascribe this to the tails of the giant dipole resonance, but superimposed on this trend, and clearly visible also in the $M1/E2$ ratios, are very sizeable fluctuations easily amounting to 1 or 2 orders of magnitude. It is well to remember that in the pure statistical model, all three ratios should be straight horizontal lines; for the

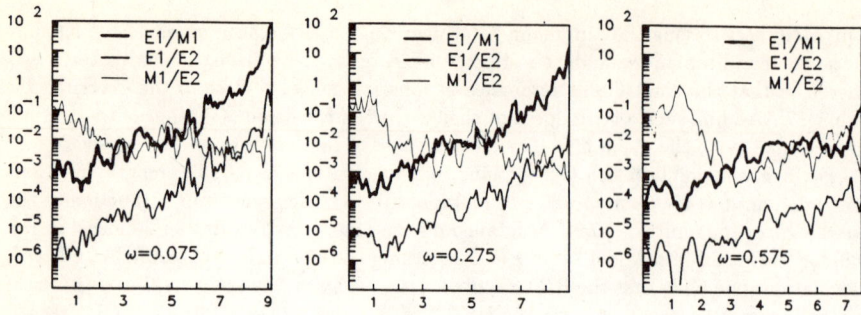

Figure 9: The RPA strength-function ratios S(E1)/S(E2), S(E1)/S(M1) and S(M1)/S(E2) versus γ-ray energy (in MeV) on a logarithmic scale for $T = 0.7$ MeV as explained in the text.

E1/M1 and E1/E2 ratios this behaviour would change if the giant E1 resonance were accounted for.

Calculations of the type shown here can be carried oit on an entire grid covering part of the (E^*, I) plane. Such calculations are needed in order to follow systematically the deexcitation by gamma emision of a compound nucleus formed by a heavy-ion-induced reaction. It is obvious from the results above that specially in a transitional nucleus, the construction of such a grid cannot be performed reliably with the help of simple models like the statistical and/or the collective model.

I am indebted to C. Esebbag, V. Martin, P. Ring, L. M. Robledo and H. A. Weidenmüller for the collaborations on which this lectures are based.

I would like to express my gratitude to J.F. Berger, M. Girod and D. Gogny of the Centre d'Etudes Nucleaires de Bruyeres le Chatel for their kindness in allowing us to use their original Hartree–Fock code and also to thank them for helpful discussions and for sending me relevant preprints on superdeformation.

REFERENCES

[1] A. Bohr and B. Mottelson, *Nuclear Structure. Vol II* (Benjamin, Reading, 1975), 19

[2] J. F. Sharpey-Shafer and J. Simpson. Prog. Part. Nucl. Phys. **21**(1988)293

[3] S. Aberg, H. Flocard and W. Nazarewicz, Ann. Rev. Nucl. Part. **40**(1990)

[4] Proceedings of the Conference on Nuclear Structure in the Nineties, Oack-Ridge, Tennessee, 1990, Ed. N. R. Johnson, Nucl. Phys. **A520**(1990)

[5] Proceedings of the Workshop and Symposium " Future directions in Nuclear Physics with 4π Gamma detection System of the New Generation", Strasbourg(1991), Eds. J. Dudek et al.

[6] W. Kurcewicz et al. Nucl. Phys. **A356**, 15(1981)

[7] P. Moller and J. R. Nix. Nucl. Phys. **A361**, 117 (1981)

[8] R.R. Chasman. Phys. Lett. **96B**, 7 (1980)

[9] G. A. Leander et al. Nucl. Phys.**A388**, 452 (1982)

[10] W. Nazarewicz, P. Olanders, I. Ragnarsson, J. Dudek, G.A. Leander, P. Moller and E. Rochowska. Nucl. Phys. **A429**, 269 (1984)

[11] W. Nazarewicz et al. Phys. Rev. Lett. **52**, 1272 (1984); **53**, 2060 (1984)

[12] G.A. Leander, W. Nazarewitcz, P. Olanders, I. Ragnarsson and J. Dudek. Phys. Lett. **152B**, 284 (1985)
S. Frauendorf and V.V. Pashkevich. Phys. Lett. **B141**,23 (1984); **B152**, 289 (1985).

[13] P. Bonche, P.H. Heenen H. Flocard and D. Vautherin. Phys. Lett **B175**, 387 (1986)

[14] P. Bonche. *The Variety of Nuclear Shapes*. World Scientific Press. J.D. Garret et al. eds. p 302

[15] J.L. Egido and L.M. Robledo. Nucl. Phys. **A494**, 85 (1989)

[16] J.L. Egido and L.M. Robledo. Nucl. Phys. **A518**, 475 (1990)

[17] D. Gogny. *Nuclear Selfconsistent fields*. Eds. G. Ripka and M. Porneuf (North Holland 1975).

[18] P. Ring and P. Shuck. *The Nuclear Many Body Problem* . Springer–Verlag Edt. Berlin 1980.

[19] P.G. Reinhard and K. Goeke. J. of Phys. **G4**, 245 (1978)

[20] M. Girod and B. Grammaticos. Nucl. Phys. **A330**, 40 (1979)

[21] L.M. Robledo, J.L. Egido, B. Nerlo–Pomorska and K. Pomorski. Phys. Lett. **B201**, 409 (1988)

[22] J.F. Berger, M. Girod and D. Gogny. Nucl. Phys. **A428**, 23c (1984)

[23] J.L. Egido and L.M. Robledo. *Proceedings of the XII Workshop on Nuclear Physics*, Iguazu Falls, Argentina (World Scientific 1990 Eds. M.C. Cambiaggio and A.J. Kreiner)

[24] B. Nerlo–Pomorska, K. Pomorski, M. Brack and E. Werner. Nucl. Phys. **A462**, 252 (1987)

[25] P.A. Butler and W. Nazarewicz. University of Liverpool preprint (1991) and references therein.

[26] R.H. Spear. Atomic data and Nuclear Data Tables **42** , 55-104 (1989)

[27] P.J. Twin et al., Phys. Rev. Lett. **57** (1986) 2141.

[28] C.C. Anderson et al, Phys. Scr. **24** (1981) 266.

[29] S. Aberg, Phys. Scr. **25** (1982) 25.

[30] J. Dudek and W. Nazarewicz, Phys. Rev. **C31** (1982) 298.

[31] R.R. Chasman, Phys. Lett. **187 B** (1987) 219.

[32] R.R. Chasman, Phys. Lett. **219 B** (1989) 232.

[33] M. Girod, J.P. Delaroche, D. Gogny, and J.F. Berger. Phys. Rev. Lett. **62** (1989) 2452.

[34] J. P. Delaroche, M. Girod, J. Libert and I. Deloncle. Phys. Lett. **232** (1989) 145.

[35] P. Bonche et al., Nucl. Phys. **A510** (1990) 466, Nucl. Phys. **A519** (1990) 509.

[36] J. K. Johansson et al. Phys. Rev. Lett. **63** (1989) 2200.

[37] W. Nazarewicz, R. Wyss, A. Johnson, Nucl. Phys. **A503** (1989) 80.

[38] T. Byrski et al, Phys. Rev. Lett. **64** (1990) 1650.

[39] F. S. Stephens et al. Phys. Rev. Lett. **65** (1990) 301.

[40] J. Dudek, in "The Variety of Nuclear Shapes", Proc. Int. Conf. on Nuclear Shapes, Crete, ed. J. D. Garret. Would Scientific, Singapore, 1988.

[41] J. C. Bacelar, G. B. Hagemann, B. Herskind, A. Holm, J. C. Lisle and P.O. Tjom Phys. Rev. Lett. **55** (1985) 1858

[42] J. E. Draper, E. L. Dines, M. A. Deleplanque, R. M. Diamond and F. S. Stephens, Phys. Rev. Lett. **56** (1986) 309.

[43] I. Y. Lee *et al*, Proceedings of the Conference on High-Spin Nuclear Structure and Novel Nuclear Shapes, April 1988, ANL-PHY-88-2.

[44] G. A. Leander, Phys. Rev. **C25** (1982) 2780

[45] B. Lauritzen, T. Dossing and R. A. Broglia, Nucl. Phys. **A447**(1986)61

[46] F. S. Stephens and R. S. Simon Nucl. Phys. **A183**(1972)257

[47] A. Bohr and B. R. Mottelson, Nuclear Structure, p. 302, Benjamin, Reading, Mass., 1975.

[48] J. L. Egido and A. Faessler Z. Phys. **A339** (1991) 115.

[49] R. A. Broglia, Frontiers and Borderlines in Many-Particle Physics, CIC Corso, Soc. Italiana di Fisica, Italy, 1988.

[50] S. Vydrug-Vlasenko, R. A. Broglia, T. Dossing and W. E. Ormand, Proceedings of the Conference on High-Spin Nuclear Structure and Novel Nuclear Shapes, April 1988, ANL-PHY-88-2.

[51] J. L. Egido, Report of the Workshop on Nuclear Structure, Niels Bohr Institute, May 1986.
Phys. Lett. **B232** (1989) 1.

[52] I. Hamamoto and B. R. Mottelson in Contemporary Topics in Nuclear Structure Physics, Ed. R. F. Casten, A. Frank, M. Moshinsky and S. Pittel, World Scientific, 1988.

[53] R. A. Broglia, T. Dossing, B. Lauritzen and B. R. Mottelson, Phys. Rev. Lett. **58** (1987) 326.

[54] F. S. Stephens, J. E. Draper, M. A. Deleplanque, R. M. Diamond and A. O. Macchiavelli, Phys. Rev. Lett. **60**(1988)2129

[55] F. S. Stephens, in Contemporary Topics in Nuclear Structure Physics, Ed. R. F. Casten, A. Frank, M. Moshinsky and S. Pittel, World Scientific, 1988.

[56] R.J. Liotta and R.A. Sorensen, Nucl. Phys. **A297**(1978) 136.

[57] G. Leander, Y.S. Chen and B.S. Nilsson. Phys. Ser. **24** (1981) 164.

[58] M. Wakai and A. Faessler, Nucl. Phys. **A307** (1978) 349.

[59] A.L. Goodman, Advances in Nuclear Physics, Vol. 11, eds. J. Negele and E. Voigt (Plenum, New York, 1979).

[60] J.L. Egido, H.J. Mang and P. Ring, Nucl. Phys. **A339**(1980) 390.

[61] A.L. Goodman, Nucl. Phys. **A369**(1981)365, **A370**(1981) 90.

[62] K. Tanabe, K. Sugawara-Tanabe and H.J. Mang, Nucl. Phys. **A357** (1981) 20, 45.

[63] J.L. Egido P. Ring and H. J. Mang. Nucl. Phys. **A451** (1986) 77.

[64] J.L. Egido and J. O. Rasmussen, Nucl. Phys. **A476** (1988) 48.

[65] J.L. Egido and H. A. Weidenmueller, Phys. Lett. **208**(1988)58

[66] J.L. Egido and H. A. Weidenmueller, Phys. Rev. **C39**(1989)2398

[67] P. Ring. L.M. Robledo, J.L. Egido and M. Faber, Nucl. Phys. **A419** (1984) 261.

[68] B. Baranger and K. Kumar, Nucl. Phys. **A110** (1968) 490.

[69] A. Goodman, Phys. Rev. **C34**(1986) 1942.

Recent Spectroscopy of Exotic Nuclear Shapes at High Spin

M.A. Bentley

SERC Daresbury Laboratory, Daresbury
Warrington WA4 4AD, U.K.

Abstract : Recent advances in experimental high spin physics are discussed. The advent of arrays of high performances Compton-suppressed Germanium detectors has allowed weakly populated exotic structures in nuclei to be observed. The observation of discrete-line superdeformed bands at the highest known spins has posed many physics questions which are currently being addressed, and some of these aspects of superdeformation are discussed. Rotational bands associated with superdeformed shapes have been observed in the neighbouring nuclei, and in some cases the transition energies are identical to 1 part in 1000. This implies that the nuclear moment of inertia can remain unchanged with respect to the addition or removal of a particular nucleon. The next generation of γ-ray detector arrays for the 1990's is discussed.

1 Introduction

In these lectures, I will give an experimental overview of the study of nuclei at high spin, and in the spirit of the title of the Summer School, will discuss the work currently underway at the "borderlines" of high spin physics.

It was suggested at the beginning of the century that nuclei could adopt a variety of shapes, and that the study of gamma-ray transitions between rotational states in nuclei could yield valuable information on fundamental properties such as the nuclear shape. Since these original suggestions, huge advances have been made, and a wealth of new physics has been uncovered.

Probably the most significant advances in experimental high spin work have been made in the last decade with the development of large arrays of high resolution, low signal-to-noise gamma-ray detectors, in the form of arrays of Escape Suppressed Spectrometers (or ESS's) [1,2]. This has enabled measurement of structures in nuclei which are weakly populated in the reaction, and has resulted in the observation of exotic shape effects such as superdeformed rotating nuclei with a major-to-minor axis ratio of 2:1 [3,4,5]. Simultaneously, development of detailed models of the nucleus, such as deformed mean-field calculations which can be compared with experimental observations, has also played a major role in the recent renaissance of low temperature nuclear structure.

In the following sections I shall give a flavour of some of the recent experimental developments in high spin physics, and demonstrate what we have learnt about nuclear shapes. Latest results from the "borderlines" of nuclear structure will be discussed, and in particular I will describe the fascinating effect of identical superdeformed bands in nuclei. This phenomenon has demonstrated that under certain conditions nuclei can have underlying symmetries which govern their overall behaviour. The future of gamma-ray spectroscopy lies in attempting to observe new effects in nuclei which may be populated only very weakly in the reaction, and hence much more sensitive detector systems are required. In the final section I will describe the experimental development currently taking place world-wide which will open up research into effects in nuclei way beyond our present borderlines.

2 Recent advances in experimental high spin physics

One of the most effective ways of populating high spin states in nuclei is the fusion evaporation reaction, which produces states in the compound nucleus which are both highly excited in energy and in states of high angular momentum. Rich nuclear structure information can be derived from the study of the gamma-radiation emitted in these reactions, although the efficient and accurate measurement of these gamma-rays can be difficult. Germanium semi-conductor detectors are used for accurately measuring the energy of the gamma-ray. However, when a gamma-ray interacts in the Ge crystal, it is most likely that it Compton scatters from one of the atomic sites within the crystal, depositing only part of the full energy of the gamma-ray. This results in a huge background in the gamma-ray spectrum, which must be rejected. This is achieved by surrounding the Ge crystal with a dense scintillation crystal [6] (such as bismuth germanate - BGO) which records the Compton scattered gamma-ray, and rejects the event when a signal is recorded in both the Ge and BGO.

This composite detector - the Escape Suppressed Spectrometer (ESS [1]) - results in a major improvement in the quality of the gamma-ray spectrum, shown in figure 1, in which the size of the background is considerably reduced with no loss of intensity in the peak. In a $\gamma - \gamma$ coincidence experiment, this results in a factor of 10 improvement in the ratio of useful events to background events. Arrays of ESS's measuring $\gamma - \gamma$ coincidences which were pioneered during the early 1980's have revolutionised gamma-ray spectroscopy, and have culminated in the current highly sucessful large ESS arrays. Examples of this are the Tessa3 [7] and PolyTessa (or ESSA30) arrays at Daresbury, shown in figure 2. Tessa3 consists of 16 ESS's and a 50 element BGO "crystal ball" [7] for measuring the total energy and multiplicity of the gamma-rays emitted (this helps select the nucleus from which the gamma-rays were emitted). Many other large ESS arrays exist today [2], all of which are extremely successful in their own right.

The $\gamma - \gamma$ coincidence technique enables the energy level scheme of a nucleus to be constructed, and the relatively high efficiency of the ESS arrays for measuring two γ-rays

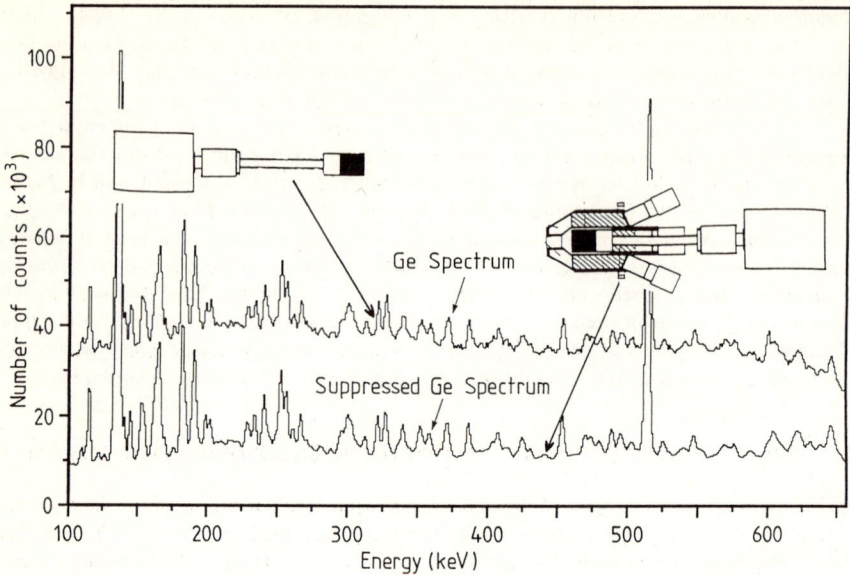

Figure 1: A typical spectrum from a heavy ion fusion reaction measured using a single Ge detector. The upper spectrum was recorded using a bare detector, and in the lower spectrum, a suppression shield was used to record and reject events in which the γ-ray Compton scatters in the Ge crystal.

simultaneously has enabled weak structures in nuclei to be observed. One of the most important consequences of this is that it is now possible to observe structures towards the highest values of spin that the nucleus can tolerate before fissioning.

One of the most important questions that must be addressed by studies of low temperature nuclear structure is - what shapes do nuclei adopt to achieve their lowest state of energy? This question has been addressed in some depth using ESS arays, and the answers are surprising complex, depending in subtle ways on the numbers of protons and neutrons. The power of the ESS arrays in elucidating nuclear shape effects at high spin can perhaps be best demonstrated by considering the case of a "classic nucleus", ^{158}Er.

The spectrum of gamma-rays emitted from the excited states of ^{158}Er is shown in figure 3, in which gamma-rays are observed between states of spin values ranging from 2 to $46\hbar$. The spectrum for gamma-decays near to the ground state shows a regular rotational structure (ie $E_{rot} \propto I(I+1)$) up to around spin $12\hbar$, suggesting that at low spins, the nucleus is rotating and has a prolate deformation. At these low spins the individual protons and neutrons are fully paired - ie they interact through the monopole pairing interaction, and occupy time reversed orbits with each pair coupled to zero spin.

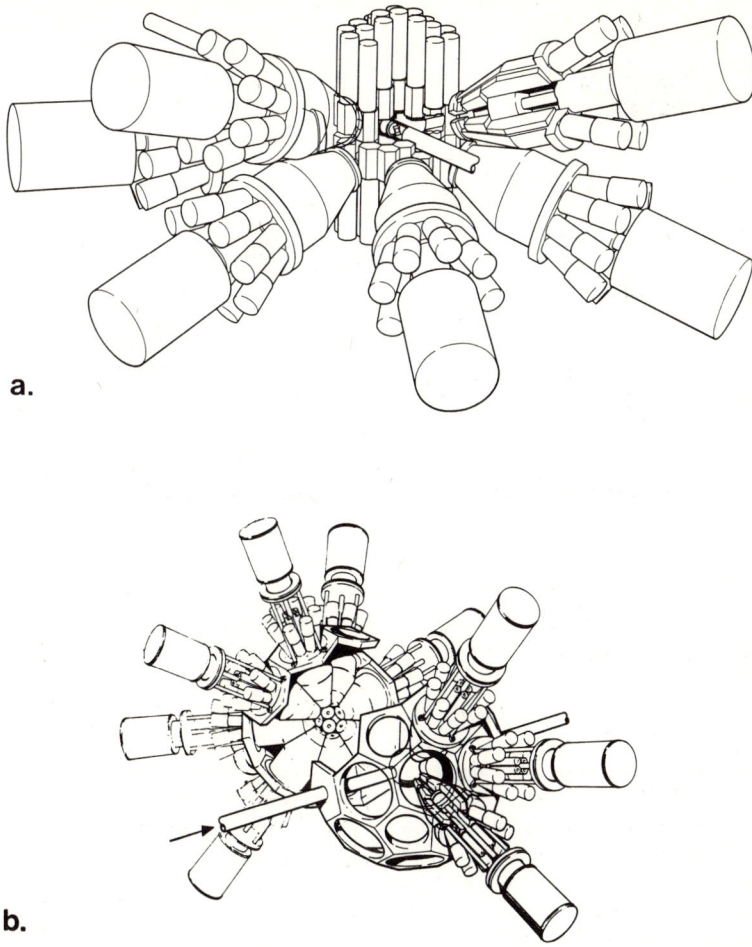

Figure 2: (a) The Tessa3 γ-ray detector array at Daresbury Laboratory, consisting of 16 Escape Suppressed Spectrometers (ESS) and 50 element BGO crystal ball. (b) The Daresbury PolyTessa array, with a framework capable of taking 30 ESS's, which was achieved in the ESSA30 collaboration during 1987.

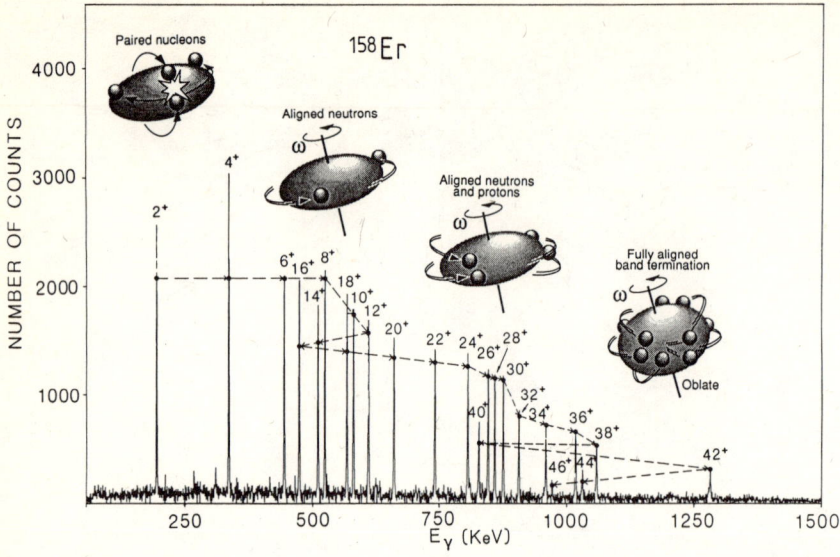

Figure 3: A spectrum of the yrast states in the nucleus ^{158}Er. The spin and parity of the state emitting the γ-rays is marked.

All the angular momentum of the nucleus is generated by the collective rotation of the entire system. At spin $12\hbar$, a discontinuity appears, and the spectrum no longer follows the rotational pattern, and the gamma-ray energy briefly decreases with increasing spin. This effect is known as backbending, and is now known to be caused by the breaking of a pair of specific nucleons. This occurs when the Coriolis force exerted on the nucleons due to the collective rotation is greater than the pairing force which would tend to keep the nucleons in time-reversed orbits. When this happens, the angular momenta of the two nucleons are aligned with the rotational axis of the nucleus contributing to the total angular momentum, and hence causing the backbending phenomeon. As the Coriolis force is greatest for a particle with large angular momentum, the first pair of particles to align is always a pair of high-j nucleons lying near the Fermi surface - in this case a pair of $i_{13/2}$ neutrons. Above spin $16\hbar$, the spectrum again returns to the rotational pattern, until a second backbend occurs - this time caused by the alignment of a pair of $h_{11/2}$ protons at spin $26\hbar$ [2,8]. Moving to yet higher spins the spectrum once again settles down into a regular rotational sequence until at spin $38\hbar$, a different effect occurs. Above this point the spectrum becomes extremely irregular, no longer exhibiting behaviour expected of a rotational nucleus. This effect is now known to be due to a drastic change in shape from a prolate to an oblate shape. In these oblate states, no collective rotation occurs, and

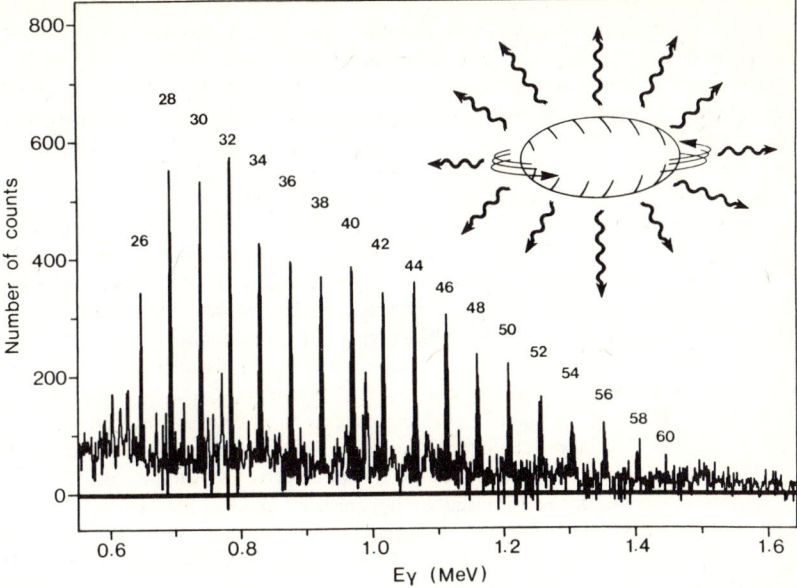

Figure 4: The superdeformed band in the nucleus ^{152}Dy [3-5]. The spectrum corresponds to a rapidly rotating nucleus with a major-to-minor axis ratio of around 2:1. The γ-rays are marked with the estimated spins of the emitting states.

the angular momentum is generated simply by occupation of many different single particle levels. The spin $46\hbar$ state, above which no more γ-rays are observed, corresponds to a complete alignment of all the available nucleons and is known as the band termination.

The case of ^{158}Er demonstrates that not only is it possible to infer properties of the bulk nucleus from gamma-ray techniques (such as shapes, shape coexistence, collective rotation etc.) but also to gain significant insight into how individual nucleons or pairs of nucleons affect the behaviour of the entire nucleus through Coriolis and alignment effects.

3 Superdeformed shapes at high spin

The increased sensitivity of the present generation of gamma-ray detector arrays has allowed weakly populated structures at high spin to be investigated (eg ^{158}Er above). This has resulted in one of the most startling discoveries in low temperature nuclear physics - the observation of superdeformed shapes at high spin in heavy nuclei. In an experiment [3,5] at Daresbury on the Tessa3 array, a rotational band was observed in the nucleus ^{152}Dy, which displays some remarkable properties (see figure 4). An extremely regular rotational sequence of 19 γ-rays was observed extending from spin $22\hbar$ to spin $60\hbar$ (some $15\hbar$ higher in spin than had previously been observed). The spectrum displays the properties of a rigid rotational nucleus, and the spacing between the gamma-rays implies an extremely high moment of inertia - hence a large deformation. The quadrupole moment of the nucleus for these rotational states has been measured at (18 ± 3)eb [4,5] which corresponds to a superdeformed prolate shape with an axis ratio of 2:1. The case of the superdeformed band in ^{152}Dy is now well documented, and although many cases of superdeformed bands have subsequently been discovered in this mass region (eg [9-12]), there are still many properties of this remarkable phenomenon that are not well understood. Some of these will be discussed later in this section. In fact, before the discovery of the superdeformed band in ^{152}Dy, a rotational band had been observed in ^{132}Ce [13] (also at Daresbury using the Tessa3 array). This band has many of the properties of the superdeformed band in ^{152}Dy, and although the deformation is somewhat smaller - an axis ratio of 3:2 - this was in fact the first observation of a discrete line highly-deformed band.

Superdeformed nuclei were in fact well known before the discovery of these superdeformed bands. Almost 30 years ago in the A = 240 region, superdeformed prolate shapes were observed in the fission isomers [14]. These superdeformed states are favoured in energy due to the fact that large "shell-gaps" appear in the single particle spectrum of states for axis ratios around 2:1 [15], causing a low energy relative to other states. These shell effects cause a minimum in the potential energy surface - the so called second minimum - which stabilises the superdeformed shape. Transitions have been observed [16] between the superdeformed states at very low spins prior to fission. Similarly, around A = 150, there are also shell gaps for protons and neutrons which tend to give a negative shell energy for the superdeformed shape. However, unlike the fission isomer region, there is a significant fission barrier at large deformations which entirely dominates the potential energy surface at low spins. This is demonstrated in figure 5(a) which shows a portion of the potential energy surface calculated from a Woods-Saxon model [17] for ^{152}Dy at low spin ($10\hbar$). Only one minimum is evident at low spins corresponding to a low deformation prolate shape. However, it is well known that if a nucleus is rotated, the total energy is lowered for large deformations, effectively pushing the fission barrier to a larger deformation. In this case, the low shell energy begins to have an effect, and a second minimum appears in the potential energy surface for a superdeformed shape. This is demonstrated in figure 5(b) which shows the same calculations for spin ($60\hbar$). The fact that the second minimum appears suggests that if the nucleus can be populated at very high spins, then one may expect to see effects associated with superdeformation. Indeed the range of angular momenta over which the superdeformed band is observed corresponds well to the potential energy calculations of figure 5. However, whilst the calculations show that ^{152}Dy is indeed a favourable case in which to see such a superdeformed shape, many of the remarkable features of the band remain unexplained.

Measurements [5] of the intensity of the gamma-ray transitions between superdeformed states have shown that the superdeformed states are populated with an intensity of around 1% of the total population of ^{152}Dy - the relative intensities of the gamma-rays are plotted in figure 6. The data show that the superdeformed band is populated only at very high spins between 50 and $60\hbar$, implying that the superdeformed states are yrast only in this spin range. At lower spins the intensity of the band remains constant at around 1% down to around spin $30\hbar$ at which point the band disappears. One of the most fascinating features

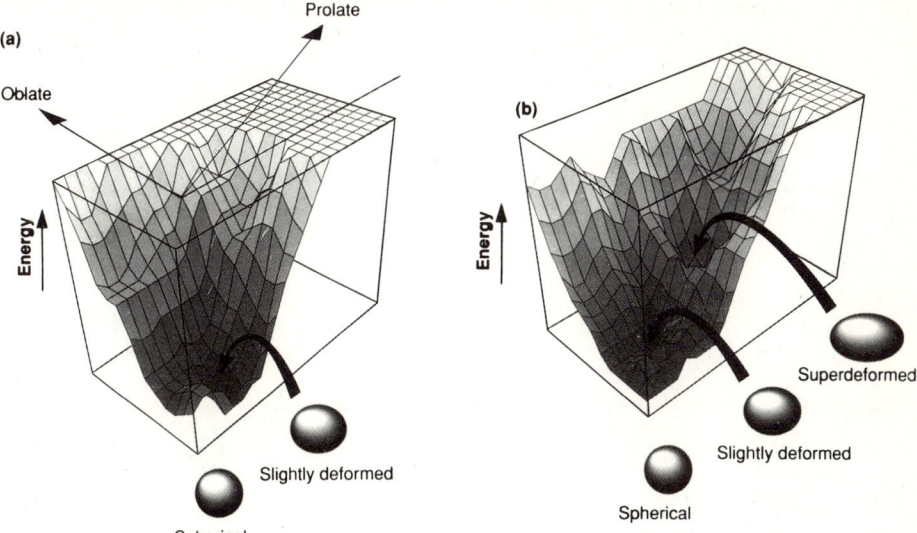

Figure 5: A portion of the potential energy surface calculated from a cranked Woods-Saxon code [17] for ^{152}Dy as a function of deformation. The near left hand corner corresponds to spherical shapes and other points on the surface represents deformed shapes for prolate and oblate nuclei as marked. (a) shows the calculation for spin $10\hbar$ and (b) for $60\hbar$.

of the data in figure 6 is the intensity with which the states are populated at the highest spins. For comparison, the relative intensities of normal deformation rotational bands in two other nuclei (^{160}Er [18] and ^{168}Yb [19]) are also plotted.

The comparison shows that although the superdeformed states are populated weakly relative to other states, they are populated at the highest values of spin with an intensity at least an order of magnitude larger than one would expect for normal deformed structures. This is even more remarkable when one considers that the superdeformed states are populated at spins which are very close to the limit that the nucleus can tolerate before fissioning. This implies that there is some mechanism which is unique to superdeformed shapes which, following the fusion reaction, "cools" the nucleus rapidly and populates the yrast line at very high spins. This phenomenon alone has been the subject of much theoretical and experimental work, and is still not well understood. One of the early ideas about this effect is that the transition probability for high energy E1 decays (which cool the nucleus at high spins) is enhanced for the superdeformed shape. It was suggested [20,21] that this could be caused by combination of effects due to the splitting (and lowering of one component) of the giant dipole resonance, and the relatively low level density for a superdeformed shape. The fact that the superdeformed states are suddenly depopulated at around $30\hbar$ is also rather surprising, and is again the subject of much speculation (eg [20,22]). It has been suggested [20] that a sudden onset of pairing in the second minimum could account for this effect. Indeed, the different rotational frequencies at which the bands

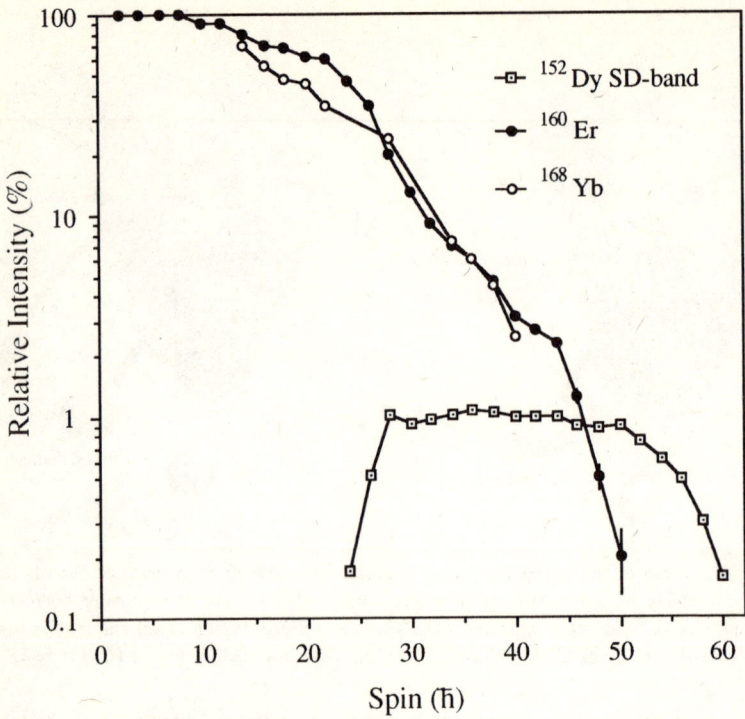

Figure 6: The intensities as a function of spin of the superdeformed band in ^{152}Dy, and the yrast low deformation rotational bands in ^{160}Er [18] and ^{168}Yb [19]. The intensity is plotted relative to the total population intensity of the particular nucleus.

in different nuclei depopulate have been explained in terms of different pairing strengths [23] for superdeformed shapes.

It is known that following the depopulation of the superdeformed states, the band decays into the normal deformation states in ^{152}Dy. However, the linking transitions between the two structures have not been observed. Therefore, it is not possible to make unabiguous spin assignments to the superdeformed states (the spins plotted in figure 4 are reasonable estimates), and hence the information available from the data is rather limited. One quantity which can be readily deduced even in the absence of spin assignments is the dynamic moment of inertia, $\mathcal{J}^{(2)}$, given by

$$\mathcal{J}^{(2)} = \frac{4\hbar^2}{\Delta E_\gamma} \qquad (1)$$

where ΔE_γ is the difference in gamma-ray energy between two subsequent transitions. The values of $\mathcal{J}^{(2)}$ for four cases of superdeformed bands around A = 150 [5,9,11] are plotted in figure 7 as a function of rotational frequency ($= E_\gamma/2$), along with the kinematic moments of inertia $\mathcal{J}^{(1)}$ (defined in equation (3) later), deduced from estimated spin assignments. Although the value of $\mathcal{J}^{(2)}$ depends on the deformation of the nucleus, it is also extremely sensitive to single particle effects such as rotational alignments of specific particles. Indeed as the values for $\mathcal{J}^{(2)}$ vary strongly from one nucleus to the next, this suggests that it may be possible to gain some information on the single particle configurations by considering the differences between the moments of inertia of the different superdeformed bands.

In order to attempt to discover which single particle orbits are involved in these superdeformed bands, it is desirable to calculate the dynamic moments of inertia for specific configurations, and compare these with experimental data. For a specific orbital, ν, the dynamic moment can be calculated as follows

$$\mathcal{J}_\nu^{(2)} = \frac{di_x^\nu}{d\omega} \tag{2}$$

where i_x^ν is the alignment of the single particle angular momentum on the rotational axis x, and ω is the rotational frequency. The total $\mathcal{J}^{(2)}$ for the nucleus can then be calculated by summing the contributions from all the orbitals [24]. It is clear that the orbits with the largest alignment effects (ie high angular momemtum, or high-j, orbits) will have the largest influence on the moment of inertia, and the differences between the values of $\mathcal{J}^{(2)}$ for the different superdeformed nuclei will arise from the different high-j configurations involved. The dynamic moments of inertia calculated [24] in this way are plotted in figure 7 assuming the configurations which the models predict to be most likely. The trends of the moments of inertia are indeed reproduced quite well by the calculations, and it would appear that the high-j configurations suggested by the mean-field models are correct. (The agreement in the case of ^{150}Gd is not so good, but when the effects of static pairing are taken into account [23], the agreement is considerably better).

In this way, details of the microscopic configurations involved in superdeformed states are emerging, and in particular the role of specific high-j orbits is now quite well understood. These comparisons are an excellent test of the different mean-field theories, and the systematic studies of superdeformed bands allow us to test the validity of such models at the extremes of both deformation and spin.

Since the original discovery of the superdeformed band in ^{152}Dy, similar structures have been observed in 9 other nuclei around A = 150, as well as in several nuclei around A = 130 [25]. These structures have revealed some new and exciting phenomena (described in the next section). A third region of superdeformation, around A = 190, has recently been discovered (eg refs [26-37]). Many of the features of the superdeformed bands are quite similar between all these mass regions (eg. no linking transitions have been observed in any of these nuclei), although in the mass 190 region the bands are observed to much lower spins (around $10\hbar$). Again, rich nuclear structure information has been derived from these studies, and some details are presented in the next section.

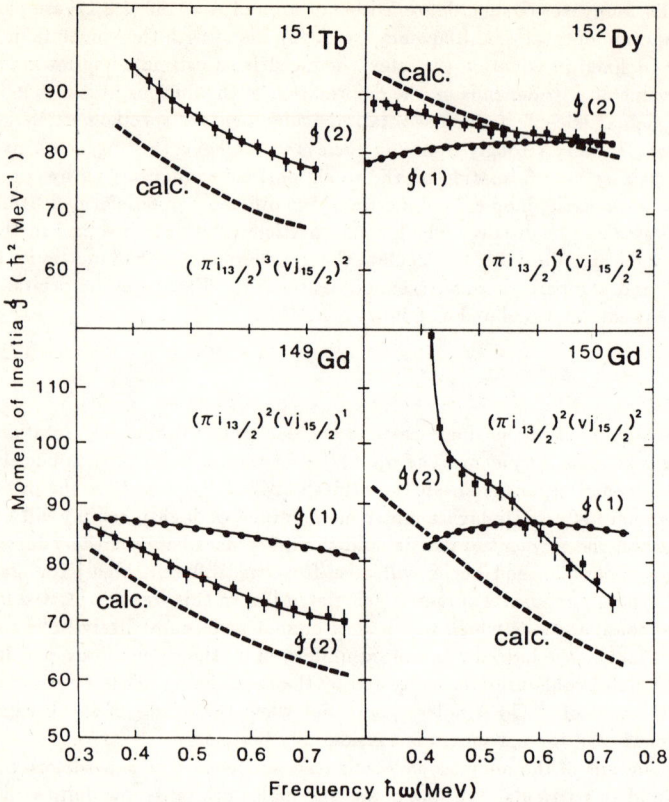

Figure 7: The dynamic and kinematic moments of inertia for four superdeformed bands in the A=150 region [5,9,11]. The kinematic moments are plotted using estimated spin assignments. Calculations (from [24]) of the dynamic moments of inertia are also plotted, assuming the single particle configurations as shown.

4 Excited and identical superdeformed bands

So far in this discussion, only one superdeformed band in any given nucleus has been mentioned, and indeed for the first four years of the work on superdeformation, there were no examples of second superdeformed bands in any nuclei. This suggests that only the lowest energy, or yrast, superdeformed states have been populated in every case. In order to gain more detailed information on the structure of states in the second minimum, it is clearly of interest to attempt to find excitations of the superdeformed shape - ie. excited or non-yrast superdeformed bands. In this way, it may be possible to establish details of the distribution of single particle states in the second minimum.

The first example of a second superdeformed band was found in ^{151}Tb [38] following a re-examination of a previous data-set on this nucleus. The same group subsequently observed a second band in ^{150}Gd, and these two bands are plotted in figure 8. The first thing which is immediately apparent from the spectra is that the intensity of the gamma-rays is rather weak, with the intensities of the bands measured at about 30% of the intensity of the yrast superdeformed states. Hence it would seem reasonable to describe these structures as excited states.

Once again, no linking transitions to either the yrast superdeformed band or the normal deformed structures were observed, as hence the dynamic moment of inertia $\mathcal{J}^{(2)}$ is one of the few quantities that can be derived directly from the experimental data. The values of $\mathcal{J}^{(2)}$ for the excited bands are plotted as a function of rotational frequency in the inserts in figure 8. The data points are shown along with the values of $\mathcal{J}^{(2)}$ for the yrast superdeformed bands in these two nuclei (dashed lines - ^{151}Tb and ^{150}Gd respectively), along with the values of $\mathcal{J}^{(2)}$ for the yrast superdeformed bands of the Z+1 isotopes (solid lines - ^{152}Dy and ^{151}Tb respectively). The first striking feature of these data is that the $\mathcal{J}^{(2)}$ for the excited band does not follow the $\mathcal{J}^{(2)}$ for the yrast superdeformed band in the same nucleus. In fact, the data for the excited bands follows almost exactly the values of $\mathcal{J}^{(2)}$ for the yrast bands of the respective Z+1 N=86 isotones. This is perhaps not as surprising as it may seem at first. As discussed earlier, the $\mathcal{J}^{(2)}$ is influenced very strongly by the high-j configuration of the superdeformed band, and similar values of $\mathcal{J}^{(2)}$ for neighbouring Z, Z+1 isotopes implies that the high-j configurations are the same. It would suggest that the extra proton occupies an orbital with no influence on $\mathcal{J}^{(2)}$ - such as a low-j orbital.

A closer examination of the data reveals a rather more startling effect. The transition energies of the excited band in ^{151}Tb, plotted in figure 8(a), are in fact extremely similar to the transition energies of the superdeformed band in ^{152}Dy - certainly an effect which had not been seen before. The difference between the γ-ray energies for the excited band in ^{151}Tb and the band in ^{152}Dy is plotted in figure 9(a), and shows that the transition energies are identical to around 1 part in 1000! A closer examination of the γ-ray energies of the excited band in ^{150}Gd reveals that these transitions are also identical to 1 part in 1000 to the γ-ray energies of the yrast superdeformed band in ^{151}Tb. It is worth mentioning at this point that these identical gamma-ray sequences cannot be in the same nuclei (ie the same bands) as it is not experimentally possible to form ^{152}Dy in the reaction used to populate the excited band in ^{151}Tb (similarly ^{151}Tb cannot be formed in the reaction used to populate the excited band in ^{150}Gd).

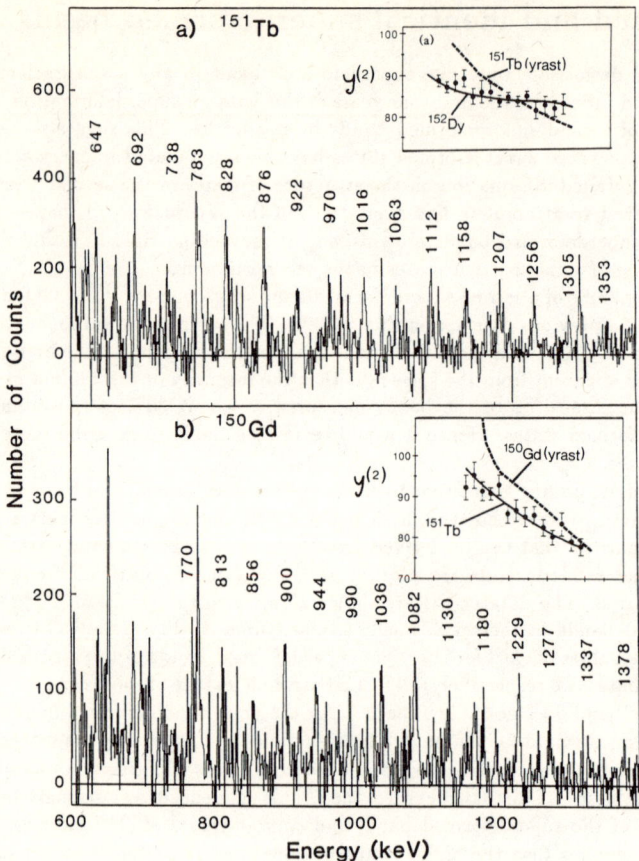

Figure 8: Excited superdeformed bands in the nuclei (a) ^{151}Tb and (b) ^{150}Gd [38] measured using the Tessa3 array. The measured dynamic moments of inertia are plotted in the inserts, along with the moments of inertia for the yrast superdeformed band in the same nucleus and the yrast superdeformed band in the Z+1 isotone.

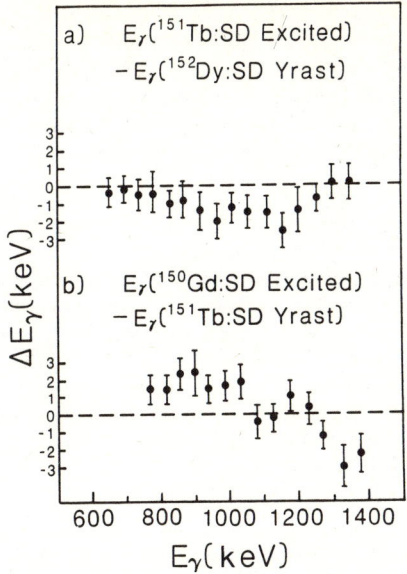

Figure 9: The difference in γ-ray energy between transitions in (a) the excited superdeformed band in ^{151}Tb and the yrast superdeformed band in ^{152}Dy, and (b) the excited band in ^{150}Gd and the yrast band in ^{151}Tb.

In order to analyse this extraordinary effect, it is useful to consider the parameters which affect the γ-ray energy. The transition energy between states of a rotational band (from spin I to I-2) can be written as follows

$$E_\gamma = \frac{\hbar^2}{2\mathcal{J}^{(1)}}(4I-2) \qquad (3)$$

where I is the spin. This means that in the case of identical superdeformed bands in neighbouring nuclei, the kinematic moments of inertia of the two nuclei are also identical to 1 part in 1000. This is somewhat surprising as $\mathcal{J}^{(1)}$ is a function of (among other things) the size and shape of the nucleus. Indeed, if one considers the nucleus purely classically, the change in mass alone should cause a change in the moment of inertia of the order of 1 part in 100. Thus it would seem that under certain circumstances, a superdeformed nucleus can be rigid with respect to the removal or addition of an individual nucleon.

It is also clear from equation (3) that the spins of the superdeformed states in the two neighbouring nuclei would also need to be the same in order for the γ-ray energies to be the same. Of course, the spins of the odd and even nuclei cannot be the same, and at first this would seem to present a problem. In order to understand this, it is necessary to consider the particle-rotor model [39] which describes an odd nucleus in terms of a rotating core of even mass and a valence particle (or hole), with the two degrees of freedom coupled through the Coriolis interaction. Let us consider the specific limiting case in which the angular momemtum of the odd nucleon is strongly coupled to the deformation of the core.

In this case the angular momentum of this deformation aligned orbital precesses around the symmetry axis of the core, and usually has an average zero projection on the rotational axis. However, for the specific case of a $K = 1/2$ orbital (K is the projection of the angular momentum of the nucleon on the symmetry axis) the projection on the rotational axis can be non-zero, adding an extra term to equation (3). In this strong coupling limit of the particle rotor model, equation (3) becomes for the odd nucleus

$$E_\gamma = \frac{\hbar^2}{\mathcal{J}^{(1)}} \left(2I - 1 + a(-)^{I+1/2}\delta_{K,1/2}\right) \quad (4)$$

where a is called the decoupling parameter. For a particle occupying a $K = 1/2$ orbit, it is trivial to show from equation (4), with a decoupling parameter of $a = 1$, that the γ-ray energies for the even nucleus with integer spins are equal to the γ-ray energies of the odd nucleus with spins $I = I_{even} + 1/2$ and $I_{even} - 1/2$. It is possible to understand, therefore, from this simple picture how identical γ-ray sequences can be obtained for odd and even neighbouring nuclei. However, the observation of identical bands still requires the moments of inertia to be identical to 1 part in 1000. It would also appear that these excited structures are almost perfect examples of $a = 1$ bands (or positivley decoupled) bands.

It is perhaps useful to consider which proton orbitals are involved in this process, and some of the ideas are presented in figure 10. Calculations based on a Woods-Saxon model [40] give some indication as to which orbitals are involed in these nuclei, and there appear to be three proton orbits which lie near the Fermi surface for the superdeformed shape. Two of these originate from the $i_{13/2}$ shell ([660]1/2 and [651]3/2 Nilsson orbits) and the third comes from the $p_{1/2}$ shell ([301]1/2 Nilsson orbit). When a proton is removed from the superdeformed structure in ^{152}Dy, it can either form the excited superdeformed band in ^{151}Tb or the yrast superdeformed band. It is clear from the moment of inertia data that the high-j configurations of ^{152}Dy and the excited band of ^{151}Tb are the same, and hence the proton must be removed from the low-j, hence $p_{1/2}$, orbital. Removal of the proton from the highest $i_{13/2}$ orbit will result in the yrast superdeformed band - known to have a different high-j configuration. Hence it would appear that the removal of a proton from the $p_{1/2}$ orbit has little or no effect on the rest of the nucleus. Is there some special property of this orbital when in the superdeformed shape? Indeed, as shown schematically in figure 10, it has a very different spatial distibution in the nucleus than the other orbits near the Fermi surface.

About the same time as these discoveries, a closer look at the data in ^{153}Dy [41] revealed a similar phenomenon. Two superdeformed bands had previously been observed which appeared to be signature partners with no signature splitting (ie two bands with odd and even spins with interleving energy levels). Whilst the energies of the γ-rays in the two bands are not similar to any sequence in a neighbouring nucleus, the average γ-ray energies of the two signature partner bands are identical to the energies of the ^{152}Dy sequence to around 1 part in 1000. This feature is exactly what would be expected if the moments of inertia were once again identical, but in this case with a decoupling parameter of $a = 0$. So now we have seen the same effect again (identical moments of inertia) but this time a neutron orbital is involved - the calculations predict a [514]9/2 Nilsson orbit. It is clear therefore that this phenomenon is not simply a curious feature of one specific orbital.

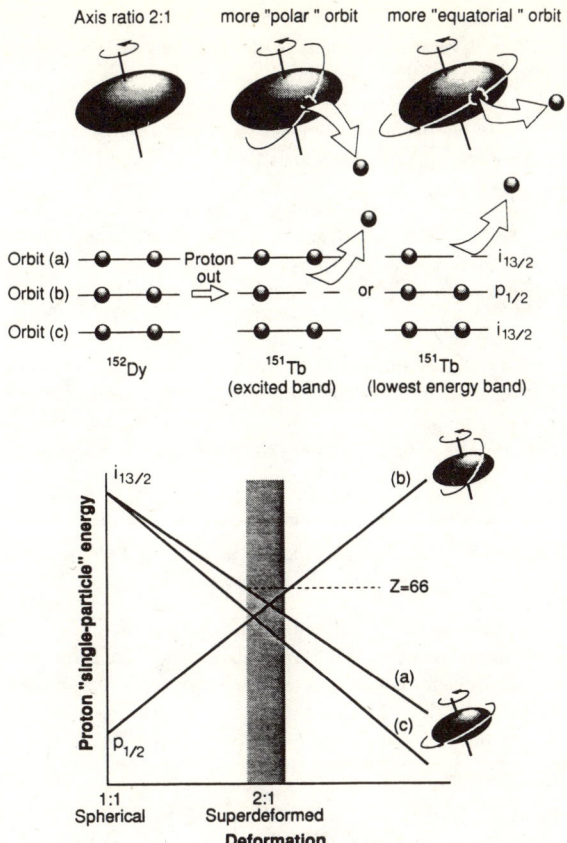

Figure 10: The upper part of the figure shows the expected proton configurations of the two superdeformed bands in ^{151}Tb and the band in ^{152}Dy. The lower part shows schematically the behaviour as a function of deformation of these three orbitals which lie near the Fermi surface for a superdeformed shape.

Since these investigations, excited superdeformed bands have been found in seven nuclei in this mass region, and in several of these cases identical or near-identical sequences have been found - particularly in the $^{146-150}$Gd isotopes, eg [42].

It is useful at this point to look at the different effects which can contribute to the moment of inertia $\mathcal{J}^{(1)}$. It has been pointed out in recent discussions [43] that there at four effects to which the $\mathcal{J}^{(1)}$ is particularly sensitive.

- Orbital alignment: A single particle orbital in a rotating nucleus should be affected by the rotation. As the rotational frequency is increased, the alignment of the single particle angular momentum on the rotational axis will change. As $\mathcal{J}^{(1)}$ is very sensitive to alignment effects, its value should change if a particle is removed.

- Deformation-driving effects: Unless a specific single particle energy level is completely unaffected by deformation (an orbit with no slope in a Nilsson-type diagram), the removal or addition of a particle to a nucleus will have some deformation driving effect. For example, the $p_{1/2}$ orbit described above rises sharply in energy with increasing prolate deformation. The removal of this particle should tend to increase the deformation of the nucleus, and this would ultimately affect the moment of inertia.

- Mass: As the moment of inertia classically is proportional to $A^{5/3}$, a change in mass should significantly affect the moment of inertia. Clearly one would not expect $\mathcal{J}^{(1)}$ to scale exactly as $A^{5/3}$, but it generally follows this trend.

- Pairing: This effect has a major influence on the moment of inertia through rotational alignment effects (see the first item above). It is well known that pairing is significantly lower in odd mass nuclei than in even-even nuclei (caused by the blocking effect of the odd particle). If pairing plays a role in these nuclei, alignment effects should be different between odd and even nuclei - and hence the $\mathcal{J}^{(1)}$ should be different.

Hence identical moments of inertia in neighbouring nuclei would seem to imply that all of these effects make no contribution to the $\mathcal{J}^{(1)}$ for particular superdeformed nuclei. Under certain circumstances, one may expect one or two of these effects to be negligible (eg there is evidence to suggest that pairing plays no significant role in high spin superdeformed bands), although it seems extremely difficult to understand how they can all be negligible when they seem to affect the moments of inertia in normal deformed bands. It is of course possible that some of these effects may cancel each other out - it has been suggested by Stephens et al [44] that under certain conditions the influence of orbital alignment and deformation may indeed cancel.

Turning to the A = 190 region, superdeformed bands have now been observed in thirteen nuclei (eg [26-37]), with these structures showing many of the features of the A = 150 superdeformed bands. Excited bands have now been observed in many of these cases. As an example, the nucleus ^{194}Hg was investigated simultaneously at Daresbury [34] (again on the Tessa3 array) and at Berkeley [33] in which three superdeformed bands were observed, and a closer examination of the data revealed a now familiar effect. The γ-ray energies of one of the excited bands in ^{194}Hg were identical to around 1 part in 500 to the γ-ray energies of a superdeformed band previously observed in ^{192}Hg [28-30] (some of these data

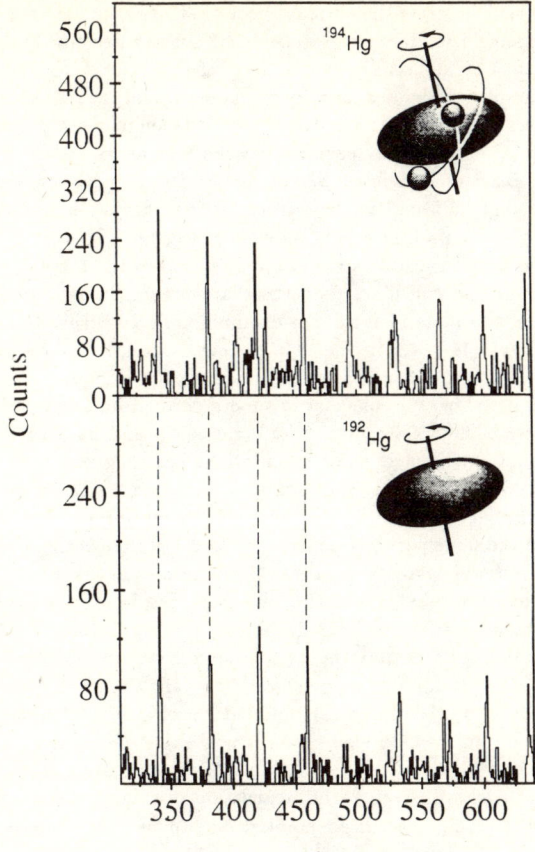

Figure 11: Spectra of the second superdeformed band in ^{194}Hg and the band in ^{192}Hg (data from [54]). The dashed lines indicate that the γ-ray energies are identical.

[54] are shown in figure 11). Therefore the curious effect of identical moments of inertia for neighbouring superdeformed nuclei does not appear to be constrained to one mass region. Indeed, in this case identical bands have been seen in nuclei separated by *two* nucleons.

Once again, many examples of identical or very similar γ-ray energy sequences have now been observed for superdeformed bands around A = 190 between nuclei separated by both one *and* two nucleons (eg [43-45]). The agreement between the energies of the gamma-rays in these bands is not as exact as in a few cases in the A=150 region, and there is some speculation [46] that the slight differences come from pairing effects (these structures are seen to much lower spins where pairing effects may be stronger). The fact that identical structures have now been seen in a different mass region makes these results even more remarkable especially when one considers that there are many differences between the details of the bands in the two mass regions. For example, different orbitals are involved, the deformation of the nuclei is lower in the A=190 region and the ranges of spin and rotational frequency over which the bands are observed are very different.

The observation of identical γ-ray sequences in superdeformed nuclei is perhaps one of the most surprising results in high spin physics for some time, and has been the subject of much theoretical speculation. Some of these speculations have suggested [40,43-45] that some of these effects may involve pseudo-spin symmetry in nuclei. The idea of pseudo-spin [47-49] arises from the fact that in the shell model, the spin-orbit interaction causes particular energy levels of the same parity to come very close together in energy (eg $f_{7/2}$ and $h_{9/2}$). In fact, these levels remain near-degenerate as the deformation changes. The pseudo-spin model reclassifies these levels such that they are pseudo spin-orbit partners (eg the $f_{7/2}$, $h_{9/2}$ pair becomes the pseudo spin orbit pair $g_{7/2}$, $g_{9/2}$). One of the most interesting features of the pseudo spin model is that the pseudo spin-orbit coupling is rather weak and it was predicted [50] that at large rotational frequency the pseudo intrinsic spin can decouple from from the pseudo orbital angular momentum and align with the rotational axis through the Coriolis interaction. Indeed, there is now some evidence for this curious effect happening in the identical bands around A=190. If this is correct, then this would be the first observation of such an effect. A full discussion of some of these ideas can be found in references [43-45]. It has also been pointed out above that the observation of identical bands in the pairs of nuclei ^{150}Gd, ^{151}Tb and ^{151}Tb, ^{152}Dy requires a decoupling parameter of $a = 1$ for the odd particle - thought to occupy the [301]1/2 Nilsson orbit. It has been shown in reference [40] that the although calculations based on this orbit give $a = 0$, when this level is reclassified under the pseudo spin model, the calculations predict a decoupling parameter of $a = 1$.

It is clear that these observations now challenge many of the ways in which nuclei are understood in terms of the nuclear mean-field. It is hoped that with the new generation of gamma-ray spectrometers we may find more examples of these fascinating effects in an attempt to understand how they occur.

5 Future developments

During the last decade, the significant increase in the sensitivity of γ-ray detector arrays has allowed structures populated at an intensity of 0.5% of the reaction channel to be observed, such as the excited superdeformed bands described in the last section. However, many of the exciting problems raised by recent γ-ray studies require far more sensitive arrays, able to probe much weaker structures in nuclei. For example, one of the outstanding experimental problems in the field of superdeformation is the difficulty encountered in observing the linking transitions between the superdeformed and normal deformed states. If these links could be established, more complete spectroscopy of superdeformed states would be possible as the energy and spins of the states would be measurable. However, these transitions are expected to be very weak, an it is expected that two orders of magnitude increase in sensitivity of arrays is required in order to observe these linking transitions. Other exotic shapes effects are predicted to occur in some nuclei at high spins, such as the occurrance of hyperdeformed nuclei (with a 3:1 axis ratio). These states are expected to be extremely weakly populated, and would require a more powerful γ-ray spectrometer than at present. Indeed, in many areas of research using γ-rays, significant advances can now only be made by removing the limitation imposed by the observatioanl limit of the existing arrays.

It is clear from the spectra in figure 8 that excited superdeformed bands are populated with an intensity close to the limit of observation of the current arrays (the peak-to-background ratio is extremely poor). These spectra were generated using the technique of $\gamma - \gamma$ coincidence measurements, where two γ-rays from the same event were recorded simultaneously and coincidences between γ-rays of particular energies are analysed. In order to decrease the effective limit of observation even further, one technique is to increase the number of γ-rays measured from an event (known as the "fold"). In this way, extremely weak γ-rays may be observed as the technique of analysing high fold coincidences significantly improves the peak-to-background ratio.

Typically, up to 30 γ-rays may be emitted in a heavy compound nucleus reaction, and in order to measure a significant number of them, the detector array must have both a high efficiency for detection of γ-rays as well as a large number of detectors (high granularity). These two criteria are difficult to meet simultaneously, and hence the geometrical design of the individual detectors must allow for efficient packing of many detectors around a target. However, in order to be able to identify weak γ-rays above a background, it is necessary to ensure that each individual detector has a high resolving power - a function depending critically on the performance of the suppression shield and the energy resolution of the Ge detector. The resolving power of a detector and the limit of observation of a detector array may be written as (eg [51])

$$R_\gamma = \frac{SE_\gamma . PT}{\Delta E_\gamma} \qquad (5)$$

where SE_γ is the average spacing between the γ-rays, PT is the peak-to-total ratio for a detector and ΔE_γ is the energy resolution of the detector. The limit of observation of a detector array can be shown [52] to be proportional to $1/R_\gamma^n$ where n is the number of γ-rays measured, or fold. A low limit of observation will therefore only be achieved with a high fold and high resolving power.

At present, based on these ideas, the next generation of γ-ray spectrometers are currently being designed and built. Two of these arrays, EUROGAM and Gammasphere, are shown schematically in figure 12. EUROGAM [51] is a UK/France collaboration to build a large array of high efficiency Ge detectors and suppression shields. The current design

Figure 12: A schematic representation of (left) the USA "Gammasphere" array of 110 Compton suppressed Ge detectors and (right) the EUROGAM array with 70 detectors.

of EUROGAM will allow 70 detector systems to be arranged in close geometry around a target, whilst still maintaining a high resolving power for each detector [51]. A diagram of a EUROGAM Ge detector and shield is shown in figure 13, incorporating a tapered high efficiency Ge crystal (three times the efficiency of Tessa3 detector). The array utilises the technique of shared suppression, in which a Ge detector is Compton suppressed by both its own BGO shield and the neighbouring shields (this allows a smaller amount of BGO to be used in each shield).

As the EUROGAM array incorporates a large number of high efficiency Ge detectors, the total efficiency for measuring a γ-ray peak will increase by a factor of 12 (comparing a 70 detector EUROGAM array with a 16 detector Tessa3 array). This array will enable large numbers of events to be recorded with around fold 4-5 (compared with fold 2 for existing arrays). As this has been achieved without deteriorating the resolving power of the detectors, this will reduce the limit of observation by around two orders of magnitude. This would enable structures in nuclei populated with an intensity of 0.005% - 0.01% to be observed.

The limit of observation of a detector array may be decreased further if used in conjunction with another device which reduces the background (such as an inner crystal ball like Tessa3, or a recoil spectrometer). Indeed, the first phase of EUROGAM at Daresbury with 45 detector systems will be used in conjunction with the Daresbury Recoil Mass Separator [53], capable of measuring the mass and charge of recoiling nuclei. In the field of studies of nuclei very far from stability, the combination of a large detector array and a high efficiency recoil spectrometer will be extremely effective in identifying nuclear struc-

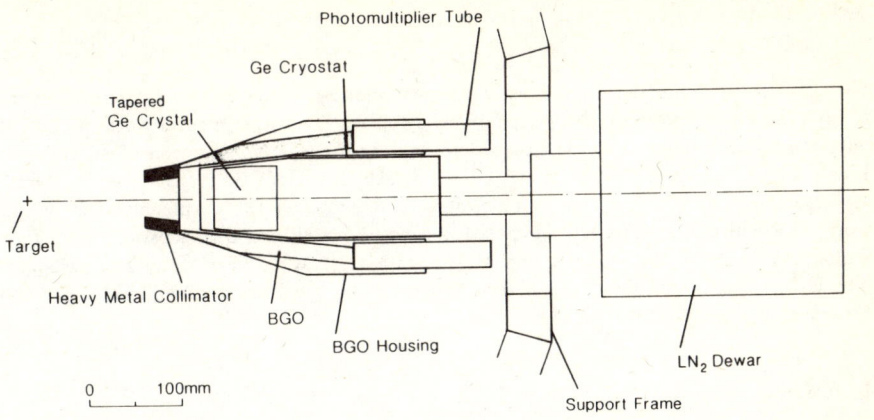

Figure 13: A cross-section through a EUROGAM detector and Compton suppression shield. The drawing shows the special design of the tapered Ge and BGO crystals.

ture effects in extremely weakly populated nuclei. EUROGAM is currently being built, and the first experiments are due to be underway early in 1992.

The USA Gammasphere project, like EUROGAM, is designed to measure efficiently high-fold coincidences in order to probe very weak nuclear structure effects. Gammasphere (figure 12) will comprise 110 high efficiency Ge detectors with BGO shielding. A third project to build a 40 detector γ-ray array, GASP, is currently underway in Italy.

The advent of these new detector arrays, and the possibility of further collaborations to build even more powerful devices (eg EUROBALL), heralds a very bright future for γ-ray spectroscopy and in particlular the study of high spin states, to which these arrays are particularly suited. Advances in detector array technology over the last decade, culminating in the current arrays such as Tessa3, have decreased the limit of observation in nuclei by around an order of magnitude. During this time, many exciting and unexpected effects have been observed as a result of these developments. The list includes the observation of discrete-line superdeformed bands, multi-shape coexistence in nuclei, shape changes and phase transitions at high spin, band terminations, the observation of heavy N=Z nuclei, octupole deformed nuclei and the observation of identical γ-ray sequences in neighbouring nuclei. With the next generation of detector arrays, a further two orders of magnitude sensitivity will be achieved. The truly exciting aspect of these new arrays is that nobody really knows what fascinating features of the nucleus these studies will reveal.

Acknowledgements

I would like to thank the organisers of this stimulating Summer School for inviting me to give these lectures on some of my favourite topics. The experimental work in the UK on superdeformation and excited superdeformed bands is a combination of efforts from Liverpool University, York University, CNRS Strasbourg, Neils Bohr Institute, Daresbury and other establishments. It is not possible to mention all the people involved in this work, so I will thank them all. Stimulating (and above all useful) discussions with Paul Fallon, David Brink, Neil Rowley and M.A.Nagarajan (Nag) are gratefully acknowledged. The UK research work described here is supported by the United Kingdom Science and Engineering Research Council.

References

1. P.J.Nolan et al, Nucl. Inst. Meth. **A236** (1985) 95
2. J.F.Sharpey-Schafer and J.Simpson, Prog. Part. Nucl. Phys. **21** (1988) 293
3. P.J.Twin et al, Phys. Rev. Lett. **57** (1986) 811
4. M.A.Bentley et al, Phys. Rev. Lett. **59** (1987) 2141
5. M.A.Bentley et al, J.Phys.G : Nucl. Part. Phys **17** (1991) 481-510
6. G.T.Ewan and A.J.Tavendale, Can. J. Phys. **42** (1964) 2286
7. P.J.Twin et al, Nucl. Phys. **A409** (1983) 343c
8. M.A.Riley et al, Phys. Lett. **B135** (1984) 275
9. B.Haas et al, Phys. Rev. Lett. **60** (1988) 503
10. M.A.Deleplanque et al, Phys. Rev. Lett **60** (1988) 1626
11. P.Fallon et al, Phys. Lett **B218** (1989) 137
12. G.E.Rathke et al, Phys. Lett. **B209** (1988) 177
13. A.J.Kirwan et al, Phys. Rev. Lett. **58** (1987) 467
14. S.M.Polikanov et al, Sov. Phys. JETP **15** (1962) 1016
15. V.M.Strutinsky, Nucl. Phys. **A95** (1967) 420
16. H.J.Specht et al, Phys. Lett. **B41** (1972) 43
17. J.Dudek, Woods-Saxon calculations, private communication.
18. J.Simpson et al, J.Phys.G : Nucl. Phys. **13** (1987) 43 and private communication
19. J.Bacelar et al, Nucl. Phys. **A442** (1985) 509
20. B.Herskind et al, Phys. Rev. Lett. **59** (1987) 2416
21. B.Herskind and K.Schiffer, Proc. Int. School of Physics 'Enrico Fermi' 1987
22. I.Ragnarsson and S.Aberg, Phys. Lett. **B180** (1986) 191
23. W.Nazarewicz et al, Nucl. Phys. **A503** (1989) 285
24. T.Bengtsson et al, Phys. Lett. **B208** (1988) 39
25. P.J.Nolan and P.J.Twin, Ann. Rev. Part. Nucl. Sci. **38** (1988) 533 and references therein
26. E.F.Moore et al, Phys. Rev. Lett. **63** (1990) 360
27. M.P.Carpenter et al, Phys. Lett. **B240** (1990) 44
28. J.A.Becker et al, Phys. Rev. **C41** (1990) 9

29. D.Ye et al, Phys. Rev. **C41** (1990) 13
30. E.F.Moore et al, Phys. Rev. Lett. **64** (1990) 3127
31. E.A.Henry et al, Z.Phys. **A335** (1990) 361
32. D.M.Cullen et al, Phys. Rev. Lett. **65** (1990) 1547
33. C.W.Beausang et al, Z.Phys. **A335** (1990) 325
34. M.A.Riley et al, Nucl. Phys. **A512** (1990) 178
35. P.B.Fernandez et al, Nucl. Phys. **A517** (1990) 886
36. F.Azaiez et al, Z.Phys **A336** (1990) 244
37. F.Azaiez et al, Phys. Rev. Lett. **66** (1990) 1030
38. T.Byrski et al, Phys. Rev. Lett. **64** (1990) 1650
39. A.Bohr and B.R.Mottelson, Nuclear Structure (Benjamin, New York, 1975) Vol. 2
40. W.Nazarewicz et al, Phys. Rev. Lett. **64** (1990) 1654
41. J.K.Johansson et al, Phys. Rev. Lett. **63** (1989) 2200
42. P.Fallon, Proc. Int. Conf. High Spin Physics and Gamma-soft Nuclei, Pittsburgh, PA USA. 17-21 Sept, 1990 and references therein.
43. R.M.Diamond, Proc. Int. Conf. High Spin Physics and Gamma-soft Nuclei, Pittsburgh, PA USA. 17-21 Sept 1990.
44. F.S.Stephens et al, Phys. Rev. Lett. **64** (1990) 2623
45. F.S.Stephens et al, Phys. Rev. Lett. **65** (1990) 301
46. P.Fallon, private communication, and to be published.
47. K.T.Hecht and A.Adler, Nucl. Phys. **A137** (1969) 129
48. A.Arima et al, Phys. Lett **B30** (1969) 517
49. R.D.Ratna-Raju et al, Nucl. Phys. **A202** (1973) 433
50. A.Bohr et al, Phys. Scr. **26** (1982) 267
51. P.J.Nolan, Nucl. Phys. **A520** (1990) 657c
52. EUROGAM proposal document, 1990.
53. A.N.James et al, Nucl. Inst. Meth. **A267** (1988) 144
54. D.M.Cullen, private communication

Nuclear Phase Transitions at Finite Temperature

Alan L. Goodman

Physics Department, Tulane University
New Orleans, Louisiana 70118, USA

> *Abstract:* Statistical mechanics is applied to small systems. Emphasis is given to statistical fluctuations in collective order parameters.

1. Statistical Mechanics of Small Systems

The traditional presentation of statistical mechanics describes macroscopic systems, where $N \to \infty$. Macroscopic systems in thermal equilibrium occupy the most probable state. Thermal fluctuations away from the most probable state are negligible, except at critical points and phase transitions. The various statistical ensembles are equivalent.

For small systems, where $N \sim 100$, these standard results of statistical mechanics do not apply. Thermal fluctuations away from the most probable state can be large, and they can significantly alter the observable properties of a small system. The different statistical ensembles are not equivalent.

To describe hot nuclei, we have to carefully apply statistical mechanics to small systems. The first task is to approximate the most probable state of hot interacting nucleons. This is normally accomplished with a mean field approximation. In this approximation hot nuclei display phase transitions in their collective order parameters: deformations, pair gaps, and density.

The second task is to determine the statistical fluctuations in the collective order parameters. These fluctuations often smooth out the phase transitions which are predicted by the mean field theories. The conventional method for calculating these fluctuations uses the grand canonical ensemble (GCE) and makes the assumption that the intensive variables (temperature T and rotational frequency ω) remain constant as the order parameters fluctuate. However since the nucleus is an isolated system, the proper constraints are to keep the extensive variables (energy E and angular momentum I) constant as the nucleus fluctuates. How can this be accomplished? Do the microcanonical ensemble and the grand canonical ensemble agree on the most probable values of the order parameters? Do the

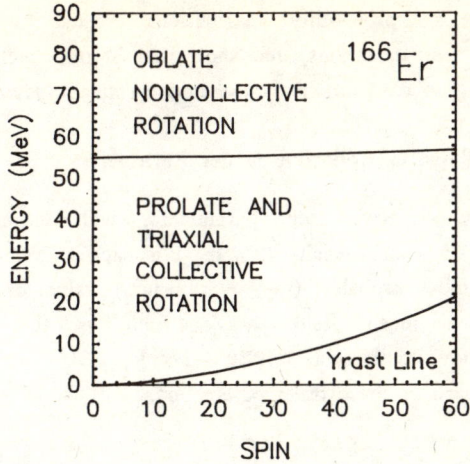

Fig. 1. Phase diagram for the shape and rotation mechanism of ^{166}Er.

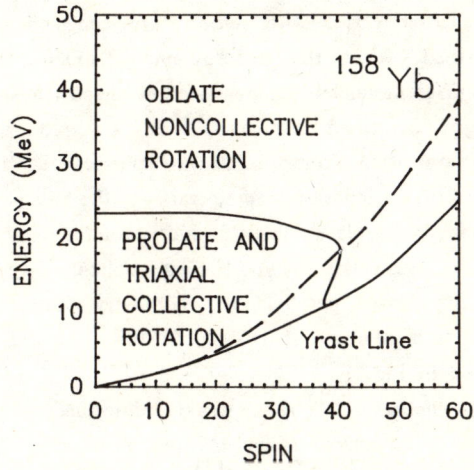

Fig. 2. Phase diagram for the shape and rotation mechanism of ^{158}Yb.

ensembles agree on the probability distributions created by the fluctuations? We hope to answer these questions, and show that atomic nuclei provide a good laboratory for learning to use statistical mechanics in small systems.

2. Most Probable Value of Collective Order Parameters

A nucleus has collective order parameters, such as multipole moments ($Q_{\lambda\mu}$), pair gaps (Δ_p, Δ_n) and density ($\rho-\rho_c$). For each combination of T and ω (or E and I), the most probable (i.e., equilibrium) value of these parameters is determined by a mean field theory, such as the finite-temperature Hartree-Fock-Bogoliubov cranking (FTHFBC) theory [1-3]. This theory predicts phase transitions in each of these order parameters.

2.1 Shape transitions

Many nuclei have ground state intrinsic shapes which are deformed. These deformations are caused by shell effects. When the temperature is non-zero, there are statistical excitations of particles (quasiparticles) which weaken the shell effects, and change the equilibrium deformation. The microscopic FTHFBC equation has been solved for various nuclei to show how the quadrupole deformation varies with temperature and spin [4-9]. This has also been accomplished with the macroscopic Landau theory of phase transitions [10-13]. Different nuclei display different shape transitions.

For the strongly deformed nucleus ^{166}Er at spin zero, raising the temperature causes a transition from prolate to spherical at $T_c = 1.74$MeV. If this nucleus rotates with a constant spin, raising the temperature induces a transition from a nearly prolate shape which rotates collectively (rotation axis perpendicular to symmetry axis) to an oblate shape which rotates noncollectively (rotation axis parallel to symmetry axis). The critical temperature decreases with spin, so that $T_c = 1.44$MeV at I=60$\hbar$. The FTHFBC phase diagram is shown in Fig. 1.

The transitional nucleus ^{158}Yb is weakly deformed. Consequently the transition from prolate to spherical shape at spin zero occurs at $T_c = 1.04$MeV, which is much lower than in ^{166}Er. If ^{158}Yb is cold, then rotating to spin 40$\hbar$ causes a transition from nearly prolate collective rotation to oblate noncollective rotation. For a constant spin between 0 and 40$\hbar$, raising the temperature causes the same transition. The FTHFBC phase diagram is in Fig. 2. A comparison of Figs. 1 and 2 shows how a strongly deformed nucleus and a transitional nucleus have different thermal and rotational responses.

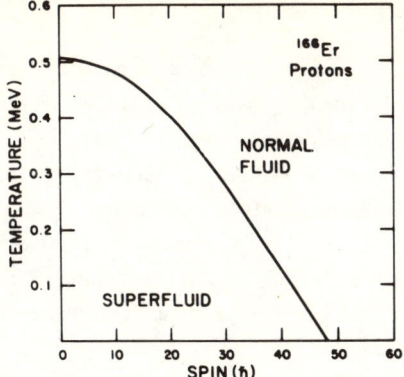

Fig. 3. Phase diagram for the proton pair correlations in ^{166}Er. The critical temperature T_c versus the spin I.

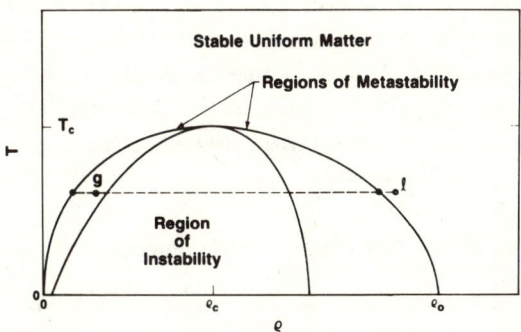

Fig. 4. Phase diagram for nuclear matter.

2.2 Pairing transitions

The ground states of many nuclei contain pair correlations, whose strength is measured by the pair gap Δ. Heating a nucleus creates statistical quasiparticle excitations, which break the correlated pairs, and weaken the pair gap. At a critical temperature $T_c \sim 0.57\Delta(T=0)$, there is a transition from a superfluid ($\Delta \neq 0$) to a normal fluid ($\Delta = 0$). In a rotating nucleus, the Coriolis force also weakens the pair correlations, and reduces the pair gap. Consequently the critical temperature decreases with spin, and at a critical spin I_c, $T_c=0$. Fig. 3 gives the phase diagram for the proton pair correlations in ^{166}Er [7].

If neutron-proton pair correlations occur in a specific nucleus, then the phase diagram may contain separate phases for the np correlated superfluid and the nn, pp correlated superfluid [14].

2.3 Liquid-gas transition

The nucleon-nucleon interaction contains a short range repulsion and a long range attraction. Consequently the equation of state for nuclear matter resembles the van der Waals equation of state, which indicates the presence of a nuclear liquid-gas phase transition. There is a critical point at a temperature $T_c \sim 15$MeV and a density $\rho_c \sim 0.4\rho_0$, where ρ_0 is the equilibrium density at T=0. If $T < T_c$, a liquid-gas transition occurs.

The phase diagram for nuclear matter is shown in Fig. 4 [15, 16]. For temperatures below T_c, there are regions of stable liquid and gas phases. There is a mechanically unstable region, where $\partial P/\partial \rho < 0$. The metastable region on the left is the supersaturated vapor.

3. Statistical Fluctuations in Collective Order Parameters

For a finite system at non-zero temperature, there are statistical fluctuations in the collective order parameters. These fluctuations can create large deviations from the most probable (i.e., equilibrium) values of the order parameters.

To calculate statistical fluctuations in some order parameter β, we use the fundamental postulate of statistical mechanics, which states that all accessible states are equally probable. Therefore the probability that a system will have a specific value of β is proportional to the number of accessible states which have that value of β. For a system with fixed temperature, this is the partition function $Z(T,\beta)$, so that

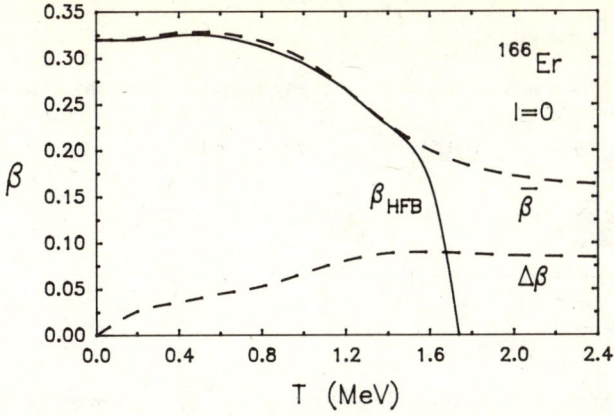

Fig. 5. The quadrupole deformation β versus the temperature T for ^{166}Er at spin zero.

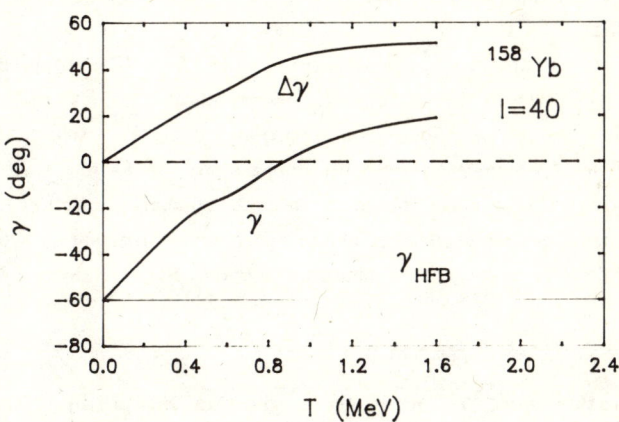

Fig. 6. The quadrupole deformation γ versus the temperature T for ^{158}Yb. The spin is $40\hbar$.

$$P(T,\beta) \propto Z(T,\beta) = e^{-F(T,\beta)/T} ,\qquad(1)$$

where F is the free energy E-TS. For any operator $O(\beta)$, the average value is

$$\overline{O}(T) = <O> = \frac{\int O(\beta)\, P(T,\beta)\, d\tau}{\int P(T,\beta)\, d\tau} ,\qquad(2)$$

where $d\tau$ is the metric. The standard deviation in O is

$$\Delta O = [<O^2> - <O>^2]^{1/2} .\qquad(3)$$

3.1 Shape fluctuations

Quadrupole deformations are characterized by the order parameters β and γ. The FTHFBC theory determines the free energy $F(T,I;\beta,\gamma)$ and the shape probability distribution $P(T,I;\beta,\gamma)$.

To see how statistical fluctuations affect phase transitions, consider ^{166}Er at spin zero. Fig. 5 shows how β varies with temperature [17]. The mean field theory predicts a shape transition from prolate to spherical at the critical temperature $T_c = 1.74$MeV. When the shape fluctuations are included, the average shape $\overline{\beta}$ does not undergo a transition from prolate to spherical. At temperatures above T_c, $\overline{\beta}$ is $\sim \frac{1}{2}\beta(T=0)$. Egido et al. found similar results in ^{158}Er [18].

The next example is ^{158}Yb at spin $40\hbar$ [8, 9]. The equilibrium phase is oblate noncollective rotation for all temperatures. However at moderate temperatures, the shape fluctuations populate collective structures. Fig. 6 shows how the average $\overline{\gamma}$ varies with temperature for $I=40\hbar$. At $T=0.9$ MeV ($E^*=12.7$MeV), the average phase is prolate collective rotation ($\overline{\gamma}=0°$), even through the most probable phase is oblate noncollective rotation ($\gamma=-60°$).

The collective strength is measured by the B(E2) value. For $I \gg 1$, this is

$$B(E2,I \to I \pm 2) \propto \beta^2 \cos^2(\gamma-30°) .\qquad(4)$$

The B(E2) value in ^{158}Yb at $I=40\hbar$ is given in Fig. 7 [9]. The mean field value is zero at all temperatures. When shape fluctuations are included, the average B(E2) value increases with excitation energy. For $T \sim 0.6$MeV ($E^* \sim 6.5$MeV), the average B(E2)$\sim$50 or 100 W.u., depending upon the choice of the metric $d\tau$. Including the omitted orbitals $\nu j_{15/2}$ and $\pi i_{13/2}$ should substantially increase the B(E2) value. The significant conclusion is that thermal shape fluctuations provide a simple mechanism for generating collective B(E2) strength at moderate temperatures, even when the equilibrium phase is noncollective. This explains the ORNL experiment [19] which shows

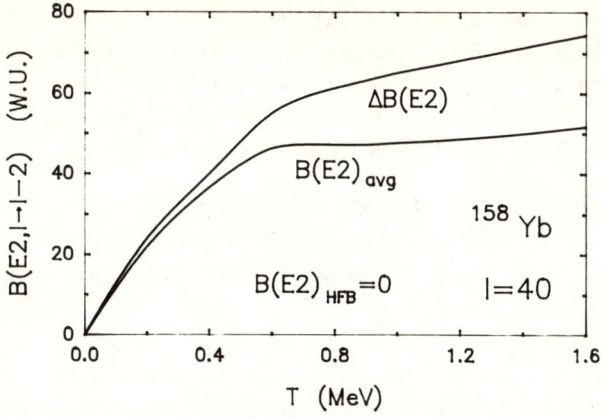

Fig. 7. The B(E2) value versus the temperature T for ^{158}Yb. The spin is 40$\hbar$.

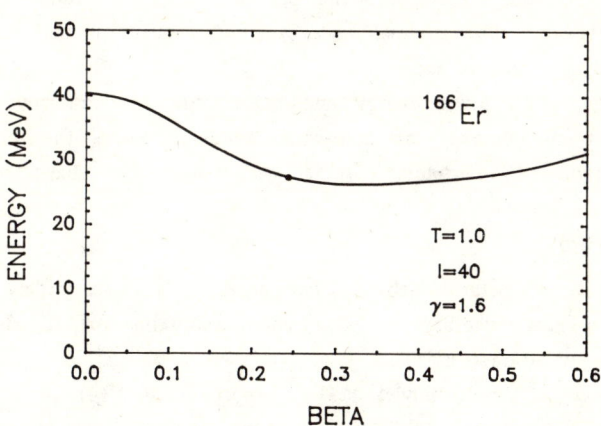

Fig. 8. The energy versus the quadrupole deformation β for the constant temperature constraint.

that although the yrast line of ^{158}Yb is noncollective for spins 40-50$\hbar$, there is significant collective structure when the thermal excitation energy is raised to 5-10MeV.

3.2 Constant energy versus constant temperature constraints

In the previous section we assumed that a nucleus has a constant temperature as its shape fluctuates. Is this a justifiable assumption? The energy of a warm non-rotating nucleus should contain a thermal component and a deformation component, as in

$$E = aT^2 + V(\beta) , \qquad (5)$$

where the thermal energy aT^2 is taken from the Fermi gas model, and a is the level density parameter. If the temperature is the same for all shapes, then it is obvious from eq. (5) that each shape will have a different energy. However the fluctuating nucleus is an isolated system, whose energy should remain constant, until it de-excites. So the assumption that the fluctuating nucleus has a constant temperature violates the conservation of energy. The correct constraint is to let the energy remain constant as the shape fluctuates [20-24]. Eq. (5) shows that each shape will have a different temperature. As the shape fluctuates, the constant total energy is repartitioned between the thermal energy and other forms of energy (deformation, pairing, rotation, etc.).

For a system with given energy and deformation, the entropy $S(E,\beta)$ is the logarithm of the number of accessible states. From the fundamental postulate of statistical mechanics, it follows that the shape probability distribution is

$$P(E,\beta) \propto e^{S(E,\beta)} . \qquad (6)$$

Since the constant temperature shape distribution (1) is so widely used, it is important to know whether it is a good approximation to the correct constant energy distribution (6).

First consider a simple model [23]. Approximate $V(\beta)$ in eq. (5) by $b(\beta-\beta_0)^2$, where β_0 is the equilibrium deformation, and b measures the stiffness of the shape. Then it can be shown that the constant temperature constraint gives

$$\frac{P(T,\beta)}{P(T,\beta_0)} = e^{-b(\beta-\beta_0)^2/T} , \qquad (7)$$

and the constant energy constraint gives

$$\frac{P(E,\beta)}{P(E,\beta_0)} = e^{-b(\beta-\beta_0)^2/T - b^2(\beta-\beta_0)^4/(4aT^3) + \ldots} , \qquad (8)$$

where T is the temperature of the equilibrium shape, defined by $E=aT^2$. The

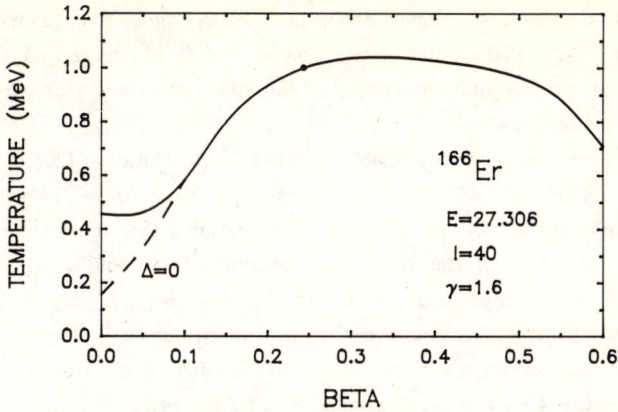

Fig. 9. The temperature versus the quadrupole deformation β for the constant energy constraint.

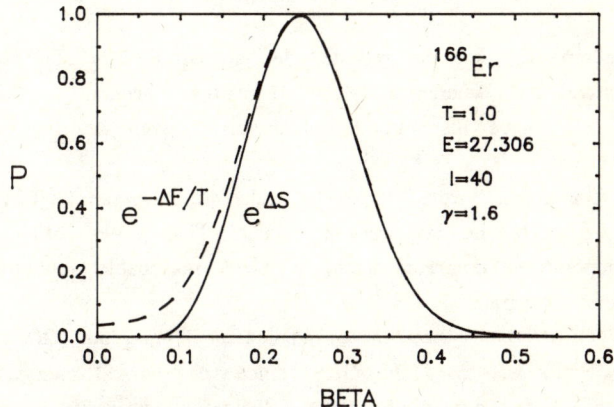

Fig. 10. The relative shape probability distribution P versus the quadrupole deformation β.

distributions (7) and (8) have exponents which agree to quadratic order in $(\beta-\beta_0)$, but they differ in quartic order. Furthermore, for $\beta \neq \beta_0$, the relative shape probability at constant energy is smaller than the probability at constant temperature.

Next we consider microscopic FTHFBC calculations [22]. Choose the equilibrium state of ^{166}Er which has the properties: $I=40\hbar$, $T=1$MeV, $E=27.306$MeV, $\beta_0=0.243$ and $\gamma_0=1.6°$. First assume that the temperature remains constant at $T=1$MeV as the shape fluctuates about its equilibrium value. Then the energy of each shape is given in Fig. 8. For example the $\beta=0.1$ shape has an energy which is 8.8MeV above that of the equilibrium shape. The constant T constraint violates energy conservation by 8.8MeV when $\beta=0.1$. Nevertheless this shape occurs with a relative probability of $P=0.14$.

Next assume that the energy remains constant at $E=27.306$MeV as the shape fluctuates about this equilibrium state. Then the temperature of each shape is given in Fig. 9. As β decreases below β_0, the temperature rapidly falls. This again indicates that constant T is not equivalent to constant E.

Fig. 10 compares the shape probability distributions given by the two constraints. In the neighbourhood of the equilibrium shape, the two constraints give similar probabilities. For small deformations, the constant T constraint has larger relative probabilities than the constant E constraint.

This comparison can be repeated for an equilibrium state located at the critical temperature where a shape transition occurs. Then the two constraints give shape distributions which differ even in the vicinity of the equilibrium shape.

Both constraints were used to calculate average B(E2) values and quadrupole moments in hot rotating nuclei [24]. For this purpose, the constant temperature constraint seems to be a reasonable approximation to the constant energy constraint.

The FTHFBC theory uses the grand canonical ensemble (GCE). The simple model of eq. (5) also uses the GCE, since the thermal energy aT^2 is taken from the ideal Fermi gas in the GCE. Therefore whenever *energy* appears in this section, this implies the *average energy* $\overline{E} = <H>$ evaluated in the GCE. The microcanonical ensemble will be considered in Sect. 3.4.

3.3 Orientation fluctuations: constant angular momentum versus constant rotational frequency constraints

If a hot rotating even-even nucleus is in an equilibrium state, then the rotation axis is usually parallel to a principal axis of the intrinsic shape.

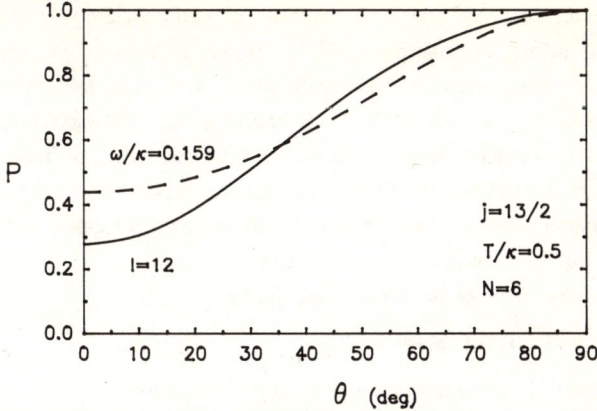

Fig. 11. The orientation probability distribution P versus the orientation angle θ. The constant spin constraint is compared to the constant rotational frequency constraint.

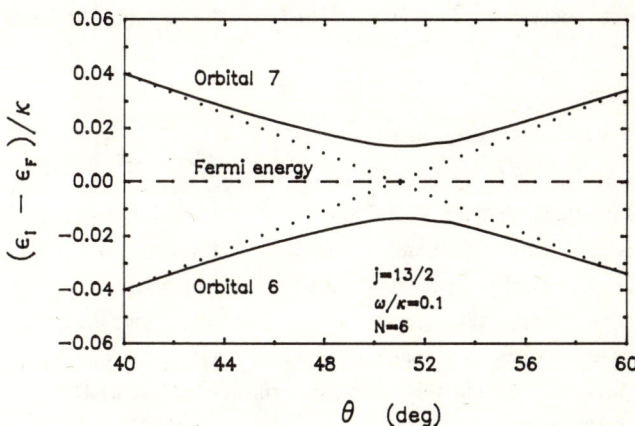

Fig. 12. Single particle energies in the rotating frame versus the orientation angle θ.

However thermal fluctuations can generate non-equilibrium states in which the rotation axis does not coincide with a principal axis. In such states, the orientation of the intrinsic principal axes with respect to the rotational frequency vector ω is defined by two angles (θ,ϕ). Alhassid and Bush (AB) [25, 26] have shown that the statistical fluctuations in these angles alter the observed properties of the giant dipole resonance.

AB assume that the rotational frequency remains constant as the orientation angles fluctuate. For given ω and T, the equilibrium state minimizes the free energy in the rotating frame

$$F'(T,\omega) = F'(T,\omega,\theta,\phi) = E - TS - \omega \cdot I . \tag{9}$$

For given ω and T, the orientation probability distribution is

$$P(\theta,\phi) \propto e^{-F'(\theta,\phi)/T} . \tag{10}$$

We are considering the fluctuation in the orientation of an isolated nucleus. Therefore the variable which is conserved is the angular momentum, not the rotational frequency. So the orientation fluctuations should be calculated with a constant spin constraint, rather than a constant frequency constraint. Mean field theories provide an average spin. For each orientation (θ,ϕ), the frequency ω should be adjusted so that the average spin I remains constant. For given I and T, the equilibrium state minimizes the free energy

$$F = E - TS , \tag{11}$$

and the orientation probability distribution is

$$P(\theta,\phi) \propto e^{-F(\theta,\phi)/T} . \tag{12}$$

Is the constant frequency distribution (10) a good approximation to the correct constant spin distribution (12)? These two distributions are compared for an axially symmetric quadrupole potential acting in a $j=13/2$ shell [27]. Since the potential is axially symmetric, all possible orientations are defined by one orientation angle θ, which is the angle between the frequency ω and the symmetry axis of the potential.

Consider the equilibrium state at $T/\kappa=0.5$, $\omega/\kappa=0.159$, $I=12\hbar$, and $\theta=90°$. (The constant κ is proportional to the deformation β. If $\kappa=2$MeV, then $\beta=0.24$.) The orientation fluctuations around this equilibrium state are calculated (a) with constant $I=12\hbar$ using distribution (12), and (b) with constant $\omega/\kappa=0.159$ using distribution (10). The two distributions are compared in Fig. 11. For orientations near $\theta=0°$ (where the rotation axis is parallel to the symmetry axis), the constant frequency constraint gives larger relative probabilities than the constant spin constraint. For orientations near $\theta=90°$ (where the rotation axis is perpendicular to the symmetry axis), the reverse occurs. The differences between the constant spin and constant frequency constraints may be significant. Further comparisons with more realistic microscopic models should be pursued.

The AB investigation made the implicit assumption that the principal moments of inertia are independent of the orientation angles (θ,ϕ). This is obviously correct for a rigid classical system. But is it justified for a rotating nucleus? This question is addressed with the j=13/2 model described above. The cranking Hamiltonian is

$$H' = H - \omega_x J_x - \omega_z J_z \qquad (13)$$
$$= H - \omega \sin\theta \, J_x - \omega \cos\theta \, J_z .$$

The eigenvalues of H' are the single particle energies in the rotating frame ε_i. Fig. 12 shows how ε_i depends upon the orientation angle θ, for $\omega/\kappa=0.1$ and N=6. At the angle $\theta_c \sim 51°$, there is a level crossing at the Fermi surface. If the temperature is small, there is a transition at θ_c from one set of occupied orbitals to another set. Since the particle configuration changes at θ_c, it is natural that the inertial properties should change at θ_c. It should be emphasized that this transition is caused by varying the orientation angle θ, while ω and $\beta_{potential}$ remain constant. The conclusion is that the principal moments of inertia are strongly dependent upon the orientation of the shape, if the temperature is small. However at moderate temperatures, the effect of the level crossing on the inertial properties is dampened by the thermal single particle excitations. Therefore the principal moments of inertia are almost independent of the shape orientation at moderate temperatures.

3.4 Microcanonical ensemble versus grand canonical ensemble

In the limit $N \to \infty$, the various statistical ensembles are equivalent. But this equivalence is not guaranteed for small systems. In previous sections we used the GCE to calculate statistical fluctuations,. Now we will use the microcanonical ensemble (ME) to calculate the fluctuations. The two ensembles will be compared [24].

For the constant energy constraint, the shape probability distribution $P(E,\beta)$ is the exponential of the entropy $S(E,\beta)$, as in eq. (6). Previously we used the GCE, where the energy is the average $\overline{E}=<H>$ and the entropy is

$$S_{GC}(\overline{E},\beta) = - \sum_i [f_i \ln f_i + (1-f_i) \ln(1-f_i)] , \qquad (14)$$

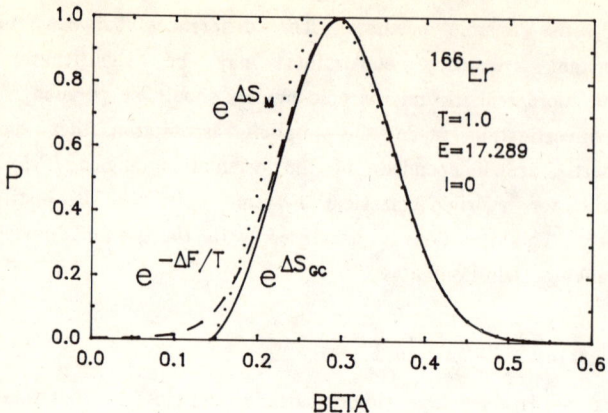

Fig. 13. The relative shape probability distribution P versus the quadrupole deformation β.

$$f_i = \frac{1}{1 + e^{E_i/T}} , \qquad (15)$$

where E_i is a quasiparticle energy.

For a mean field approximation, the number of states with energy between E and E+δE is

$$\rho(E,\beta) \, \delta E = e^{S_M(E,\beta)} , \qquad (16)$$

where ρ is the nuclear level density and S_M is the entropy in the microcanonical ensemble. In the saddle point approximation, the FTHFB level density is

$$\rho(E,\beta) = \frac{e^{S_{GC}(\overline{E},\beta)}}{(2\pi)^{1/2} \Delta E_{HFB}} , \qquad (17)$$

where $\overline{E}$=E and the energy fluctuation is

$$\Delta E_{HFB} = [\sum_i E_i^2 f_i (1-f_i)]^{1/2} . \qquad (18)$$

From eqs. (16) and (17) it follows that

$$\ln \rho(E,\beta) = S_M(E,\beta) - \ln(\delta E) \qquad (19)$$
$$= S_{GC}(\overline{E},\beta) - \ln(\Delta E_{HFB}) - \ln(2\pi)^{1/2} ,$$

where $\bar{E}=E$. This equation enables us to compare the microcanonical entropy S_M with the grand canonical entropy S_{GC}. For specified energy $\bar{E}=E$, the FTHFB self-consistent shape β_0 maximizes the entropy $S_{GC}(E,\beta)$. Therefore β_0 is the equilibrium (i.e., most probable) shape for the GCE. Since the fluctuation ΔE_{HFB} does not appear in the FTHFB variational principle, ΔE_{HFB} does not have an extremum at β_0. Therefore the entropy $S_M(E,\beta)$ and the level density $\rho(E,\beta)$ are not maximized at β_0. The conclusion is that the microcanonical entropy and the grand canonical entropy predict different equilibrium shapes.

This effect is illustrated in Fig. 13, which gives the β fluctuations around the ^{166}Er equilibrium state with I=0, T=1.0MeV, and E=17.289 MeV. The GCE distributions which use the grand canonical entropy and the free energy are both maximized at the self-consistent shape $\beta_0=0.294$. In contrast the distribution which uses the microcanonical entropy peaks at $\beta=0.284$. This shows how the microcanonical ensemble and the grand canonical ensemble can differ for finite systems.

Acknowledgement

This work was supported in part by the National Science Foundation.

References

1. A. L. Goodman, Nucl. Phys. A352, 30 (1981).
2. K. Tanabe, K. Sugawara-Tanabe, and H. J. Mang.,Nucl. Phys. A357, 20 (1981).
3. M. Sano and M. Wakai, Prog. Theor. Phys. 48, 160 (1972).
4. A. L. Goodman, Phys. Rev. C33, 2212 (1986).
5. A. L. Goodman, Phys. Rev. C34, 1942 (1986).
6. A. L. Goodman, Phys. Rev. C35, 2338 (1987).
7. A. L. Goodman, Phys. Rev. C38, 977 (1988).
8. A. L. Goodman, Phys. Rev. C38, 1092 (1988).
9. A. L. Goodman, Phys. Rev. C39, 2008 (1989).
10. S. Levit and Y. Alhassid, Nucl. Phys. A413, 439 (1984).
11. Y. Alhassid, S. Levit, and J. Zingman, Phys. Rev. Lett. 57, 539 (1986).
12. Y. Alhassid, J. Zingman, and S. Levit, Nucl. Phys. A469, 205 (1987).
13. Y. Alhassid, J. Manoyan, and S. Levit, Phys. Rev. Lett. 63, 31 (1989).
14. A. L. Goodman, Nucl. Phys. A402, 189 (1983).
15. A. L. Goodman, J. I. Kapusta, and A. Z. Mekjian, Phys. Rev. C30, 851 (1984).
16. D. H. Boal and A. L. Goodman, Phys. Rev. C33, 1690 (1986).
17. A. L. Goodman, Phys. Rev. C37, 2162 (1988).
18. J. L. Egido, C. Dorso, J. O. Rasmussen, and P. Ring, Phys. Lett. B178, 139 (1986).
19. C. Baktash, in Proceedings of the Workshop on Nuclear Structure at Moderate and High Spin, Berkeley, 1986.

20. L. G. Moretto, Nucl. Phys. A182, 641 (1972).
21. J. Dudek, B. Herskind, W. Nazarewicz, Z. Szymanski, and T. R. Werner, Phys. Rev. C38, 940 (1988).
22. A. L. Goodman, Nucl. Phys. A520, 567c (1990).
23. A. L. Goodman, in Understanding the Variety of Nuclear Excitations, ed. by A. Covello (World Scientific, Singapore, 1991) p. 639.
24. A. L. Goodman, in High Spin Physics and Gamma-Soft Nuclei, ed. by J. Saladin, R. Sorensen, and C. M. Vincent (World Scientific, Singapore, 1991) p. 1.
25. Y. Alhassid and B. Bush, Phys. Rev. Lett. 65, 2527 (1990).
26. Y. Alhassid and B. Bush, preprint YCTP-N9-90.
27. A. L. Goodman, in Future Directions in Nuclear Physics, ed. by J. Dudek, in press.

Algebraic Approach to Molecular Spectra

A. Frank
Instituto de Ciencias Nucleares, UNAM
Apdo. Postal 70-543, México, D.F., 04510 México.

Abstract: Algebraic methods for the description of molecular excitations are presented, starting with a schematic $U(2)$ model which only requires angular momentum theory. I then describe the $U(4)$ based vibron-electron model, which treats rotation, vibration and electronic excitations in diatomic molecules in a unified way.

1. Introduction

Symmetry methods have found numerous applications in different branches of physics, from solid state and molecular physics where obvious geometric symmetries are present, to nuclear and particle physics where these symmetries appear in more subtle ways and where group theory has become the standard language. These applications often involve continuous groups, such as the $SO(3)$ group of rotations, where the local properties of the corresponding transformations are described by a set of generators that conform an algebra under the commutator operation, hence the term "algebraic" associated to such group-theoretical techniques.

Among the better-known examples of algebraic methods are the early ones by Heisenberg and Wigner concerning isospin [1] and spin-isospin [2] symmetries in Nuclear Physics, followed by Racah's work on atomic and nuclear configurations [3], Elliot's $SU(3)$ description of rotational spectra [4] and Gel-Mann and Ne'eman's $SU(3)$ hadron model [5]. A more recent example is provided by the Interacting Boson Model of nuclear structure [6], where algebraic techniques play a central role, not only due to the elegance and simplification they bring about in calculations but because of the physical insights they provide for dealing with more complex situations. It is important to emphasize these features against the background of the very large numerical calculations arising in the usual integro-differential techniques.

Similar methods have been recently proposed for the description of molecular spectra [7]. Although the point groups associated to discrete geometrical symmetries in molecules have been used for many years in the study of rotation, vibration and electronic spectra and electromagnetic transitions, continuous groups have not played an equally important role for these systems. The usual approach is to start from the exact non-relativistic Schrödinger equation for the nuclei and electrons in the molecule and carry out successive simplifications, including the Born-Oppenheimer approximation, corresponding to the adiabatic motion of the nuclei with respect to the electrons. The remaining problem is, however, still difficult to solve, involving the diagonalization of very large matrices, particularly for polyatomic molecules.

The alternative route followed by the vibron model was devised following the interacting boson model paradigm. Instead of the s and d bosons of the IBM, the building blocks of the vibronic Hamiltonian are s and p bosons, the latter reflecting the dipole character of the local modes in each bond [7]. The model was first applied to rotation-vibration spectra in diatomic [8] and polyatomic [9] molecules and was recently supplemented with the introduction of electronic degrees of freedom, thus providing a unified algebraic framework for the analysis of molecular spectra [10].

In the next section a simplified algebraic model based on the $U(2)$ algebra is studied in order to introduce the algebraic language and techniques. In section 3 the vibron model is discussed, concentrating on the electron-vibron model and its application to the hydride diatomic molecules, while in section 4 a method to extract Born-Oppenheimer potential curves from the algebraic Hamiltonian is presented. Finally, in the last section I present the conclusions.

2. $U(2)$ Model

We start by considering the well-known $SU(2)$ algebra

$$[\hat{J}_i, \hat{J}_j] = i\epsilon_{ijk}\hat{J}_k \ , \qquad i,j = x,y,z \ , \tag{2.1}$$

where the $\hat{J}_i$ are cartesian components of the angular momentum operators [11]. Defining creation (annihilation) opertors $s^\dagger, t^\dagger (s,t)$ satisfying

$$[s, s^\dagger] = [t, t^\dagger] = 1 \ , \tag{2.2}$$

with all other commutators being zero, the $\hat{J}_i$ may be realized by

$$\hat{J}_x = \frac{1}{2}(s^\dagger s - t^\dagger t) \ , \qquad \hat{J}_y = \frac{1}{2}(t^\dagger s + s^\dagger t) \ , \qquad \hat{J}_z = \frac{i}{2}(t^\dagger s - s^\dagger t) \ , \tag{2.3}$$

which satisfy (2.1). In addition, we define the boson number operator

$$\hat{N} = s^\dagger s + t^\dagger t \ , \tag{2.4}$$

which commutes with the $\hat{J}_i$. The four operators together constitute a $U(2)$ algebra, isomorphic to $SU(2) \times U(1)$. We note that the total angular momentum operator becomes

$$\hat{J}^2 = \frac{1}{4}\hat{N}(\hat{N}+2) \ , \tag{2.5}$$

which compared with the eigenvalue $j(j+1)$ gives $j = N/2$. The eigenstates are thus classified by either N or j and we may realize them in terms of polynomials in the creation operators $s^\dagger$ and $t^\dagger$.

In analogy to the nuclear IBM, we consider a model of N interacting s and t bosons, in this case both with zero angular momentum. The most general one- and two-body, hermitean Hamiltonian, which preserves the total number of bosons is then given by

$$\hat{H} = E_0 + \epsilon_s s^\dagger s + \epsilon_t t^\dagger t + u_1 s^\dagger s^\dagger ss + v_1(s^\dagger s^\dagger tt + t^\dagger t^\dagger ss)$$
$$+ u_2 t^\dagger t^\dagger tt + v_2 t^\dagger s^\dagger ts \ , \tag{2.6}$$

where E_0 is constant for a fixed boson number N. In this space we find two symmetry limits, which correspond to the two chains

$$\begin{array}{ccc} U(2) & \supset & U(1) \\ \{\hat{N}\} & & \{t^\dagger t\} \end{array} \ , \tag{2.7a}$$

$$\begin{array}{ccccc} U(2) & \supset & SU(2) & \supset & SO(2) \\ \{\hat{N}\} & & \{\hat{J}^2\} & & \{\hat{J}_z\} \end{array} \ , \tag{2.8}$$

where below each group the operators that label their representations are indicated. In (2.7a) both $\hat{n}_t = t^\dagger t$ and $\hat{n}_s = s^\dagger s = \hat{N} - \hat{n}_t$ are diagonal, while in (2.8) it is $\hat{J}_z$ which is chosen to be diagonal. Note that chain (2.7a) is isomorphic to the chain

$$\begin{array}{ccccc} U(2) & \supset & SU(2) & \supset & \overline{SO(2)} \\ \{N\} & & \{\hat{J}^2\} & & \{\hat{J}_x\} \end{array} \ , \tag{2.7b}$$

since $\hat{J}_x = \frac{1}{2}(\hat{n}_s - \hat{n}_t)$. The branching rules are simply given by

$$n_t = 0, 1, 2, \ldots N \tag{2.9}$$

and

$$m = j, j-1, \ldots -j \ , \tag{2.10}$$

where m is the projection quantum number associated to $\hat{J}_z$. We denote by σ the $\overline{SO(2)}$ label in chain (2.8b), which also satisfies the $SU(2) \supset SO(2)$ rule (2.10). Because of the explicit form of $\hat{J}_z$ we find

$$\sigma = \frac{1}{2}(N - 2n_t) = j - n_t \ . \tag{2.11}$$

The eigenstates associated to (2.7a) are found as a product of two one-dimensional oscillator states

$$|Nn_t> = \frac{(s^\dagger)^{N-n_t}(t^\dagger)^{n_t}}{\sqrt{n_t!(N-n_t)!}}|0> \ , \tag{2.12a}$$

which in terms of the σ quantum number (2.11) are given by

$$|j\sigma> = \frac{(s^\dagger)^{j+\sigma}(t^\dagger)^{j-\sigma}}{\sqrt{(j+\sigma)!(j-\sigma)!}}|0> \ . \tag{2.12b}$$

From (2.12b)

$$|\tfrac{1}{2}\ \tfrac{1}{2}> = s^\dagger|0> \ , \qquad |\tfrac{1}{2}\ -\tfrac{1}{2}> = t^\dagger|0> \ , \tag{2.13}$$

which identifies the $s^\dagger$ and $t^\dagger$ single boson states as components of a $j = \frac{1}{2}$ spinor in the $SU(2) \supset \overline{SO(2)}$ basis. To find the $SO(2)$ states (2.9) we carry out an $SU(2)$ rotation in the $s-t$ space [11]

$$|j\mu> = \sum_\sigma D^j_{\sigma\mu}(-\frac{\pi}{2}, -\frac{\pi}{2}, 0)|j\sigma> \ , \tag{2.14}$$

which takes the single boson states $(j = \frac{1}{2})$ into

$$s^\dagger \to \frac{e^{i\pi/4}}{\sqrt{2}}(s^\dagger + it^\dagger) \ , \quad t^\dagger \to \frac{e^{i\pi/4}}{\sqrt{2}}(s^\dagger - it^\dagger) \ , \tag{2.15}$$

and transforms the $SU(2)$ generators (2.3) cyclically: $\hat{J}_x \to \hat{J}_z$, $\hat{J}_y \to \hat{J}_x$, $\hat{J}_z \to \hat{J}_y$. It follows that the transformation (2.16) leads to the desired states (2.8)

$$|jm> = \frac{(-)^j(s^\dagger - it^\dagger)^{j-m}(s^\dagger + it^\dagger)^{j+m}}{2^j\sqrt{(j-m)!(j+m)!}}|0> \ . \tag{2.16}$$

The $SU(2)$ rotation matrix (2.14) constitutes the transformation bracket between the two chains. The Hamiltonian (2.6) may be conveniently expressed in terms of the Casimir invariants in chains (2.7) and (2.8) and the operations in the model carried out in the $|jm>$ basis (or equivalently in the $|j\sigma>$ basis) using the well known $SU(2) \supset SO(2)$ angular momentum algebra, its Clebsch-Gordan coefficients, Wigner-Eckart theorem, etc.

We now consider the application of the $U(2)$ model to the stretching vibrations in A-B-A triatomic molecules, constituted by two identical and one (in general) unlike atom, such as H_2O and CO_2 [12]. These vibrations are usually described in terms of normal modes which are either symmetric or antisymmetric with respect to a plane of symmetry of the molecule perpendicular to the A-B-A plane and passing through the B atom. Normal modes occur in pure form only in the ideal case of exactly harmonic motion, arising from quadratic interaction terms in the molecular Hamiltonian. Another possibility is to consider the motion in terms of local anharmonic modes, through the weak coupling of two independent oscillators, while in real cases it is a mixture of these two idealized schemes [13]. The symmetric and antisymmetric stretching vibrations in A-B-A molecules occur in both linear and non-linear molecules and a full description of their excitation requires the use of algebras which contain the angular momentum subalgebra, as discussed in the next section. However, in contrast to the bending vibrations, they correspond to non-degenerate intrinsic excitations of the molecule [12]. By restricting ourselves to the $L = 0$ excitations, we may carry out a simpler analysis based on the $U(2)$ algebra [13]. To see why this is so consider the one-dimensional Morse Hamiltonian,

$$H_M = \frac{-\hbar^2}{2\mu} \frac{d^2}{dx^2} + D[exp(-2x/d) - 2exp(-x/d)] \; . \tag{2.17}$$

The eigenstates of (2.17) may be put in a one to one correspondence with the $U(2) \supset SO(2)$ states (2.16), as long as we restrict the value of m to non-negative values. To show how this connection comes about, take the radial equation

$$\frac{1}{2}(-\frac{1}{r}\frac{d}{dr}r\frac{d}{dr} + \frac{m^2}{r^2} + r^2)\phi(r) = (N+1)\phi(r) \; , \tag{2.18}$$

which corresponds to a two-dimensional harmonic oscillator and thus to a $U(2)$ symmetry algebra [14]. Now carry out the transformation [15]

$$r^2 = (N+1)exp(-\rho) \; ,$$

which implies

$$\frac{d}{dr} = \frac{d\rho}{dr}\frac{d}{d\rho} = \frac{-2}{\sqrt{N+1}}e^{\rho/2} \; .$$

Substituting in (2.18)

$$\left[-\frac{d^2}{d\rho^2} + \left(\frac{N+1}{2}\right)^2 (e^{-2\rho} - 2e^{-\rho})\right]\phi(\rho) = -m^2\phi(\rho) \; , \tag{2.19}$$

defining $x \equiv \rho \cdot d$ and multiplying (2.19) by $\hbar^2/2\mu d^2$ leads to (2.17), provided that

$$N + 1 = \sqrt{\frac{8\mu d^2 D}{\hbar^2}} \;, \qquad (2.20)$$

$$E = -\frac{\hbar^2}{2\mu d^2} m^2 \;. \qquad (2.21)$$

Since $N = 0, 1, 2, \ldots$ and $m = \pm N/2, \pm(N-2)/2 \ldots$, we see that the Morse spectrum is reproduced twice and that one has to restrict the values of m to non-negative values, as stated above. The Morse eigenstates can thus be classified by the $|jm>$ states (2.16). Note that relation (2.21) implies that the Morse Hamiltonian has the algebraic realization

$$H_M = -\frac{\hbar^2}{2\mu d^2} J_z^2 \;. \qquad (2.22)$$

From (2.20) we also note that the $U(2)$ eigenvalue N is related to the potential depth, as expected from the fact that N fixes the number of bound states m.

Given the connection of the $U(2) \supset SO(2)$ chain with the one-dimensional Morse oscillator, for A-B-A molecules we use the chains [13]

$$\begin{array}{cccccc} SU_1(2) & \times & SU_2(2) & \supset & SO_1(2) & \times & SO_2(2) & \supset & SO(2) \\ j_1 & & j_2 & & m_1 & & m_2 & & m \end{array} \;, \qquad (2.23)$$

and

$$\begin{array}{cccccc} SU_1(2) & \times & SU_2(2) & \supset & SU(2) & \supset & SO(2) \\ j_1 & & j_2 & & j & & m \end{array} \;, \qquad (2.24)$$

to consider the interactions among two Morse oscillators describing the A-B bonds. While in the first coupling the nature of the individual Morse oscillators is preserved, in the second one it is not.

In (2.23) and (2.24) we have written the $SU(2)$ groups instead of the $U(2)$ ones, where $j_i = N_i/2$, $i = 1, 2$, as explained before. States classified by chain (2.23) correspond to the "local modes" basis, since both Morse oscillators are well defined and weakly interact. It is convenient to define the number of quanta in the local modes basis by the relation

$$n_i \equiv j_i - m_i \qquad i = 1, 2 \;, \qquad (2.25)$$

which take the values

$$n_i = 0, 1, \ldots j_i \quad \text{or} \quad j_i - 1/2 \quad \text{(for } N_i \text{ even or odd)} \;.$$

Recalling that we only keep non-negative values of m,

$$n_1 + n_2 = \frac{N}{2} - m = 0, 1, \ldots N/2 , \qquad (2.26)$$

and we may use $n_1 + n_2$ in (2.23) as a label to classify the $SO(2)$ states instead of m. The local modes basis is then classified by the chain

$$\begin{array}{ccccccc} SU_1(2) & \times & SU_2(2) & \supset & SO_1(2) & \times & SO_2(2) & \supset & SO(2) \\ j_1 & & j_2 & & n_1 & & n_2 & & n_1 + n_2 \end{array} . \qquad (2.27)$$

On the other hand, states (2.24) involve a strong coupling of the local modes basis, through the nondiagonal $\hat{J}^2$ interaction associated to the coupled $SU(2)$. To see whether these states correspond to normal modes, we write the $SU(2) \supset SO(2)$ states $|j_1 j_2 j m>$

$$|j_1 j_2 j m> = \sum_{m_1 m_2} (j_1 m_1 j_2 m_2 | j m) |j_1 m_1> |j_2 m_2> , \qquad (2.28)$$

and interchange the j_1 and j_2 quantum numbers

$$\begin{aligned}|j_2 j_1 j m> &= \sum_{m_1 m_2} (j_2 m_2 j_1 m_1 | j m) |j_1 m_1> |j_2 m_2> \\ &= (-)^{j_1 + j_2 - j} |j_1 j_2 j m> = (-)^{\frac{N}{2} - j} |j_1 j_2 j m>\end{aligned} \qquad (2.29)$$

where we used symmetry properties of the C.G. coefficients [11]. This exchange is equivalent in the algebraic framework to the reflection operation on the symmetry plane passing through the B atom. In (2.29), $N = N_1 + N_2$ is the total number of quanta. Since for A-B-A molecules the two bonds are identical, $N_1 = N_2 = N/2$, where N is an even number. Denoting by $\hat{\sigma}_v$ the reflection operator, eq.(2.29) reads

$$\hat{\sigma}_v |N j m> = (-)^{N/2 - j} |N j m> , \qquad (2.30)$$

so we verify that the basis (2.24) naturally describes symmetric and antisymmetric vibrational states.

We now define the labels for the number of quanta in the normal modes basis, as we did in (2.25) for the local modes basis. Due to (2.30), the number of quanta in the antisymmetric mode is given by

$$v_3 \equiv N/2 - j = 0, 1, \ldots N/2 , \qquad (2.31)$$

so that we obtain symmetric or antisymmetric states under $\hat{\sigma}_v$ for v_3 even or odd, respectively. To define the number of quanta in the symmetric mode, v_1, we note that it must depend on the quantum number $m = m_1 + m_2$, since only N and j appear in the definition (2.31). The $SO(2)$ subgroup is common to the two bases, (2.23) and (2.24), so we may fix v_1 by requiring $n_1 + n_2 = v_1 + v_3$, since the local modes basis (2.27) was chosen to have $n_1 + n_2 = \frac{N}{2} - m$ as $SO(2)$ label. Using (2.26) and (2.31) we find

$$v_1 = j - m \; , \tag{2.32}$$

which is analogous to relations (2.25) for the coupled $SU(2)$ system. Equation (2.32) is thus consistent with the interpretation of v_1 as related to the symmetric modes, which are analogous to one-dimensional vibrations.

Summing up, we have shown that the basis states (2.23) and (2.24) may be written in the form $|Nn_1n_2>$ and $|Nv_1v_3>$ and correspond to the local and normal modes bases, respectively.

The most general one- and two-body Hamiltonian associated to the system is

$$\hat{H} = E_0 + A[\hat{J}_{z_1}^2 + \hat{J}_{z_2}^2] + B\hat{J}_z^2 + C\hat{J}^2 \; , \tag{2.33}$$

where the coefficient of $\hat{J}_{z_1}^2$ and $\hat{J}_{z_2}^2$ was set equal due to the symmetry of the A-B-A molecule and E_0 only depends on N, which is a fixed number determined for a given molecule by the potential depth. The Hamiltonian is diagonal in the local modes basis when $C = 0$, while it is diagonal in the normal modes basis for $A = 0$. In the general case we may evaluate the matrix elements of J^2 in the local modes basis and carry out the diagonalization of the Hamiltonian. In order to compute these matrix elements it is convenient to rewrite (2.23) in the form

$$\hat{H} = E_0' + \alpha_1[\hat{J}_{z_1}^2 + \hat{J}_{z_2}^2] + \alpha_2 \hat{J}_{z_1} \cdot \hat{J}_{z_2} + \alpha_3 \mathbf{J}_1 \cdot \mathbf{J}_2 \; , \tag{2.34}$$

where $\mathbf{J}_1 \cdot \mathbf{J}_2 = \hat{J}_{z_1}\hat{J}_{z_2} + \hat{J}_{y_1}\hat{J}_{y_2} + \hat{J}_{z_1}\hat{J}_{z_2}$, $E_0' = E_0 + CN(N+4)/8$, $\alpha_1 = A+B$, $\alpha_2 = 2B$ and $\alpha_3 = 2C$. In this form we find that the first three terms are again diagonal in the local modes basis, while $\mathbf{J}_1 \cdot \mathbf{J}_2$ gives the matrix elements [11]

$$< Nn_1'n_2'|\mathbf{J}_1 \cdot \mathbf{J}_2|Nn_1n_2> = (\frac{N}{4} - n_1)(\frac{N}{4} - n_2)\delta_{n_1 n_1'}\delta_{n_2 n_2'}$$
$$+ \sqrt{n_1(N/2 - n_1 + 1)(N/2 - n_2)(n_2 + 1)}\delta_{n_1 n_1' - 1}\delta_{n_2 n_2' + 1}$$
$$+ \sqrt{(N/2 - n_1)(n_1 + 1)n_2(N/2 - n_2 + 1)}\delta_{n_2 n_2' + 1}\delta_{n_2 n_2' - 1} \; . \tag{2.35}$$

Note that the matrix turns out to be tridiagonal. An alternative approach is to compute the matrix elements of $\hat{J}_{z_1}^2$ and $\hat{J}_{z_2}^2$ in the normal modes basis $|Nv_1v_3>$. This can be done using the transformation brackets (2.14), which leads to a sum over Clebsch-Gordan coefficients.

In Table 2.1 we show the results of a least square fit to the stretching modes of water [13]. We identify the levels by the normal modes states $|Nv_1v_3>$ which give maximum overlap with the resulting eigenstates.

Table I

Stretching Modes of Water. The parameters in (2.34) are $\alpha_1 = -20.1134$, $\alpha_2 = -2.2639$, $\alpha_3 = -.5131$, $N = 88$. The energy is in units of cm^{-1}. The mean square deviation is 3.99 cm^{-1}.

$v_1\, v_3$	Calc.	Expt.	$v_1\, v_3$	Calc.	Expt.
(10)	3 658.82	3 657.05	(04)	14 546.49	14 536.87
(01)	3 749.13	3 755.93	(32)	16 895.22	16 898.40
(20)	7 205.41	7 201.54	(41)	16 895.34	16 898.83
(11)	7 247.04	7 249.82	(50)	17 459.57	17 458.20
(02)	7 438.49	7 445.05	(23)	17 492.63	17 495.52
(30)	10 604.46	10 599.66	(14)	17 746.26	17 748.07
(21)	10 614.69	10 613.41	(05)	17 971.71	17 970.90
(12)	10 865.28	10 868.86	(42)	19 795.67	...
(03)	11 031.58	11 032.40	(51)	19 795.67	...
(22)	13 832.77	13 828.30	(60)	20 529.98	...
(31)	13 834.12	13 830.92	(24)	20 535.80	...
(40)	14 220.69	14 221.14	(15)	20 897.17	...
(13)	14 316.17	14 318.80	(06)	21 046.45	...

The $SU(2) \times SU(2)$ model has been applied successfully to other A-B-A molecules [13] and has recently been extended to polyatomic molecules by Iachello and Oss [16].

3. The $U(4)$ Model

Having discussed in detail the $U(2)$-based model, we briefly indicate the way to describe molecular rotation-vibration degrees of freedom. In the vibron model, instead of the two $L = 0$ bosons of last section we introduce an $s^+(L = 0)$ boson and a $p_m^+(L = 1)$ boson, the latter carrying negative parity, being associated to the dipole character of the molecular bond. In this case the vibron algebra for each bond is generated by the operators

$$C_{lm}^{l'm'} = b_{lm}^{\dagger} b_{l'm'} \, , \qquad l, l' = 0, 1, \tag{3.1}$$

where $b_{lm}^\dagger \equiv p_m^\dagger$, $b_{00}^\dagger \equiv s^\dagger$. These generators satisfy the $U(4)$ commutation relations and again the Hamiltonian and other observables are expressed in terms of them [7,8]. In this case there are two chains of algebras (or dynamical symmetries) incorporating the angular momentum subalgebra:

$$U(4) \supset U(3) \supset SO(3) , \tag{3.2}$$

$$U(4) \supset SO(4) \supset SO(3) , \tag{3.3}$$

the latter being closely associated to the spectrum of a three-dimensional Morse oscillator [8], while the former corresponds to a harmonic oscillator. This is analogous to the situation with the two chains in the $U(2)$ model of the last section. All calculations in the $U(4)$ model can again be carried out using algebraic techniques, for which one can exploit the isomorphism [17]

$$\begin{array}{cccc} SO(4) & \approx & SU(2) & \times & SU(2) \\ (\sigma_1 \sigma_2) & & j & & j' \end{array} . \tag{3.4}$$

where the $SO(4)$ labels (σ_1, σ_2) are related to the $SU(2)$ ones by

$$\sigma_1 = j + j' , \qquad \sigma_2 = j - j' . \tag{3.5}$$

This connection implies that one can carry out all calculations in chain (3.3) again in terms of the well-known $SU(2)$ algebra [8,18,19]. For polyatomic molecules, it is necessary to introduce an independent $U(4)$ algebra for each molecular bond [9]. For lack of space I shall not discuss the vibron model any further, but refer the reader to the original publications [7, 8, 9, 18]. It should be pointed out, however, that it constitutes a very successful technique for the description of molecular rotation-vibration excitations and an active area of research [9].

In spite of its remarkable success, the model does not incorporate the molecular electronic degrees of freedom and is restricted to the description of rotation-vibration spectra associated to the ground electronic configuration, which in addition should be an $L = 0$ state, i.e., a Σ^+ state in the case of linear molecules. A full description of molecular spectra requires the explicit introduction of the electronic excitations. We need to supplement the bosonic degrees of freedom with fermionic ones, representing the $D = \Sigma_i 2(2l_i + 1)$ electronic single particle states. We shall discuss the simplest possible molecular system, that of heteronuclear diatomic molecules [10].

For simplicity we take the united atoms limit and use an $O(4)$ electronic basis. These states should be coupled with the N-boson vibron model states to construct a complete set, classified by the $[1^m] \times [N]$ irreps of $U^{rv}(4) \otimes U^e(D)$, where m is the number of active electrons and the superscripts (rv) and (e) refer to the vibron and electronic dynamical groups, respectively. Specifically, if we consider as a basis for the united atoms limit that of hydrogenic levels, $D = 2n^2$, where n is the principal quantum number. As an example, we consider the case of electronic states originating from the $2s$, $2p$ levels, for which $D = 8$. In this case the basis states carry representations of the

$$U^{rv}(4) \otimes U^e(8) \tag{3.6}$$

direct product group. Besides the bilinear products $b_{l\mu}^\dagger b_{l'\mu'}$ that generate the unitary group $U^{rv}(4)$, there is a set of creation $a_{lm\frac{1}{2}\sigma}^\dagger$ and annihilation $a_{lm\frac{1}{2}\sigma}$ fermion operators which satisfy the anticommutation relations

$$\begin{aligned} \{a_{lm\frac{1}{2}\sigma}, a_{l'm'\frac{1}{2}\sigma'}^\dagger\} &= \delta_{ll'}\delta_{mm'}\delta_{\sigma\sigma'}, \\ \{a_{lm\frac{1}{2}\sigma}^\dagger, a_{l'm'\frac{1}{2}\sigma'}^\dagger\} &= \{a_{lm\frac{1}{2}\sigma}, a_{l'm'\frac{1}{2}\sigma'}\} = 0, \end{aligned} \tag{3.7}$$

where l and m correspond to the orbital angular momentum and projection, and σ is the projection of the spin $s = 1/2$. For the $n = 2$ manifold $l = 0, 1$, which correspond to the $2s - 2p$ hydrogenic levels. The products $a_{lm\frac{1}{2}\sigma}^\dagger a_{l'm'\frac{1}{2}\sigma'}$ generate the unitary group $U^e(8)$ and we may separate the electronic degrees of freedom in their orbital and spin parts by means of the reduction

$$U^e(8) \supset U^e(4) \otimes SU_s(2), \tag{3.8}$$

which corresponds to the $L-S$ coupling scheme. With this decomposition the group $U(4)$ appears for both the vibrons and the electrons and the chains (3.2) and (3.3) are present for both systems. The algebraic problem is reduced to that of the coupling of two $U(4)$ algebras. (Note again the analogy with the description of A-B-A stretching modes). The different decompositions of the dynamical group (3.6) followed by (3.8) is given by the different ways in which the chains (3.2) and (3.3) can be coupled, preserving the $O^{rv+e}(3)$ group. The latter is isomorphic to the spin group $SU_s(2)$ and thus can be coupled to it to yield the total angular momentum group $SU_J(2)$.

The full spectrum of rotation-vibration-electronic states is obtained by diagonalization of the molecular Hamiltonian

$$\hat{\mathcal{H}} = \hat{\mathcal{H}}^{rv} + \hat{\mathcal{H}}^e + \hat{V}^{rv-e}, \tag{3.9}$$

where $\hat{H}^{rv}$ is the vibron Hamiltonian [7], $\hat{H}^e$ is the electronic contribution (formally equivalent in this case to an atomic Hamiltonian) and $\hat{V}^{rv-e}$ is the interaction term [10, 19]. The interaction term is fundamental for the ordering of the electronic states and for spin-zero interactions may be written in the multipole expansion form

$$\hat{V}^{rv-e} = V^0 \hat{n}_p \hat{n}_F + V_1^1 \hat{D}_{rv} \cdot \hat{D}_e + V_2^1 \hat{\bar{D}}_{rv} \cdot \hat{\bar{D}}_e + V_3^1 \hat{L}_{rv} \cdot \hat{L}_e + V^2 \hat{Q}_{rv} \cdot \hat{Q}_e , \qquad (3.10)$$

where

$$\begin{aligned}
\hat{n}_p &= \sqrt{3}(p^\dagger \times \tilde{p})^{(0)}, & \hat{n}_F &= -\sqrt{6}(a^\dagger_{1\frac{1}{2}} \times \tilde{a}_{1\frac{1}{2}})^{(00)}, \\
\hat{L}_{rv,\mu} &= \sqrt{2}(p^\dagger \times \tilde{p})^{(1)}_\mu, & \hat{L}_{e,\mu} &= -2(a^\dagger_{1\frac{1}{2}} \times \tilde{a}_{1\frac{1}{2}})^{(10)}_\mu, \\
\hat{Q}_{rv,\mu} &= (p^\dagger \times \tilde{p})^{(2)}_\mu, & \hat{Q}_{e,\mu} &= -\sqrt{2}(a^\dagger_{1\frac{1}{2}} \times \tilde{a}_{1\frac{1}{2}})^{(20)}_\mu, \\
D_{rv,\mu} &= (p^\dagger s - s^\dagger \tilde{p})^{(1)}_\mu, & D_{e,\mu} &= -\sqrt{2}(a^\dagger_{1\frac{1}{2}} \times \tilde{a}_{0\frac{1}{2}} - a^\dagger_{0\frac{1}{2}} \times \tilde{a}_{1\frac{1}{2}})^{(10)}_\mu, \\
\bar{D}_{rv,\mu} &= i(p^\dagger s + s^\dagger \tilde{p})^{(1)}_\mu, & \bar{D}_{e,\mu} &= -i\sqrt{2}(a^\dagger_{1\frac{1}{2}} \times \tilde{a}_{0\frac{1}{2}} + a^\dagger_{0\frac{1}{2}} \times \tilde{a}_{1\frac{1}{2}})^{(10)}_\mu .
\end{aligned} \qquad (3.11)$$

While $\hat{n}$ and $\hat{Q}$ correspond to monopole and quadrupole operators, the other three are $L=1$ operators, with $\hat{D}$ being the physical dipole operator. A complete set of states for the system is a combination of the rotation-vibration states [7,8,18]

$$\begin{array}{cccc}
U^{rv}(4) \supset & SO^{rv}(4) \supset & SO^{rv}(3) \supset & SO^{rv}(2) \\
\downarrow & \downarrow & \downarrow & \downarrow \\
\mid [N] , & \omega , & l , & m >,
\end{array} \qquad (3.12)$$

and the electronic wave functions [10,19]

$$\begin{array}{cccc}
U^e(8) \supset U^e(4) \times SU(2) \supset & SO^e(4) \times SU(2) \supset & SO^e(3) \times SU(2) \supset \\
\mid [1^m] , & S , & (\tau_1 \tau_2) , & l_e , \\
& & SO^e(2) \times SU(1) & \\
& & m_e , \quad \sigma >,
\end{array} \qquad (3.13)$$

where we indicate below each group the quantum numbers that label their representations. The full wave functions are then classified by

$$\begin{array}{cccc}
U^{rv}(4) \times U^e(8) \supset & U^{rv}(4) \times U^e(4) \times SU(2) \supset & SO^{rv}(4) \times SO^e(4) \times \\
\downarrow \quad \downarrow & & \downarrow \quad \downarrow \\
\mid [N], \quad [1^m], & & \omega, \quad (\tau_1 \tau_2), \\
SU(2) \supset SO(4) \times SU(2) \supset & SO(3) \times SU(2) \supset & SU_J(2) \supset \quad SU_J(1) \\
\downarrow & \downarrow \quad \downarrow & \downarrow \quad \downarrow \\
(\sigma_1 \sigma_2) & L, \quad S, & J, \quad M > .
\end{array} \qquad (3.14)$$

Explicitly, the coupling (2.9) is given by:

$$|[N][1^m]\omega(\tau_1\tau_2)(\sigma_1\sigma_2)SLJM> = \sum_{l,l_e,\forall m} \left\langle \begin{matrix} \omega & (\tau_1\tau_2) \\ l & l_e \end{matrix} \middle| \begin{matrix} (\sigma_1\sigma_2) \\ L \end{matrix} \right\rangle C(ll_eL;mm_em_L)$$
$$C(LSJ;m_L\sigma M)|[N]\omega lm>|[1^m](\tau_1\tau_2)l_em_e;S\sigma>, \qquad (3.15)$$

where $<|>$ is an isoscalar factor corresponding to the $SO(4) \supset SO(3)$ reduction. Taking again advantage of the $SO(4) \simeq SU(2) \times SU(2)$ isomorphism, the isoscalar factors can be expressed in terms of $SU(2)$ 9-j symbols. Explicit expressions for the states (3.12)-(3.15) as well as the reduction rules for the quantum labels are known [10,19]. Although the quantum numbers in (3.14) completely characterize the molecular states, we need to correlate them with the usual spectroscopic notation. The vibrational quantum number v is given by [7]

$$v = \frac{N-\omega}{2}, \qquad (3.16)$$

while for the electronic part the usual notation is based on the Born-Oppenheimer approximation, where the projection of the orbital angular momentum on the symmetry axis is conserved and denoted by $\Lambda = \Sigma, \Pi, \Delta, \ldots$. The Σ terms are denoted by Σ^+ or Σ^- depending on their behavior under reflections on a plane passing through the molecular axis. The $\Lambda \neq 0$ states are doubly degenerate and contain both signs, $\Pi^\pm$, $\Delta^\pm$, etc. The reflection parity (π_σ) is related to the total parity (p) by [19]

$$\pi_\sigma = (-)^L p. \qquad (3.17)$$

Using the $SO(4) \supset SO(3)$ reduction rule $L = |\sigma_2|, |\sigma_2|+1, \ldots, \sigma_1$, we can identify Λ with the quantum number σ_2

$$\Lambda = |\sigma_2|. \qquad (3.18)$$

The quantum numbers (τ_1, τ_2) and σ_1 have no counterparts in the standard classification, which is incomplete. The identifications (3.16)-(3.18), together with the S, L, J and M quantum numbers, which are already in standard form, allow us to compare the algebraic model results with experimental data.

In Figure 1, we show the corresponding electronic and rotation-vibration spectrum for the $SO(4)$ dynamical symmetry for $m = 1$ and 7.

Using the model Hamiltonian

$$\hat{H} = E_O + \beta_6 \hat{C}_2(O^{rv}(4)) + f_3 \hat{n}_F + f_4 \hat{Q}_e^2 + f_6 C_2(O^e(4)) \\ + A_2 \hat{L}_e^2 + A_1 \hat{C}_2(O(4)) + 2k\hat{Q}_B \cdot \hat{Q}_F, \qquad (3.19)$$

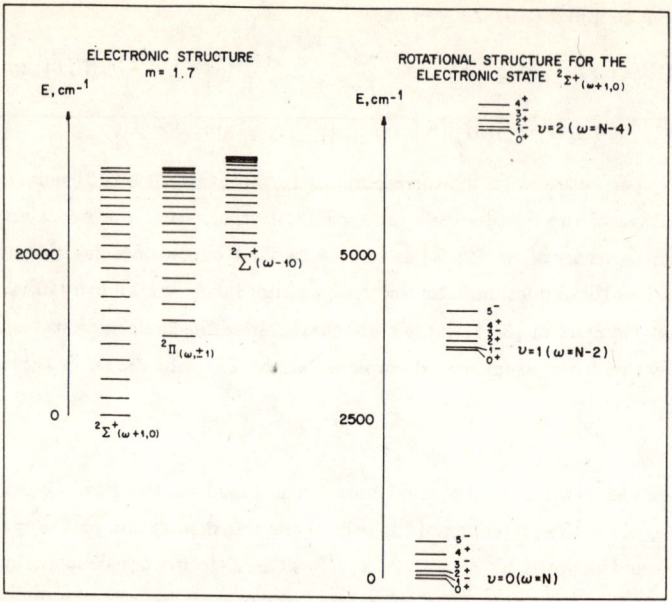

Figure 1. Schematic structure of the electronic vibration-rotation spectrum provided by an $O(4)$ dynamical symmetry for one and seven electrons.

we have applied the model to the hydride molecules LiH, BeH, BH, CH, NH and OH. In fig. 2 we show the results for NH and OH.

The r.m.s. deviation for the states in the 6 molecules is of the order of 4000 cm^{-1} which is comparable to the accuracy of abinitio calculations, but including simultaneously $\simeq 25$ electronic levels. An interesting characteristic of this calculation is the fact that most parameters are determined by atomic fits in the united atoms limit and that all parameters vary smoothly as a function of the electron number m [10]. There are other interesting features of the calculation, like the fact that the model produces naturally all electronic states originating from the $2s - 2p$ united atomic levels, including states not known experimentally.

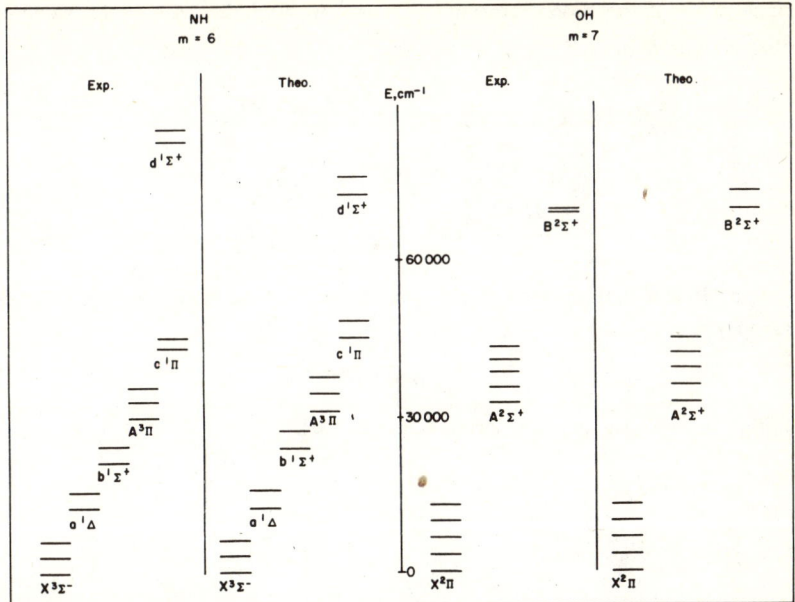

Figure 2. Comparison between calculated and experimental electronic energy levels in NH and $OH (m = 6$ and $7)$.

4. Potential Energy Curves

Even if the algebraic approach is useful in correlating a large amount of experimental data and in predicting states that have not been observed, it is desirable to correlate this method with the more traditional procedures, including the Born-Oppenheimer approximation, which leads to the evaluation of potential energy surfaces associated to the electronic states. We proceed in analogy to the interacting boson model coherent state method [20]. The normalized vibron intrinsic states are given by [21]

$$|N;\ r> = \frac{1}{\sqrt{N!\,(1+r^2)^N}} (s^+ + r p_0^+)^N |0>_v\ , \qquad (4.1)$$

where r is a radial coordinate and $|0>_v$ is the vibron vacuum state. The expectation value of the $O^{rv}(4)$ vibron Hamiltonian in the intrinsic states (4.1), and the change of variable

$$r = \sqrt{\frac{e^{-\rho}}{2 - e^{-\rho}}}\ , \qquad (4.2)$$

leads to a Morse potential in the ρ coordinate [21]. Using

$$[b_i, f(b^+)] = \frac{\partial}{\partial b_i^+} f(b^+)\ , \qquad (4.3)$$

for a function f of the boson creation operators, we find the expectation value of the molecular Hamiltonian (3.19):

$$\begin{aligned}< N; r|\hat{H}|N; r > = & E_0 + (\beta_6 + A_1)[3N + \frac{4N(N-1)r^2}{(1+r^2)^2}] \\
& + f_3 \hat{n}_F + (f_6 + A_1)\hat{C}_2(O^e(4)) + f_4 \hat{Q}_e \cdot \hat{Q}_e + A_2 \hat{L}_e^2 \\
& + A_1 \frac{4Nr}{(1+r^2)} \hat{D}_{e,0} - 2k\sqrt{\frac{2}{3}} \frac{Nr^2}{(1+r^2)} \hat{Q}_{e,0} \,,
\end{aligned} \qquad (4.4)$$

which is an electronic Hamiltonian with r-dependent interactions. We then diagonalize (4.4) in the $O^e(4)$ basis

$$|[1^m](\tau_1 \tau_2) l_e m_e; S\sigma > \,, \qquad (4.5)$$

identifying $|m_e|$ with the standard projection $\wedge$ [19,22].

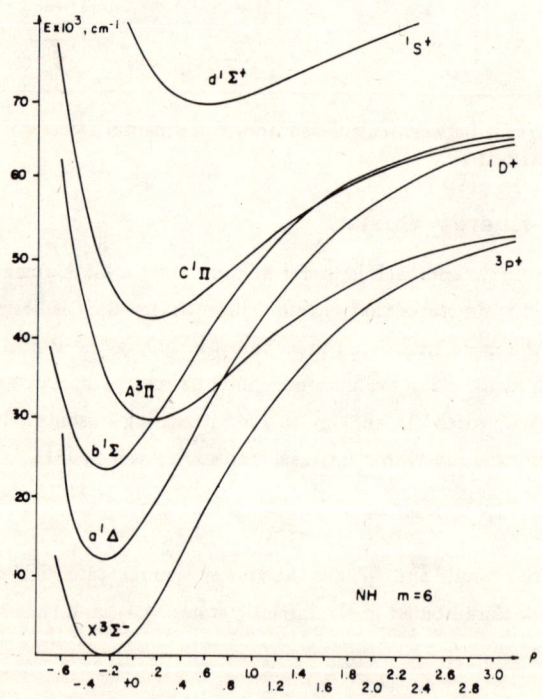

Figure 3. Theoretical potential energy curves for NH according to (4.4).

In Fig. 3 we show the result for NH. The position of the minimum in all the potentials coincides to a close approximation with the energy of the ground state configurations in the exact calculations. The Morse-like potentials are distorted in different proportions by the V^{rv-e} interaction and the minima are displaced. We are not able to reproduce the large ρ behavior of the potentials, however, since we recover the united atoms limit for $\rho \to \infty (r \to 0)$. At smaller values of the radial coordinate, we do find results similar to the standard potentials. To get the correct asymptotic result it is necessary to use the separated atoms limit for the model [23].

5. Summary and Conclusions

In this paper I have attempted to illustrate the usefulness of the algebraic methods for molecular excitations, by means of $U(2)$- and $U(4)$- based models. The $U(2)$ framework is very schematic but is valuable as a tool to introduce the algebraic language and techniques, which appear in similar (though more complicated) form in higher-dimensional models. By means of the well known properties of the $SU(2)$ angular momentum algebra one is able to carry out a full analysis of the corresponding dynamical symmetries, branching rules, coupling of representations and other algebraic operations, and evaluate the matrix elements for functions of the dynamical group generators. The model turns out to be useful for the description of stretching modes in A-B-A molecules and one can see in a transparent way how different subgroup chains are related to physically meaningful coupling schemes for the A-B bonds.

I have also briefly described the $U(4)$ vibron model for molecular rotation-vibration spectra [7,8,9] and pointed out that its application to three- and polyatomic molecules is now a fruitful area of investigation, which may play a significant role in the sistematization of molecular spectroscopic data including spectra and electromagnetic transition probabilities. The inclusion of the electronic degrees of freedom leads to the vibron-electron model, which allows a full description of molecular excitations and the simultaneous analysis of series of molecules [10,19]. The coherent-state method can be used to derive potential energy curves associated to the algebraic Hamiltonian in the Born-Oppenheimer approximation [21,22].

The algebraic approach can correlate the spectroscopic properties of different molecules and their ions and make predictions of states not yet observed [9,10]. Since the wave functions turn out to be of manageable dimensions (at least compared with the ones stemming from other methods) one may compute other properties or use them in the study of scattering data [24].

We have recently developed a way to introduce the point symmetries in the algebraic scheme, which turns out to give rise to a remarkable dynamical symmetry description for the hydride molecules [23] and which is a first step in the direction of generalizing the vibron-electron model to polyatomic molecules. Finally, we are presently working on the application of algebraic models to atom-molecule scattering systems [25].

I thank M. Lozano and J.M. Arias for their invitation to La Rábida and my collaborators in this work R. Lemus and F. Iachello, as well as O. Castaños, P. Van Isacker and A. Leviatan for their many useful comments. This work was supported in part by UNAM-DGAPA IN10-1889.

References

1. W. Heisenberg, Z. Phys. 77, 1 (1932).
2. E.P. Wigner, Phys. Rev. 51, 106 (1937).
3. G. Racah, Phys. Rev. 61, 186 (1942); 62, 438 (1942); 63, 367 (1943); 76, 1352 (1949).
4. J.P. Elliot, Proc. Roy. Soc. A245, 128 (1958).
5. M. Gell-Mann, Phys. Rev. 125, 1067 (1962); Y. Ne'eman, Nucl. Phys. 26, 222 (1961).
6. A. Arima and F. Iachello, Ann. Phys. 99, 253 (1976); 111, 201 (1978); 123, 468 (1979).
7. F. Iachello, Chem. Phys. Lett. 78, 581 (1981).
8. O.S. Van Roosmalen, et.al., Chem. Phys. Lett. 85, 32 (1982).
9. O.S. Van Roosmalen, et.al., J. Chem. Phys. 79, 2515 (1983); F. Iachello and S. Oss, J. Mol. Spect. 142, 85 (1990); F. Iachello, et.al., J. Mol. Spect. 146, 56 (1991).
10. A. Frank, et.al., Chem. Phys. Lett. 131, 380 (1986); J. Chem. Phys. 91, 29 (1989).
11. M.E. Rose, "Elementary Theory of Angular Momentum", Wiley, New York, (1957).
12. G. Herzberg, "Infrared and Raman Spectra of Polyatomic Molecules", Van Nostrand, New York, (1950).
13. O.S. Van Roosmalen, et.al., J. Chem. Phys. 81, 5986 (1984).
14. M. Moshinsky, "The Harmonic Oscillator in Modern Physics: From Atoms to Quarks", Gordon and Breach, (1969).
15. S. Levit and U. Smilansky, Nucl. Phys. A389, 56 (1982).
16. F. Iachello and S. Oss, Phys. Rev. Lett. (in press).
17. B.G. Wybourne, "Classical Groups for Physicists", Wiley, New York, 1974.
18. A. Frank and R. Lemus, J. Chem. Phys. 84, 2698 (1986).
19. R. Lemus and A. Frank, Ann. Phys. 206, 122 (1991); A. Frank, et.al., in "Symmetries in Science V", Edited by B. Gruber, et.al., Plenum Press, New York, 1991, p. 173.
20. J.N. Ginocchio and M.W. Kirson, Phys. Rev. Lett. 44, 1744 (1980); Nucl. Phys. A350, 31 (1980).

21. A. Leviatan and M.W. Kirson, Ann. Phys. $\underline{188}$, 142 (1988).
22. R. Lemus, et.al., to be published.
23. R. Lemus and A. Frank, in "Symmetries in Science V", Edited by B. Gruber, et.al., Plenum Press, New York, 1991, p. 429; Phys. Rev. Lett. (June 1991).
24. R. Bijker, et.al., Phys. Rev. $\underline{A33}$, 871 (1986); Phys. Rev. $\underline{A34}$, 71 (1986).
25. A. Frank, et.al., to be published.

Algebraic Model of Hadronic Structure

F. Iachello

Center for Theoretical Physics, Yale University
New Haven, Connecticut 06511, USA

Abstract : A model of hadronic structure based on the algebra $U(4) \otimes SU_s(2) \otimes SU_f(n) \otimes SU_c(3)$ is reviewed. Applications to the study of $(q\bar{q})$ configurations are presented.

1. INTRODUCTION

Algebraic methods have been extensively used in the last 15 years in nuclear [1] and molecular [2] physics. Although these methods were originally developed for applications to hadronic structure [3], their use in this area of physics has, in the last few years, considerably decreased. In these lectures, I briefly review previous applications of algebraic methods to hadronic structure and introduce a new model which can accomodate previous results but, in addition, can describe other possibilities. These new possibilities can eventually be tested at new experimental facilities, such as the CEBAF facility presently under construction.

Albebraic methods make use of two basic concepts:

A. **Spectrum generating algebra, SGA**

This is the set of elements, G_α, belonging to an algebra $\mathcal{G}$, onto which all operators are expanded

$$0 = f(G_\alpha) \quad , \quad G_\alpha \in \mathcal{G} \quad . \tag{1.1}$$

The algebra $\mathcal{G}$ is usually a Lie algebra, although recently more complicated algebras have been considered, such as graded Lie algebras, q-algebras, etc.

B. Dynamic symmetry, DS

If the energy operator describing the system is expanded not into all elements G_α but into a subset (or combination of subsets) of operators of $\mathcal{G}$, the so-called invariant, or Casimir operators, G_i, one has a dynamic symmetry. For relativistic systems, a convenient operator to describe the energy of the system is the mass squared operator, M^2. Thus, if

$$M^2 = f(G_i) \qquad , \tag{1.2}$$

one has a dynamic symmetry. The importance of a dynamic symmetry is that, when such a symmetry occurs, matrix elements of all operators can be written in explicit analytic form, thus making a comparison with experiment straightforward.

2. SPECTRUM GENERATING ALGEBRA OF HADRONIC PHYSICS

The first step in the construction of an algebraic model is the choice of the appropriate spectrum generating algebra. In hadronic physics this choice is made more difficult for the fact that hadrons leave both "internal" and "geometric" degrees of freedom. Thus $\mathcal{G}$ must be written as

$$\mathcal{G} = \mathcal{G}_c \otimes \mathcal{G}_f \otimes \mathcal{G}_s \otimes \mathcal{R} \qquad , \tag{2.1}$$

where $\mathcal{G}_c, \mathcal{G}_f, \mathcal{G}_s$ denote the color, flavor and spin algebras and $\mathcal{R}$ the "geometric" algebra. In some treatments, the spin and "geometric" algebras are combined together and $\mathcal{G}$ is written as

$$\mathcal{G} = \mathcal{G}_c \otimes \mathcal{G}_f \otimes \mathcal{R}' \qquad . \tag{2.2}$$

Incidentally, Eqs. (2.1) and (2.2) actually mean direct sums of the corresponding algebras. The product sign, $\otimes$, is used conventionally since the wave functions are products functions

$$\Psi = \Psi_c \otimes \Psi_f \otimes \Psi_s \otimes \Psi_R \qquad (2.3)$$

Most applications of algebraic methods to hadronic physics have concentrated on the "internal" part. This part can be written as $SU_c(3) \otimes SU_f(n) \otimes SU_s(2)$, where n is the number of flavors (n=6 in the standard model). Indeed, most applications consider the subalgebra $SU_f(3)$ of $SU_f(n)$ (Gell-Mann,Ne'eman SU(3)) [4], and combine it with $SU_s(2)$ to give SU(6). This SU(6), either in the original Gürsey-Radicati form [5] or in the so-called $SU(6)_W$ form [6], has been extensively employed to study masses and decay widths of hadrons. In another line of research [7,8] SO(4,2) and SL(3,R) have been suggested as spectrum generating algebras R'.

Recently, a somewhat different point of view has been taken. Consider the case of hadronic configurations composed by a certain number, ν, of quarks and antiquarks, Fig. 1. The number of "geometric" degrees of freedom of these configurations is $3(\nu-1)$, or $\nu-1$ vectors. It has been suggested [9] that

$$R = \prod_{i=1}^{\nu-1} U_i(3,1) \qquad , \qquad (2.4)$$

be taken as spectrum generating algebra. I shall call this choice "minimal", since it contains, as it can be seen in the next sections, the "geometric" excitations of hadrons once. Furthermore, since the use of non-compact algebras is rather cumbersome, I will compactify U(3,1) and use its compact form, U(4). Thus, I will take

$$R = \prod_{i=1}^{\nu-1} U_i(4) \qquad . \qquad (2.5)$$

as "geometric" spectrum generating algebra of hadronic configurations with ν quarks and antiquarks.

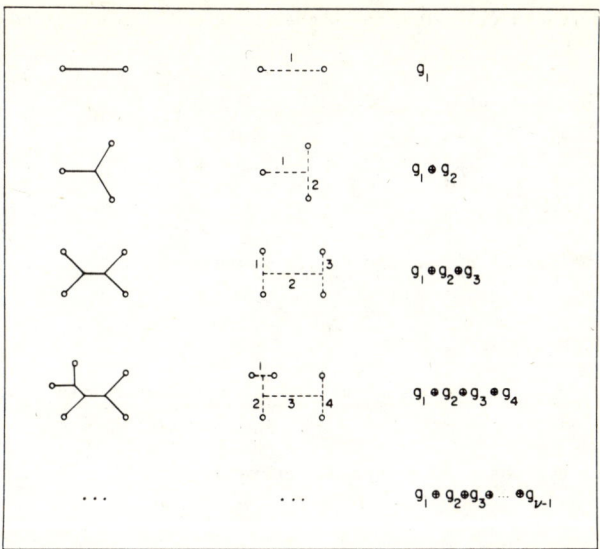

Fig.1. Hadronic configurations composed of ν quarks and antiquarks and their description in terms of spectrum generating algebras, g_i (i=1,2,...,ν-1).

3. PROPERTIES OF U(4)

In view of the central role played by U(4) it is convenient to describe its properties. The algebra U(4) admits two subalgebra chains that contain the physical angular momentum $SO_L(3)$,

$$U(4) \begin{cases} U(3) \supset SO_L(3) \supset SO_L(2) & , \quad (I) \\ SO(4) \supset SO_L(3) \supset SO_L(2) & . \quad (II) \end{cases} \quad (3.1)$$

In order to describe hadronic states we need to consider only the symmetric representation of U(4) characterized by the Young tableau

$$[N] = \overbrace{\square\square\square\ldots\square}^{\text{N-times}} \quad . \quad (3.2)$$

States in the chain (I) can then be characterized by the quantum numbers

$$\left| \begin{array}{cccc} U(4) & \supset & U(3) & \supset & SO_L(3) & \supset & SO_L(2) \\ \downarrow & & \downarrow & & \downarrow & & \downarrow \\ N & & n_\pi & & L & & M_L \end{array} \right\rangle \quad . \quad (I) \qquad (3.3)$$

The allowed values of n_π, L, and M_L are given by

$$n_\pi = N, N-1, N-2, \ldots, 0 \quad ;$$
$$L = n_\pi, n_\pi-2, n_\pi-4, \ldots, 1 \text{ or } 0 \quad , \quad n_\pi = \text{odd or even} ; \qquad (3.4)$$

and $-L \leq M_L \leq +L$. For chain (II), one has instead

$$\left| \begin{array}{cccc} U(4) & \supset & SO(4) & \supset & SO_L(3) & \supset & SO_L(2) \\ \downarrow & & \downarrow & & \downarrow & & \downarrow \\ N & & \omega & & L & & M_L \end{array} \right\rangle \quad . \quad (II) \qquad (3.5)$$

The allowed values of ω, L, and M_L are given by

$$\omega = N, N-2, \ldots, 1 \text{ or } 0 \quad , \quad N = \text{odd or even};$$
$$L = \omega, \omega-1, \ldots, 0 \quad ; \qquad (3.6)$$

and $-L \leq M_L \leq +L$.

It is possible to make a bosonic realization of U(4) in terms of 4 boson creation and annihilation operators $b^\dagger_\alpha, b_\alpha$ ($\alpha=1,\ldots 4$). The 16 generators of U(4) are

$$\mathcal{G}: \quad G_{\alpha\alpha'} = b^\dagger_\alpha b_{\alpha'} \quad , \qquad (3.7)$$

and the states are

$$\mathcal{B}: \frac{1}{\sqrt{N!}} \overbrace{b^\dagger_\alpha b^\dagger_{\alpha'}, \ldots}^{N\text{-times}} \mid 0 > \qquad (3.8)$$

The correct transformation properties under rotations can be insured by further dividing the four boson operators into a scalar, σ, and a vector, $\pi_\mu (\mu=0,\pm1)$. The 16 operators (3.7) can then be written as in Table I.

TABLE I. Transformation properties of the elements of U(4) under rotations. σ and π are boson operators with $J^\pi = 0^+, 1^-$.

Element	Explicit Form[a]	Tensorial Rank Under Rotations
$\hat{n}_\sigma$	$[\sigma^\dagger \times \tilde{\sigma}]^{(0)}_0$	Scalar
$\hat{n}_\pi$	$-\sqrt{3}[\pi^\dagger \times \tilde{\pi}]^{(0)}_0$	Scalar
$\hat{J}$	$-\sqrt{2}[\pi^\dagger \times \tilde{\pi}]^{(1)}_\kappa$	Pseudo-vector
$\hat{D}$	$[\pi^\dagger \times \tilde{\sigma} + \sigma^\dagger \times \tilde{\pi}]^{(1)}_\kappa$	Vector
$\hat{D}'$	$i[\pi^\dagger \times \tilde{\sigma} - \sigma^\dagger \times \tilde{\pi}]^{(1)}_\kappa$	Vector
$\hat{Q}$	$[\pi^\dagger \times \tilde{\pi}]^{(2)}_\kappa$	Symmetric traceless tensor or rank two

[a]The subscript κ denotes the z-component.

In the general case in which there is more than one U(4), one can introduce boson operators $b^\dagger_{\alpha i}$ where $\alpha=1,\ldots,4$; $i=1,\ldots,\nu-1$, and repeat the same construction of algebras and representations. Particularly important is the case of $\nu=3$ (qqq configurations). The algebraic structure of this configuration is $U_1(4) \otimes U_2(4)$ with two important chains

$$U_1(4) \otimes U_2(4) \diagup \begin{matrix} U_1(3) \otimes U_2(3) \supset U(3) \supset SO_L(3) \supset SO_L(2) \quad , \quad (I) \\ \\ SO_1(4) \otimes SO_2(4) \supset SO(4) \supset SO_L(3) \supset SO_L(2) \quad . \quad (II) \end{matrix}$$

(3.9)

If the coordinates i=1,2 are associated with Jacobi coordinates, it is convenient to replace the index 1 by ρ and the index 2 by λ. States in chain (I) are then characterized by

$$\left| \begin{matrix} U_\rho(4) \otimes U_\lambda(4) \supset U_\rho(3) \otimes U_\lambda(3) \supset U(3) \supset SO_L(3) \supset SO_L(2) \\ \downarrow \quad\quad \downarrow \quad\quad \downarrow \quad\quad \downarrow \quad\quad \downarrow \quad\quad \downarrow \quad\quad \downarrow \\ [N_\rho] \quad [N_\lambda] \quad n_\rho \quad n_\lambda \quad (n_1,n_2) \quad L \quad M_L \end{matrix} \right\rangle \quad , \quad (I) \quad ,$$

(3.10)

and in chain (II) by

$$\left| \begin{matrix} U_\rho(4) \otimes U_\lambda(4) \supset SO_\rho(4) \otimes SO_\lambda(4) \supset SO(4) \supset SO_L(3) \otimes SO_L(2) \\ \downarrow \quad\quad \downarrow \quad\quad \downarrow \quad\quad \downarrow \quad\quad \downarrow \quad\quad \downarrow \quad\quad \downarrow \\ [N_\rho] \quad [N_\lambda] \quad \omega_\rho \quad \omega_\lambda \quad (\omega_1,\omega_2) \quad L \quad M_L \end{matrix} \right\rangle \quad , \quad (II) \quad .$$

(3.11)

In this case, one U(4) refers to the Jacobi coordinate, $\vec{\rho}$, and the other to the Jacobi coordinate, $\vec{\lambda}$. These coordinates are related to the three quark coordinates by

$$\vec{\rho} = \frac{1}{\sqrt{2}} (\vec{r}_1 - \vec{r}_2) \quad ,$$

$$\vec{\lambda} = \frac{1}{\sqrt{6}} (\vec{r}_1 + \vec{r}_2 - 2\vec{r}_3) \quad .$$

(3.12)

The allowed values of (n_1,n_2) and L in (3.10) and (ω_1,ω_2) and L in (3.11) can be obtained by the usual rules of tensor products and their decompositions.

4. PROPERTIES OF THE FLAVOR ALGEBRA

The total number of flavors is at present unknown. In the standard model, n=6. The top quark has not been observed so far, indicating that it has a mass greater than 90 GeV. Of the remaining five quarks, the b and c quarks have comparable masses. Similarly the u, d and s quark have comparable masses. This situation suggests the following decomposition of the flavor algebra

$$SU_f(6) \supset SU_f(5) \otimes U_{Y'''}(1) \supset SU_{f'}(3) \otimes SU_{f''}(2) \otimes U_{Y'''}(1) \supset$$
$$\supset SU_I(2) \otimes U_Y(1) \otimes U_{Y'}(1) \otimes U_{Y''}(1) \otimes U_{Y'''}(1) \supset$$
$$\supset SO_I(2) \otimes U_Y(1) \otimes U_{Y'}(1) \otimes U_{Y''}(1) \otimes U_{Y'''}(1) \quad . \quad (4.1)$$

For historic reasons, the u-d quarks are usually denoted by the isospin, I, and its third component, I_3, rather than by the flavor, i.

The flavor algebra is useful for configurations with several quarks or antiquarks, for example qqq or $q^2\bar{q}^2$. It is not particularly useful for $q\bar{q}$ configurations. In the latter case, it is more convenient to use directly the quark basis, denoted by $|q_i\bar{q}_j\rangle$, where i and j represent quark flavors.

The spin-flavor algebra $\mathcal{G}_s \otimes \mathcal{G}_f$ can be realized, for quark configurations, by means of fermion creation and annihilation operators. Denoting by $a^\dagger_{\kappa,i}$ ($\kappa=\pm1/2, i=1,\ldots,6$) the creation operator for a quark of flavor i and spin component κ, and by $\bar{a}^\dagger_{\kappa,i}$ the same operator for an antiquark, one can write the generators of $SU_s(2) \otimes SU_f(6)$ as

$$\mathcal{G}_{sf}: \begin{cases} G_{\kappa i, \kappa' i'} = a^\dagger_{\kappa,i} a_{\kappa',i'} & , \\ \\ \overline{G}_{\kappa i, \kappa' i'} = \overline{a}^\dagger_{\kappa,i} \overline{a}_{\kappa',i'} & . \end{cases} \qquad (4.2)$$

States are obtained by acting with the $a^\dagger$'s on a vacuum state

$$\mathcal{F}: \quad a^\dagger_{\kappa,i} \ldots \overline{a}^\dagger_{\kappa',i'} \ldots | 0 > \qquad . \qquad (4.3)$$

When written explicity in terms of creation and annihilation operators, the wave functions of $q\overline{q}$ configurations are

$$\mathcal{BF}: \quad |q_{\kappa,i} \overline{q}_{\kappa',i'}; \alpha, \alpha', \ldots> = \frac{1}{\sqrt{N!}} \overbrace{b^\dagger_\alpha b^\dagger_{\alpha'} \ldots}^{N\text{-times}} a^\dagger_{\kappa,i} \overline{a}^\dagger_{\kappa',i'} | 0 > \quad , \qquad (4.4)$$

and similar, but more complicated, expressions hold for $qqq,\ldots$, configurations.

5. MASS FORMULAS

The second step in the implementation of the algebraic scheme is the specification of the energy operator. For relativistic problems it is more convenient to use the mass square operator, M^2. In general, this operator is a function of all elements, $G_\alpha \epsilon \mathcal{G}$. In the presence of a dynamic symmetry, it is a function of the Casimir invariants only, Eq. (1.2). In contrast with non-relativistic problems, where one usually expands the energy operator into linear functions of the Casimir operators, one must consider here more general expansions.

Consider the class of expansions

$$M^2 = \sum_i A_i [G_i]^{\alpha_i} \quad . \tag{5.1}$$

For chain (I) of U(4), the expansion (5.1) gives the "geometric" mass formula

$$M^2 = M_0^2 + A_1 [G_1(U(3))]^{\alpha_1} + A_2 [G_2(U(3))]^{\alpha_2} +$$
$$+ A_3 [G(SO_L(3))]^{\alpha_3} \quad . \tag{5.2}$$

For chain (II) it gives

$$M^2 = M_0^2 + A_1 [G(SO(4))]^{\alpha_1} + A_2 [G(SO_L(3))]^{\alpha_2} \quad . \tag{5.3}$$

It is convenient to introduce the operators

$$\begin{aligned}\hat{C} &= G(SO(4)) - N(N+2) \quad , \\ \hat{C}' &= G(SO_L(3)) + \frac{1}{4} \quad , \\ \hat{C}'' &= G_2(U(3)) + 1 \quad ,\end{aligned} \tag{5.4}$$

instead of $G(SO(4))$, $G(SO_L(3))$ and $G_2(U(3))$. Taking the expectation values of (5.2) and (5.3), after replacing $G(SO(4))$, $G(SO_L(3))$ and $G_2(U(3))$ by $\hat{C}$, $\hat{C}'$ and $\hat{C}''$, one obtains

$$M^2(N, n_\pi, L, M_L) = M_0^2 + A_1 n_\pi^{\alpha_1} + A_2 (n_\pi + 1)^{2\alpha_2} + A_3 (L + \tfrac{1}{2})^{2\alpha_3} \quad , \quad (I)$$

$$M^2(N, \omega, L, M_L) = M_0^2 + A_1 [\omega(\omega+2) - N(N+2)]^{\alpha_1} + A_2 (L + \tfrac{1}{2})^{2\alpha_2} \quad , \quad (II) \quad . \tag{5.5}$$

A particular case of Eq. (5.5), (II), is that in which $\alpha_2=1/2, \alpha_1=1$. Adding the quantity $A_2/2$ to M_0^2 and introducing the quantum number $v=(N-\omega)/2$, one can then rewrite (5.5) as

$$M^2(N,v,L,M_L) = M_0^2 - 4A_1(N+1)(v - \frac{1}{N+1} v^2) + A_2 L \quad . \quad (5.6)$$

When $N \to \infty$ with $-4A_1(N+1)=A=\text{const}$, and $A_2=B$, one obtains

$$M^2(v,L,M_L) = M_0^2 + Av + BL \quad . \quad (5.7)$$

This is the mass formula suggested by string-like models of QCD.

Expansions similar to those of Eqs. (5.2) and (5.3) can be done for the spin and flavor algebras and for "geometric" configurations including several U(4)'s. Also, one could generalize (5.1) further by introducing terms which contain products of G_i's,

$$M^2 = \sum_{ij} A_{ij} [G_i]^{\alpha_i} [G_j]^{\alpha_j} \quad . \quad (5.8)$$

6. MATRIX ELEMENTS OF OPERATORS

The third step in the exploitation of the algebraic method is the evaluation of matrix elements of operators, in particular transition operators. The basic idea here is to write the operators, $\hat{T}$, in terms of elements of the algebra, G_α,

$$\hat{T} = f(G_\alpha) \quad , \quad G_\alpha \in \mathcal{G} \quad . \quad (6.1)$$

When a dynamic symmetry exists, states are representations of $\mathcal{G}$. Thus, the matrix elements are of the type

$$\langle \text{Irrep of } \mathcal{G} \mid f(G_\alpha) \mid \text{Irrep of } \mathcal{G} \rangle , \qquad (6.2)$$

and they can be evaluated by using standard group theoretic methods.

Again, in non-relativistic physics, the operators $\hat{T}$ are usually linear functions of the G_α's and the evaluation of the matrix element in (6.2) is straightforward. In hadronic physics this is no longer the case, since the long wave-length approximation, which is at the basis of the linear expansion, no longer applies. We must therefore consider a more complicated class of operators of the type

$$\hat{T} = \sum_\alpha t_\alpha G_\alpha \, e^{i \sum_\beta k_\beta G_\beta} . \qquad (6.3)$$

In the long wave length approximation, $k_\beta \to 0$, we obtain the usual form, $\hat{T} = \Sigma_\alpha t_\alpha G_\alpha$. The evaluation of matrix elements of the operators (6.3) is very tedious. It can still be done, in the presence of a dynamic symmetry, in closed form, since G_β is a generator of $\mathcal{G}$. The matrix elements of $\hat{T}$ is related to a group element of $\mathcal{G}$.

In previous applications of algebraic methods to hadronic physics only the linear terms were considered. This approximation introduces typically errors of the order of a factor of two. The explicit introduction of the exponential term is thus necessary in cases in which the matrix elements are measured with good accuracy, as it will be the case at the new facilities.

An important new result is that the matrix elements of the exponential operator, $\exp[i\Sigma_\beta k_\beta G_\beta]$, have been explicitly evaluated in the case of U(4), both in the U(3) and in SO(4) limits. A particularly interesting case is that of the operator

$$\hat{T} = e^{i\epsilon \hat{D}_0} , \qquad (6.4)$$

where $\hat{D}$ is the operator of Table I. In this case one obtains[10] for chain I

$$\langle [N],n,L,0|\hat{T}|[N],0,0,0\rangle =$$

$$= i^n \left[\frac{2L+1}{(n+L+1)!!(n-L)!!}\right]^{1/2} (\epsilon\sqrt{N})^n e^{-\epsilon^2 N/2} \qquad , \text{(I)} \quad . \qquad (6.5)$$

and, for chain II,

$$\langle [N],\omega=N,L,0|\hat{T}|[N],\omega=N,0,0\rangle = i^L \sqrt{2L+1}\, j_L(N\epsilon) \qquad , \text{(II)} \quad . \qquad (6.6)$$

These results are valid in the limit of large N.

Matrix elements of products of generators times exponentials can be evaluated by inserting a complete set of states and using the previous results

$$\langle\Psi|G_\alpha e^{i\epsilon G_\beta}|\Psi'\rangle = \sum_k \langle\Psi|G_\alpha|\Psi_k\rangle\langle\Psi_k|e^{i\epsilon G_\beta}|\Psi'\rangle \qquad . \qquad (6.7)$$

7. SYMMETRY BREAKING INTERACTIONS

Finally, another advantage of the algebraic method is that one can consider not only those situations that are described by a dynamic symmetry but also situations intermediate between the two limits. For small symmetry breaking, the effects of breaking interactions can be evaluated in perturbation theory by adding to the mass square operator, M^2, symmetry breaking terms

$$M^2 = M^2(DS) + \text{symmetry breaking terms.} \qquad (7.1)$$

Examples of symmetry breaking interactions are, in the $U(4) \supset SO(4) \supset SO_L(3) \supset SO_L(2)$ chain, terms involving invariant operators of the other chain,

$$M^2 = M_0^2 + A_1 \hat{C} + A_2 \left[\hat{C}' - \tfrac{1}{2}\right]^{1/2} + \eta \left[G_1(U(3))\right] \quad . \tag{7.2}$$

When $\eta=0$, one has the dynamic symmetry described in Sect. 5. As η increases, the dynamic symmetry is broken. For η small the effects can be computed in perturbation theory, while for η large one must diagonalize M^2 in the basis (3.5).

A computer program that diagonalizes a generic mass operator in $U(4) \otimes SU_s(2) \otimes SU_f(n)$ has been written. This program allows one to extend the algebraic study of hadronic structure to cases with broken symmetry. In view of the fact that the M^2 operator may contain rational functions of the Casimir operators it is convenient to separate these functions and expand only the non-diagonal pieces into linear, quadratic, ..., forms in the generators

$$M^2 = M^2(DS) + \sum_\alpha \epsilon_\alpha G_\alpha + \frac{1}{2} \sum_{\alpha\beta} u_{\alpha\beta} G_\alpha G_\beta + \dots \quad , \tag{7.3}$$

as done in Eq. (7.2).

8. APPLICATIONS

The only case for which the algebraic method, in its general form, has been fully explored is that of $(q\bar{q})$ mesons. For these configurations, there is only one $U(4)$ and the spectrum generating algebra is

$$\mathcal{G} = SU_c(3) \otimes SU_f(n) \otimes SU_s(2) \otimes U(4) \quad . \tag{8.1}$$

Furthermore, color does not play any non-trivial role and can be deleted. Also, for these configurations, it is more convenient to use the quark basis, Eq. (4.4). A detailed analysis of $(q\bar{q})$ configurations in the $SO(4)$ limit of $U(4)$ has been performed recently[11]. The basis states are of the type

$$| q_i \bar{q}_j ; \; S; \; N,v,L,; \; J,M_J > \qquad , \qquad (8.2)$$

where J denotes the total angular momentum obtained by coupling $\vec{S}$ and $\vec{L}$, i.e.

$$\left| \begin{array}{cccccc} U(4) \otimes SU_s(2) \supset SO(4) \otimes SU_s(2) \supset SO_L(3) \otimes SU_s(2) \supset SU_J(2) \supset SO_J(2) \\ \downarrow & \downarrow & \downarrow & \downarrow & \downarrow & \downarrow \\ N & S & \omega & L & J & M_J \end{array} \right\rangle \quad (8.3)$$

Here S=0,1. In Ref. [11], the following mass operator was used to analyze the masses of mesons

$$M^2 = M_0^2 + A'\left[G(SO(4))-N(N+2)\right] + B\left[\sqrt{G(SO_L(3)) + \tfrac{1}{4}} - \tfrac{1}{2}\right] +$$
$$+ C\left[\sqrt{G(SU_s(2)) + \tfrac{1}{4}} - \tfrac{1}{2}\right] + D\left[\sqrt{G(SU_J(2)) + \tfrac{1}{4}} - \tfrac{1}{2}\right] \qquad , \qquad (8.4)$$

with eigenvalues

$$M^2(v,L,S,J) = M_0^2 + Av + BL + CS + DJ \qquad . \qquad (8.5)$$

The spectrum of Eq. (8.5) is very simple and it is shown in Fig. 2 (left part, S=0; right part, S=1). It consists of a series of trajectories with v=0,1,... . Within each trajectory (Regge trajectory), L=0,1,2,... . Each trajectory is a representation of SO(4). The SO(4) dynamic symmetry thus emphasizes the physics of string-like QCD which leads to linear Regge trajectories.

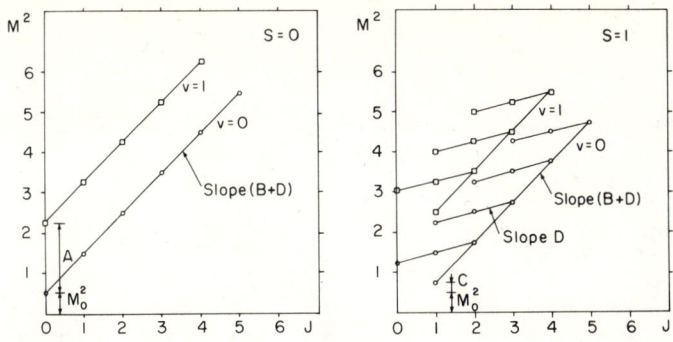

Fig.2. Spectrum of the mass sqared operator of $q\bar{q}$ mesons according to Eq.(8.5) (left panel, S=0; right panel, S=1).

The flavor dependence was introduce in Ref.[11] only through the constituent masses of quark, i, and antiquark, j, combined into $M_{ij}=M_i+M_j$,

$$M^2(v,L,S,J) = (M_0^2)_{ij} + A_{ij} v + B_{ij} L + C_{ij} S + D_{ij} J \quad . \quad (8.6)$$

Eq. (8.6) describes only the dependence of M^2 on the "geometric" and "spin" quantum numbers. One must then include the dependence on the "flavor" quantum numbers. As mentioned above, for $(q\bar{q})$ configurations for which there is only one q and one $\bar{q}$, it is not useful to go to the coupled basis

$$SU_{f,q}(n) \otimes SU_{f,\bar{q}}(n) \supset SU_f(n) \quad . \quad (8.7)$$

Furthermore, the flavor dependence comes, to a very good approximation, entirely from the quark masses, M_i and M_j. It is thus diagonal in the uncoupled basis. Introducing the quantity $M_{ij}=M_i+M_j$, one can write this dependence as

$$A_{ij} = a + a' M_{ij} \quad , \quad B_{ij} = b + b' M_{ij} \quad ,$$
$$C_{ij} = c + c' M_{ij} \quad , \quad D_{ij} = d + d' M_{ij} \quad ,$$

$$(M_0^2)_{ij} = e\, M_{ij} + (M_{ij})^2 \quad . \tag{8.8}$$

In the language of dynamic symmetry, the quantity M_{ij} can be written as an operator

$$\hat{M}_{ij} = M_i\, \hat{n}_i + M_j\, \hat{n}_j \quad , \tag{8.9}$$

where $\hat{n}_i$ and $\hat{n}_j$ are the number operators for quarks and antiquarks. They are the linear Casimir operators of the U(1) algebras of Eq. (4.1). The structure of the mass formula (8.7) is thus of the type (5.8).

The mass formula (8.6) describes accurately the masses of mesons, especially light mesons, as shown, for example, in Table II. Flavor symmetry breaking terms can be (and have been) investigated. It is convenient to write, as in the case of "geometric" degrees of freedom, the mass squared operator as

$$M^2 = M^2(DS) + \text{flavor symmetry breaking terms.} \tag{8.10}$$

For the low-lying meson monet, π, K, η, η', two symmetry breaking terms were found to play a role, a so-called "Goldstone" term and an "annihilation" term, details of which are given in Ref.[11]. The linearity of the rotational Regge trajectories appears to be well verified by the experimental data for light mesons, Fig. 3, thus supporting an interpretation of light mesons in terms of an SO(4) symmetry. The situation is not so clear for heavy mesons, for which there are no measured "rotational" trajectories. The "vibrational" trajectories of heavy mesons do not appear to be completely linear. This may be due either to a breaking of the SO(4) symmetry or to other effects not included in the mass formula (8.6).

The algebraic method has also been used to study electromagnetic decays of mesons[12]. The transition operator $\hat{T}_\gamma$ has two pieces, a magnetic and an

TABLE II. A selection of meson masses in the algebraic approach [11].

Meson	Expt.	π Family M^2(GeV^2) SO(4) Symmetry	v	L	S	J^{PC}
π	0.019±0.000	0.022	0	0	0	0^{-+}
ρ(770)	0.590±0.001	0.586	0	0	1	1^{--}
b_1(1235)	1.520±0.025	1.437	0	1	0	1^{+-}
a_1(1260)	1.588±0.076	1.636	0	1	1	1^{++}
a_2(1320)	1.738±0.002	1.680	0	1	1	2^{++}
ρ(1450)	2.103±0.023	1.884	1	0	1	1^{--}
π_2(1670)	2.772±0.067	2.531	0	2	0	2^{-+}
ρ_3(1690)	2.859±0.017	2.775	0	2	1	3^{--}
ρ(1700)	2.890±0.068	2.686	0	2	1	1^{--}

Meson	Expt.	K Family M^2(GeV^2) SO(4) Symmetry	v	L	S	J^P
K	0.246±0.000	0.231	0	0	0	0^-
K^*(892)	0.795±0.000	0.815	0	0	1	1^-
K_1(1270)	1.613±0.025	1.737	0	1	0	1^+
K^*(1370)	1.869±0.148	2.234	1	0	1	1^-
K_1(1400)	1.966±0.020	1.943	0	1	1	1^+
K^*_0(1430)	2.042±0.017	1.887	0	1	1	0^+
K^*_2(1430)	2.032±0.004	1.999	0	1	1	2^+
K^*(1680)	2.816±0.215	3.072	0	2	1	1^-
K_2(1770)	3.126±0.050	3.128	0	2	1	2^-
K^*_3(1780)	3.147±0.028	3.184	0	2	1	3^-
K^*_4(2045)	4.182±0.037	4.368	0	3	1	4^+

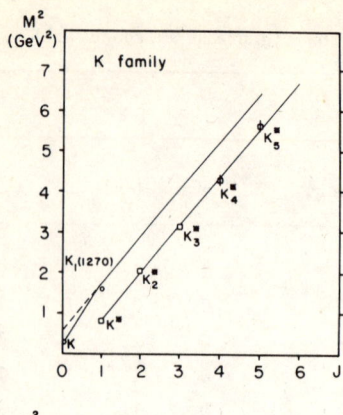
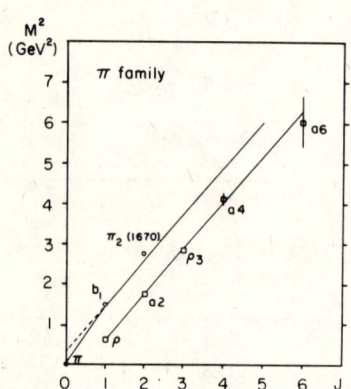

Fig.3. "Rotational" Regge trajectories for π and K families.

electric piece. The magnetic piece can be written as

$$\hat{T}_\gamma^{(M)} = \sum_i g \frac{e_i}{2M_i} \left[\vec{\sigma}^{(1)} \cdot (\vec{k} \times \vec{\epsilon}) \right] e^{-i\vec{k} \cdot \vec{r}^{(i)}} , \qquad (8.11)$$

where the sum extends over the quark and antiquark, $\vec{k}$ is the photon momentum and $\vec{\epsilon}$ its polarization. Also g is the quark g-factor. In the algebraic approach and SO(4) limit, the operator (8.11) can be written as

TABLE III. Radiative decay widths of selected light mesons

(a) Transitions with $\Delta S=1$, $\Delta L=0$

$$\Gamma(\rho^{\pm} \to \gamma\pi^{\pm}) = \left(\frac{ge}{2M_u}\right)^2 \frac{k^3}{\pi} \frac{1}{27} \left[j_0\left(\frac{k\beta}{2}\right)\right]^2$$

$$\Gamma(\rho^0 \to \gamma\eta) = \left(\frac{ge}{2M_u}\right)^2 \frac{k^3}{\pi} \frac{1}{3} \left[j_0\left(\frac{k\beta}{2}\right)\right]^2$$

$$\Gamma(\omega \to \gamma\pi^0) = \left(\frac{ge}{2M_u}\right)^2 \frac{k^3}{\pi} \frac{1}{3} \cos^2\phi_V \left[j_0\left(\frac{k\beta}{2}\right)\right]^2$$

$$\Gamma(\omega \to \gamma\eta) = \left(\frac{ge}{2M_u}\right)^2 \frac{k^3}{\pi} \frac{1}{27} \left[\cos\phi_P \cos\phi_V - 2 \frac{M_u}{M_s} \sin\phi_P \sin\phi_V\right]^2 \left[j_0\left(\frac{k\beta}{2}\right)\right]^2$$

$$\Gamma(\eta' \to \gamma\rho^0) = \left(\frac{ge}{2M_u}\right)^2 \frac{k^3}{\pi} \sin^2\phi_P \left[j_0\left(\frac{k\beta}{2}\right)\right]^2$$

$$\Gamma(\eta' \to \gamma\omega) = \left(\frac{ge}{2M_u}\right)^2 \frac{k^3}{\pi} \frac{1}{9} \left[\sin\phi_P \cos\phi_V + 2 \frac{M_u}{M_s} \cos\phi_P \sin\phi_V\right] \left[j_0\left(\frac{k\beta}{2}\right)\right]^2$$

$$\Gamma(\phi \to \gamma\pi^0) = \left(\frac{ge}{2M_u}\right)^2 \frac{k^3}{\pi} \frac{1}{3} \sin^2\phi_V \left[j_0\left(\frac{k\beta}{2}\right)\right]^2$$

$$\Gamma(\phi \to \gamma\eta) = \left(\frac{ge}{2M_u}\right)^2 \frac{k^3}{\pi} \frac{1}{27} \left[\cos\phi_P \sin\phi_V + 2 \frac{M_u}{M_s} \sin\phi_P \cos\phi_V\right]^2 \left[j_0\left(\frac{k\beta}{2}\right)\right]^2$$

$$\Gamma(\phi \to \gamma\eta') = \left(\frac{ge}{2M_u}\right)^2 \frac{k^3}{\pi} \frac{1}{27} \left[\sin\phi_P \sin\phi_V - 2 \frac{M_u}{M_s} \cos\phi_P \cos\phi_V\right]^2 \left[j_0\left(\frac{k\beta}{2}\right)\right]^2$$

$$\Gamma(K^{*0} \to \gamma K^0) = \left(\frac{ge}{2M_u}\right)^2 \frac{k^3}{\pi} \frac{1}{27} \left[j_0\left(\frac{M_s k\beta}{M_u + M_s}\right) + \left(\frac{M_u}{M_s}\right) j_0\left(\frac{M_u k\beta}{M_u + M_s}\right)\right]^2$$

$$\Gamma(K^{*\pm} \to \gamma K^{\pm}) = \left(\frac{ge}{2M_u}\right)^2 \frac{k^3}{\pi} \frac{1}{27} \left[2 j_0\left(\frac{M_s k\beta}{M_u + M_s}\right) - j_0\left(\frac{M_u k\beta}{M_u + M_s}\right)\right]^2$$

$$\hat{T}_\gamma^{(M)} = \sum_i g \frac{e_i}{2M_i} \left[\vec{\sigma}^{(i)} \cdot (\vec{k} \times \vec{\epsilon})\right] e^{-ik\beta\nu^{(i)}\hat{D}_0/N} \quad , \tag{8.12}$$

where

$$\nu^{(1)} = \frac{M_2}{M_1+M_2} \quad , \quad \nu^{(2)} = -\frac{M_1}{M_1+M_2} \quad , \tag{8.13}$$

and β is a scale of coordinates. Matrix elements of this operator have been evaluated [12], and the corresponding decay widths computed, Table III. The evaluation involves the matrix elements of the operator σ_+ (since $\vec{k}$ has been chosen in the z-direction and the photon polarization $\vec{\epsilon}$ is perpendicular to it) and of the operator $\exp(-ik\beta\nu\hat{D}_0/N)$. The former are trivial, while the latter can be obtained by means of Eq. (6.6). A comparison with the experimental decay widths is shown in Table IV. The decay widths are obtained using [12]

$$\Gamma(M \to M'+\gamma) = (\text{Phase space}) \times \sum_{m_i, m_f} |\langle M|\hat{T}_\gamma|M'\rangle|^2 \quad . \tag{8.14}$$

One can see that the agreement is good. However, the table tests mostly the flavor-spin part of the wave functions and not the SO(4) symmetry itself, since the transitions reported in Table IV are L=0 to L=0.

The electric piece can be written as

$$\hat{T}_\gamma^{(E)} = \sum_i \frac{e_i}{M_i} (\vec{p}^{(i)} \cdot \vec{\epsilon}) e^{-i\vec{k}\cdot\vec{r}^{(i)}} \quad , \tag{8.15}$$

where $\vec{p}$ is the quark momentum. In the algebraic approach and SO(4) limit, the operator (8.15) becomes

$$\hat{T}_\gamma^{(E)} = \sum_i \frac{e_i}{M_i} \zeta \, \nu'^{(i)} (\hat{D}' \cdot \vec{\epsilon}) \, e^{-ik\beta\nu^{(i)}\hat{D}_0/N} \quad , \qquad (8.16)$$

with

$$\nu'^{(1)} = 1 \quad , \qquad \nu'^{(2)} = -1 \quad , \qquad (8.17)$$

and ζ a scale of momenta. There are no data to test the electric piece for light mesons. For heavy mesons, the matrix elements of the operator (8.16) describe only very qualitately the data. This indicates that either the

TABLE IV. Comparison of light meson experimental decay widths with those computed in the U(4) ⊃ SO(4) model [12].

Decay $\Delta S=1$, $\Delta L=0$	Γ (KeV) Experiment	SO(4) Symmetry
$\rho^\pm(770) \to \gamma\pi^\pm$	68 ± 7	66
$\rho^0(770) \to \gamma\eta(549)$	62 ± 17	90
$\omega(783) \to \gamma\pi^0$	717 ± 51	609
$\omega(783) \to \gamma\eta(549)$	4.0 ± 1.9	9.9
$\eta'(958) \to \gamma\rho^0(770)$	62 ± 9	72
$\eta'(958) \to \gamma\omega(783)$	6.3 ± 1.2	9.4
$\phi(1020) \to \gamma\pi^0$	5.78 ± 0.67	5.9
$\phi(1020) \to \gamma\eta(549)$	56.4 ± 3.5	32
$\phi(1020) \to \gamma\eta'(958)$	< 1.8	0.5
$K^{*0}(902) \to \gamma K^0(498)$	117 ± 10	114
$K^{*\pm}(892) \to \gamma K^\pm(494)$	50 ± 5	55

wave functions of heavy mesons are not well described by the SO(4) symmetry, or that the operator (8.16) does not describe electric transitions. Both forms (8.11) and (8.15) are obtained from the non-relativistic expansion of the transition operator for point quarks. There is no guarantee that this is the appropriate form.

9. CONCLUSIONS

Algebraic methods provide a useful tool for the study of hadronic spectroscopy. In the past, they have been used in this area of physics mostly to deal with the spin-flavor part, SU(6) or $SU(6)_W$ [13]. Some algebraic studies of the "geometric" part, $\mathcal{R}$, have been done, in the harmonic oscillator, U(3), limit [14]. The spectrum generating algebra used is Sp(6,R)⊃U(3). Here, I have presented a more general scheme, based on the algebra U(3,1), compactified to U(4) in the large N limit. The branch of U(3,1) that goes through U(3) produces results that are identical to those of Sp(6,R). Indeed, both Sp(6,R) and U(3,1) can be used as SGA of the oscillator [15]. The algebra U(3,1) however has another branch that goes through SO(3,1) (or in the compact case SO(4)). This latter branch appears to be closer to the actual situation than U(3), as evidenced by the occurrence of Regge trajectories in light mesons. The general scheme allows one to treat also intermediate situations.

In the use of SGA and dynamic symmetries in hadronic spectroscopy, two new features appear which are not present in the non-relativistic case:
(i) the energy operator can contain non-linear terms in the invariants; in fact it is found that a good description of data is obtained by considering the expansion

$$M^2 = \sum_i A_i [C_i]^{\alpha_i} \quad ; \qquad (9.1)$$

(ii) the transition operators can contain the generators of $\mathcal{G}$ in the exponent

$$\hat{T} = \sum_\alpha t_\alpha G_\alpha \, e^{i \sum_\beta k_\beta G_\beta} \quad . \tag{9.2}$$

Both features have been fully investigated in the case of a single U(4).

When more than one U(4) is present, one has to deal with another major technical problem. For identical particles, the wave functions must have a definite overall symmetry. Since the wave functions are of the type (2.3), this implies that representations of the various pieces in (2.3) must be combined in an appropriate way. For example, for (qqq) configurations, since Ψ_c is antisymmetric, one must have

$$\Psi = \Psi_c \otimes \underbrace{\Psi_{sf} \otimes \Psi_R}_{\text{symmetric}} \quad . \tag{9.3}$$

Thus, the two pieces Ψ_{sf} and Ψ_R must be of the same symmetry character. The construction of wave functions Ψ_R of a definite symmetry under the group S_3 of quark permutations is not trivial. The problem has been solved in the U(3) limit [14] and it is being attacked at the present time in the SO(4) limit. Once solved, one will be able to complete the investigation of all U(4)⊗U(4) configurations (qqq). It is for these configurations that one expects major differences to occur between the U(3) and SO(4) limits. When the analysis of (qqq) configurations will be completed, one can then perform an investigation of more complex configurations, such as $qq\bar{q}\bar{q}$ configurations (4-body problem).

It is hoped that the development of the algebraic model based on U(4)⊗SU_s(2)⊗SU_f(n)⊗SU_c(3) will provide a general framework within which different hadronic properties can be studied. This framework is more general than that of the harmonic oscillator, U(3), and spin-flavor, SU(6) ⊃ SU_f(3)⊗SU_s(2), framework of the non-relativistic quark model, but still retains the simplicity of those approaches.

ACKNOWLEDGEMENTS

This work was supported in part by D.O.E. Grant No. DE-FG02-91ER40608.

REFERENCES

1. A. Arima and F. Iachello, Ann. Phys. (NY) $\underline{99}$, 253 (1976); $\underline{111}$, 201 (1978); $\underline{123}$, 468 (1979).
2. F. Iachello and R.D. Levine, J. Chem. Phys. $\underline{77}$, 3046 (1982); O.S. van Roosmalen, F. Iachello, R.D. Levine, and A.E.L. Dieperink, J. Chem. Phys. $\underline{79}$, 2515 (1983).
3. Y. Dothan, M. Gell-Mann and Y. Ne'eman, Phys. Lett. $\underline{17}$, 148 (1965); A.D. Barut and A. Böhm, Phys. Rev. $\underline{139B}$, 1107 (1965).
4. M. Gell-Mann, Phys. Rev. $\underline{125}$, 1067 (1962); Y. Ne'eman, Nucl. Phys. $\underline{26}$, 222 (1961).
5. F. Gürsey and L.A. Radicati, Phys. Rev. Lett. $\underline{13}$, 173 (1964).
6. H.J. Lipkin and S. Meshkov, Phys. Rev. Lett. $\underline{14}$, 670 (1965); K.J. Barnes, P. Carruthers and F. von Hippel, Phys. Rev. Lett. $\underline{14}$, 82 (1965).
7. A. Böhm, Phys. Rev. D$\underline{33}$, 3358 (1986); A. Böhm, M. Loewe and P. Magnollay, Phys. Rev. D$\underline{31}$, 2304 (1985); Phys. Rev. D$\underline{32}$, 791 (1985); A. Böhm, M. Loewe, P. Magnollay, L.C. Biedenharn, H. van Dam, M. Tarlini and R.R. Aldinger, Phys. Rev. D$\underline{32}$, 2828 (1985).
8. A.D. Barut, Phys. Lett. $\underline{26}$B, (1968); A.D. Barut, D. Corrigan and H. Kleinert, Phys. Rev. Lett. $\underline{20}$, 167 (1968); A.D. Barut and H. Beker, Phys. Rev. Lett. $\underline{50}$, 1560 (1983).
9. F. Iachello, Nucl. Phys. A$\underline{497}$, 23c (1989); A$\underline{518}$, 173 (1990).
10. R. Bijker, R.D. Amado, and D.A. Sparrow, Phys. Rev. A$\underline{33}$, 871 (1986).
11. F. Iachello, N.C. Mukhopadhyay, and L. Zhang, Phys. Lett. B$\underline{256}$, 295 (1991), and to be published in Phys. Rev. D.
12. F. Iachello and D. Kusnezov, Phys. Lett. B$\underline{255}$, 493 (1991), and to be published.
13. A.J. Hey, P.J. Litchfield, and R.J. Cashmore, Nucl. Phys. B$\underline{95}$, 516 (1975).
14. K.C. Bowler, P.J. Corvi, A.J.G. Hey, P.D. Jarvis and R.C. King, Phys. Rev. D$\underline{24}$, 197 (1981).
15. B.G. Wybourne, "Classical Groups for Physicists", Wiley, New York, 1974, Chapt. 20.

Ultrarelativistic Heavy-Ion Collisions: Present and Future

Gordon Baym

Loomis Laboratory of Physics, University of Illinois
1110 W. Green St., Urbana, Illinois 61801

> *Abstract*: These lectures describe the motivations, theoretical basis, achievements, and future goals of the program of ultrarelativistic heavy-ion collisions underway at CERN and Brookhaven National Laboratory.

1. Introduction

The ongoing and future experimental program of ultrarelativistic heavy-ion collisions goes to new frontiers of beam energies, crossing the borderlines – e.g., in the energy vs. mass-number plane – that have separated elementary particle physics and nuclear physics for the past half-century. Particle physics typically studies simple systems with very low baryon number (zero, one, or two – as at the SSC) at very high energies, while nuclear physics deals with complex systems of arbitrarily large mass number, but at low energies. Ultrarelativistic nucleus-nucleus collisions will before long be carried out with very heavy nuclei at energies per nucleon that were the state-of-the-art in particle physics only recently.

The questions asked by nuclear and particle physics are quite different, however. High energy particle physics aims to discover the fundamental building blocks of nature and their fundamental interactions. High energy nuclear physics, by looking at systems of arbitrarily large mass number, is concerned, rather, with finding out the physical phenomena that can be produced by the fundamental interactions in these complex systems with a wide variety of degrees of freedom. How, for example, does nuclear matter behave when raised to energy densities perhaps ten times that in the nuclear ground state, 0.15 GeV/fm^3 (the number density of nucleons, $\rho_0 = 0.16/\text{fm}^3$, times the rest energy of a nucleon, 0.94 GeV), in a collision of two heavy ions at center-of-mass energies greatly in excess of the rest mass of the nuclei? Many-body systems exhibit collective physical behavior that one simply cannot infer knowing only the fundamental interactions between the constituents. Although Coulomb discovered his law in 1785, the discovery of the phenomenon of superconductivity in 1911 was a great surprise, not explained in terms of the fundamental laws for another 45 years. As Hans Weidenmüller noted in his lectures in this school, one still does not predict the weather from a knowledge of the fundamental Coulomb interactions among the constituents of the atmosphere.

In nucleus-nucleus collisions, collectivity can enter at two levels. First, in what way does proton-nucleus, pA, scattering differ from a collection of A different pp collisions?

Similarly does a further level of collective behavior enter in going from pA to nucleus-nucleus, AA, scattering? Ultrarelativistic heavy-ion physics, in common with particle physics, differs from ordinary low-energy nuclear physics in the importance of particle production in scattering; in a very high-energy collision of two large nuclei, the original baryons are just a minority of the final particles seen in the collisions.

The program of ultrarelativistic heavy-ion collisions has been underway at CERN and Brookhaven since the mid-eighties.* The initial phase began in the autumn of 1986 with lighter nuclear projectiles on fixed nuclear targets. The CERN experiments have been carried out in the SPS accelerator at lab energies of 60 and 200 GeV per nucleon (GeV/A) with ^{16}O, ^{32}S, and p beams. The first phase included six major detector experiments – NA34 (Helios), NA35, NA36, NA38, WA80, and WA85 – as well as a number of passive (emulsion or plastic) experiments. The Brookhaven program has been carried out in the AGS accelerator at lab energies of 10-14.5 GeV per nucleon with ^{16}O, ^{28}Si, and p beams, with three major experiments, E802, E810, E814, plus passive experiments. The second generation of lighter projectile experiments at both laboratories is currently being implemented and will be running over the next few years. The next step is to push the frontier to larger mass numbers: The new booster ring adjacent to the AGS allows injection of fully stripped ions of arbitrarily large mass into the AGS at $\sim$12 GeV/A; similarly, CERN is designing a "Pb Injector" for the SPS with beams at $\sim$170 GeV/A. Experiments with heavy beams should begin within the next few years.

Further in the future, the Brookhaven Relativistic Heavy Ion Collider (RHIC), on which construction has recently begun, and which is expected to have beams by 1997, will provide the capability of colliding nuclei as heavy as Au on Au at 100 GeV/A in the center of mass (equivalent to 20 TeV/A fixed target). CERN is also discussing building the Large Hadron Collider (LHC) in the LEP tunnel, with the possibility of injecting heavy-ion beams from the SPS, thus enabling heavy-ion collisions at $\sim$ 4 TeV/A in the center of mass; at least one intersection region would contain heavy-ion detectors. The LHC could have heavy-ion beams as early as 1998.

The effective internal degrees of freedom that enter in nuclei depend on the spatial scale on which one looks, or equivalently on the excitation energies of the nucleus. At energy densities near those in normal laboratory nuclei, the degrees of freedom excited are nucleonic, as, e.g., in the nuclear shell model. At nuclear excitations above a few hundred MeV, further "hadronic" degrees of freedom, including internal excitations of the nucleon such as the Δ, and mesonic excitations, enter. On still higher energy scales, those being explored in ultrarelativistic heavy-ion collisions, the full quark structure of the nuclear constituents comes into play. One exciting possibility in ultrarelativistic

---

*Useful general references are the recent proceedings of the ongoing conferences on ultrarelativistic nucleus-nucleus collisions and quark matter [1,2].

collisions is the production of a new state of matter, a *quark-gluon plasma*, in which the quarks and gluons are not confined in individual hadrons, but are able to propagate over an extended region of space. The production and detection of a plasma would enable one to study quantum chromodynamics (qcd) in three dimensions over distance scales the size of the collision volume, substantially greater than the size of typical hadrons, $\sim$ 1 fm. While heavy-ion collisions to date do not appear to have produced a quark-gluon plasma, they do indicate, as we shall see, that future collisions with heavier projectiles, and at higher energies are very favorable for producing a plasma.

Information gathered from ultrarelativistic collisions is potentially very important in astrophysics. The outstanding fundamental problem in understanding the overall structure of neutron stars – e.g., the mass density profile, and the maximum possible neutron star mass – is lack of knowledge of the microscopic nature of matter under the extreme conditions deep in the stellar interiors. Present laboratory studies give information on matter only up to about twice ρ_0, while neutron star interiors can be perhaps as high as $10\rho_0$. One should be able to infer constraints on the equation of state of matter in neutron stars from heavy-ion collisions, as well as learn about the nature of matter in supernova explosions compressed by gravitational collapse, and in the first microseconds of the early universe, corresponding to temperatures, T, above several hundred MeV, where matter was in the form of a quark-gluon plasma.

2. Deconfined matter

As matter is heated or compressed its degrees of freedom change from composite to more fundamental. For example, by heating or compressing a gas of atoms, one eventually forms a plasma in which the nuclei become stripped of the electrons, which go into continuum states forming an electron gas. Similarly, when nuclei are squeezed, as happens in the formation of neutron stars in supernovae, the matter merges into a continuous fluid of neutrons and protons. One expects that a gas of nucleons, which are made of quarks, will, when squeezed or heated, turn into a gas of uniform quark matter, which at finite T should contain antiquarks and gluons as well. While it is impossible to remove an isolated quark from a hadron, in the plasma state the quarks and gluons are free to roam over macroscopic distances in the matter. Quark matter at low T consists of Fermi seas of degenerate u and d quarks.

While a very hot or dense plasma should behave in many ways as a non-interacting system, interaction effects in the plasmas expected in heavy-ion collisions, not much beyond the transition from hadronic to quark matter, are important. The only reliable approach at present to determining the properties of strongly interacting plasmas, and the deconfinement transition is through Monte-Carlo calculations of lattice gauge theory (for recent reviews see Refs. [3] and [4]). These calculations have been successful so far only for the case of zero baryon density at finite T. The results depend strongly on the masses assumed for the quarks. For infinitely heavy quarks, only gluons play a dynamic role; lattice calculations in this case predict a sharp first-order phase transition associated with deconfinement. The energy in the confined phase required to separate a test quark-antiquark pair, $\bar{Q}Q$, far apart grows linearly with separation, while it remains finite in the deconfined phase.

With finite mass quarks one can always separate a test quark-antiquark pair, $\bar{Q}Q$, with finite energy, since at sufficient separation, it becomes energetically favorable to create a $\bar{q}q$ pair in the system, which screens out the interaction between the test pair; the q binds to the $\bar{Q}$, and the $\bar{q}$ to the Q, creating effectively a pair of mesons which can be separated to infinity with finite energy. Once light quarks are in the system there no longer exists a good measure of whether the system is in a confined or deconfined state, and there need not be a sharp transition between the confined and deconfined phases. The transition between the two phases can be smooth, as in ionization of a gas being heated, where the system goes gradually from gas molecules to electrons and nuclei; the two states are qualitatively different and there is a reasonably rapid onset of ionization, but it is not sharp. Alternatively, the transition may be first order, as in the boiling of water.

For massless quarks, on the other hand, the transition is again first-order, associated now with the spontaneous breaking of the SU(3)⊗SU(3) chiral symmetry of strong interactions. In the hadronic world chiral symmetry is spontaneously broken, analogously to the breaking of rotational symmetry in a ferromagnet, while in the deconfined phase it is restored. Since chiral symmetry is not exact for finite mass quarks, the situation for realistic quarks (u and d are light with $m_u \sim m_d \sim 10$ MeV, and for the strange quark, s, $m_s \sim 150$ MeV) is not fully understood; present calculations indicate that the transition is likely first order, at a critical $T \sim 200$ MeV (or 10^{12}K), with a latent heat of order a few GeV/fm^3. This energy density sets the scale that one has to reach in heavy-ion collisions to expect to form a plasma. Whether there is a sharp transition or not is not important for reaching a plasma state in heavy-ion collisions, although the sharpness of the transition can effect the detailed dynamics.

Lattice gauge calculations at non-zero baryon density are beset by technical problems; to date we do not have a reliable estimate of the transition density at $T = 0$ from nuclear to quark matter or even compelling evidence that there is a sharp phase transition. One can roughly estimate the location of the deconfinement transition at finite baryon density by asking whether one's favorite theory of nuclear matter or quark-gluon plasmas has a lower free energy per baryon, as a function of baryon density [5]. Such an approach necessarily implies a first-order transition with a discontinuity in the baryon density; the onset of deconfinement is typically at $\rho \sim 5-10\rho_0$ at $T = 0$. Unlike in lattice gauge theory, however, phenomenological theories of nuclear matter and of quark-gluon plasmas are generally based on inequivalent physical descriptions of the two phases and thus cannot be expected to describe the transition accurately.

3. Heavy-ion collisions

The physics of ultrarelativistic heavy-ion collisions changes with increasing energy. In the lower energy AGS regime, one can picture the two colliding nuclei as effectively stopping each other (in the cm), forming, to a crude first approximation, a fireball. [In reality, of course, parts of the nuclei pass through the collision volume rather than remaining in a fireball; with light projectiles, as employed so far at the AGS and SPS, matter in the collision volume, which is small transverse to the beam direction, ought not to achieve thermal equilibration necessary for a fireball description.] Collisions between heavy nuclei in this regime may reach energy densities of order a few GeV/fm^3 – possibly high enough to reach a quark-gluon plasma – and baryon densities several times ρ_0. The high density matter produced in such collisions is relatively baryon rich.

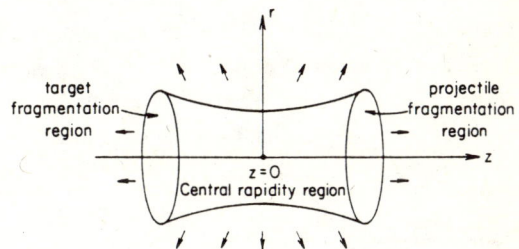

Figure 1. Nuclear fragmentation and central regions in the collision volume.

As the beam energy increases, the colliding nuclei begin to pass through each other, become highly excited internally, and at the same time, leave the vacuum between them in a highly excited state, containing matter of low baryon concentration. Such "nuclear transparency" begins to become important in the regime of the CERN fixed target experiments, and should be completely manifested at RHIC energies. The nuclear fragmentation regions, which recede from each other at the speed of light, contain essentially all the baryons of the original nuclei; the central region, to a first approximation, has no baryon excess, and, at high enough energy density, would resemble the hot vacuum of the the early universe; the matter in this region is entirely manufactured in the collision. See Fig. 1. After its production and thermalization, the matter cools and expands, and if having initially reached a quark-gluon plasma state, undergoes a hadronization transition (the deconfinement transition in reverse).

How in a collision can one learn from the emerging particles about the different regions of the collision volume? To do so requires that there be a correlation between the velocities of particles along the beam direction and their space-time positions in the collision volume. To a first approximation, the central region behaves as a rubber band stretched between the two nuclear fragementation regions, so that a given point has a determined velocity or rapidity* along the beam direction. In a collision one generally finds a range of rapidities of the final particles extending roughly from that of the target to that of the projectiles, a total rapidity spread given by $\Delta y = 2\ln(2E/m)$, where E is the beam energy per nucleon in the center-of-mass. The particles in the nuclear fragmentation regions emerge near the beam and target rapidities, and the particles of the central region emerge as a broad spread of mesons, along with baryon, antibaryon pairs, between. The correlation between space-time and rapidity can only be accurate to at most one unit of rapidity, since at any point, particles will have a spread of velocities from thermal or Fermi motion (which produces a spread $\sim 1/3$ unit of rapidity). The larger the beam energy, the greater the spread in rapidity, and the more closely one can connect regions in rapidity with space-time regions of the collision volume.

Achieving a clean separation in rapidity of the target and projectile fragmentation regions from the central regions was one of the principle factors in setting the energy of RHIC, which at $\sqrt{s}/A = 100$ GeV has a spread $\Delta y = 10.7$ units between the colliding beams. The question is to understand the widths of the fragmentation regions Δy_F in heavy-ion collisions; by how much is the nucleus spread in rapidity by the collision. Scattering of pp at the ISR at CERN showed $\Delta y_F \sim 1$. Present heavy-ion collisions indicate a Δy_F closer to 2 units [6,7]. The rapidity spread at RHIC should be adequate to separate the central region from the fragmentation regions.

4. Observables

Let us turn now to describe how one can probe the nature of ultrarelativistic heavy-ion collisions. To learn the properties of excited nuclear matter in collisions

---

*The rapidity y of a particle (either in the initial beams, or a product of the collision) is defined by $y = (1/2)\ln[(E+p_z)/(E-p_z)]$, where E is the particle energy, and p_z its momentum along the beam axis (taken to be the z direction). For the special case of motion purely along the z axis, the velocity is given by $v/c = \tanh y$. The energy and momentum along the beam direction of a particle are given by $E = m_\perp \cosh y$, and $p_z = m_\perp \sinh y$, where the transverse mass is defined by $m_\perp = \sqrt{m^2 + p_\perp^2}$, with $p_\perp$ the magnitude of the momentum transverse to the beam. Rapidities, unlike ordinary relativistic velocities, have the nice property of being additive; under a Lorentz transformation along z by velocity $u = c\tanh y_u$, rapidities transform by $y \to y + y_u$. Experimentalists often use the concept of the pseudorapidity of a particle, η, defined by $\eta = (1/2)\ln[(p+p_z)/(p-p_z)] = \tanh^{-1}\cos\theta$, where θ is the angle the particle makes with respect to the beam axis. While measurement of rapidity requires particle identification, pseudorapidity depends only on the scattering angle. For fully relativistic particles ($E \gg m$) the rapidity and pseudorapidity are equal; generally $|y| \leq |\eta|$.

is a non-trivial problem, involving a considerable interplay of many different types of measurements and theoretical simulations of collisions. There does not appear to be a well-defined signature of quark-gluon plasma production. Furthermore, unlike in solid-state experiments, the matter produced in heavy-ion collisions evolves dynamically from its initial formation, through possible equilibration, and final freezout, after which the particles no longer interact and stream freely to the detectors.

Experiments have initially concentrated on exploring the global structure of events – the degree of energy deposition and particle production, their distributions in in rapidity – as well as production of different types of particles, such as nucleons, strange particles (K mesons, Λ's), electromagnetic particles, and charm production (J/ψ mesons). In addition, Hanbury-Brown Twiss interferometry between pion pairs, and now kaon pairs, has been developed as a powerful tool for learning geometry of the collision volume [1,2]. At higher energies, as at RHIC, hard jets will become an important probe, revealing how a fast quark interacts with the matter produced.

4.1 Energy deposition in collisions

First consider measurements of the energy deposited in collisions, which are crucial to understanding whether future experiments will succeed in producing new states of nuclear matter. The degree to which colliding nuclei transfer energy and stop each other is determined most directly from measuring the transverse energy, E_T, of the event, defined as $E_T = \sum_i E_i \sin\theta_i$, where E_i is the energy from final particle i deposited in the calorimeter in which it lands (for mesons, the total energy, and for baryons, the total energy minus the nucleon rest mass); θ_i is the angle that particle i makes with respect to the beam axis, and the sum is over all final particles. [While not Lorentz invariant like the transverse momentum, the transverse energy has the great advantage of being directly measurable without needing to identify the type of particles detected.]

Measurements of E_T reveal strikingly large energy deposited in collisions. Figure 2 shows a characteristic total transverse energy distribution, $d\sigma/dE_T$, for 200 GeV/A ^{32}S beams on Al, a light target, on Ag, an intermediate mass target, and on W, Pt, Pb, and U, heavy targets, as measured by the CERN Helios collaboration (NA34) [8]. The angular coverage is essentially the entire forward hemisphere (in the beam direction) in the lab. Events in the tail of the distribution have nearly 500 GeV deposited in the collision volume!

This energy is approximately half the kinematic limit on the energy deposition, which equals the total kinetic energy in the center of mass of the nucleons participating in the collision. For a central collision of a projectile of nucleon number B on a target of nucleon number A ($>B$) the limit is $E_T^{max} = \sqrt{s_p} - m(B + A_p)$, where A_p is the number of participating target nucleons, and the participant cm energy squared is $s_p = 2mBA_pE_{lab} + (B^2 + A_p^2)m^2$. Here m is the nucleon mass and E_{lab} the lab projectile

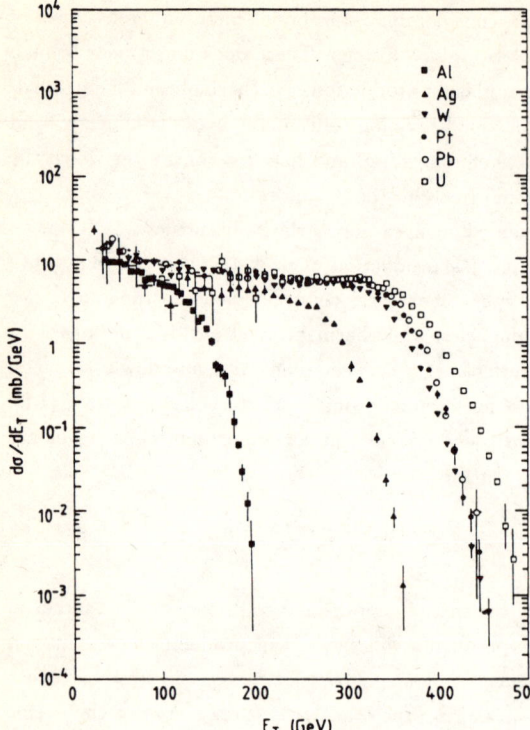

Figure 2. Transverse energy distribution $d\sigma/dE_T$ for ^{32}S at 200 GeV/A on various targets. From Ref. [8].

energy per nucleon. At ultrarelativistic energies the projectile drills a tube of radius equal to the projectile radius, $R_B \simeq 1.2 B^{1/3}$ fm, through the target nucleus, leaving the spectators in the target nucleus initially undisturbed. For a central collision the number of target participants in the tube is $A_p \simeq (3/2)B^{2/3}A^{1/3}$ when $B \ll A$. In the ^{32}S on heavy target collisions shown in Fig. 2, the available energy is $\simeq 1$ TeV, in the rare events in the tails of the distribution half this energy is deposited.

The general structure of $d\sigma/dE_T$ in Fig. 2, which is also shared by the multiplicity cross sections, $d\sigma/dm$, primarily reflects the geometry of the collision. The large cross-section at low E_T arises from peripheral interactions of the projectile and target, which occur with large probability but produce only little transverse energy. The plateau arises from events in which the entire projectile strikes the target, while the tail is produced by central collisions. The Gaussian structure of the tail arises from fluctuations in transverse energy production in central collisions; these fluctuations are, as we discuss below, a fertile probe of the production processes.

Let us turn now to estimating the energy density, ϵ, in the collision volume. The energy density decreases as the system evolves; thus, we must know the underlying space-time evolution of the collision volume. Eventually measurements from pion and kaon interferometry will be able to give detailed information on the evolving collision geometry. At present we must rely on various approximate model descriptions; the resulting estimates of the energy density are considerably model dependent.

The simplest estimate is to assume that the entire E_T is deposited uniformly in the collision volume in the participant center of mass, a cylinder with axis along the beam direction, of transverse area πR_B^2, where R_B is the projectile radius. The cylinder lies between the Lorentz contracted projectile and target nuclei, which recede from each other at the speed of light, and is thus of length ct, where t is the time since the initial collision of the nuclei, measured in the participant center of mass. This back-of-the-envelope estimate of the energy density gives $\epsilon = E_T/(\pi R_B^2 t)$, which for ^{32}S depositing 500 GeV in collision with a heavy target is $\sim (10/t_{\text{fm}/c})$ GeV/fm^3, decreasing in time.

We are interested in the energy density only after the excitations in the collision volume, whether hadrons or quarks and gluons, have had time to form. Characteristic formation times, measured in the frame comoving in the beam direction with the excitation, are, as discussed below, $\lesssim 1$ fm/c. Furthermore, we would like to study the system after it has had time to approach local thermodynamic equilibrium. A measure of the equilibration time is the viscous relaxation time, τ_η, which has recently been calculated exactly [9] to leading logarithmic order in the weak coupling limit in a quark-gluon plasma; the result is $\tau_\eta = 0.24/T\alpha^2 \log(1/\alpha) \gtrsim 1$ fm/c, where T is the plasma temperature and $\alpha = g_c^2/4\pi$ is the qcd fine structure constant. Stopping times for two interpenetrating plasmas are very similar [9,10]. The equilibration time in a strongly interacting plasma, for which we do not yet have a good description, could be considerably shorter than the weak coupling result.

Taking $t \sim 1$ fm/c in the estimate above, we would infer that we are seeing events with energy densities an order-of-magnitude beyond that in equilibrium nuclear matter, $\epsilon_0 \simeq 0.15$ GeV/fm^3 (essentially the rest mass of the nucleons). Such large values augur well for producing locally equilibrated matter at high energy and baryon densities in future experiments with heavier, more energetic beams.

An alternative method of estimating the energy density is to use numerical simulations of collisions that keep track of the space-time geometry. Representative simulations are that of Toneev and coworkers [11] and the Venus simulation of Werner [12], based on the dual parton model, the Frankfurt Relativistic Quantum Molecular Dynamics simulation [13], and extensions [14] of the Lund Fritiof model [15] for heavy-ion collisions to follow the space-time evolution of the produced particles. These models predict energy densities in light nucleus on heavy targets of order 1 GeV/fm^3, consistent with the back-of-the-envelope estimate, but with considerable variation from model to model.

Transverse energy and particle multiplicity distributions show strong evidence, both theoretically [16] and experimentally [17], for rescattering or "cascading" of secondaries in the collision volume, a crucial step towards thermalization of the matter. WA80 parametrized their measured average charged particle multiplicity, n_{ch}, distributions as a function of pseudorapidity and target mass, in the form $dn_{ch}/d\eta \propto A^{\alpha(\eta)}$. In the absence of rescattering one would expect the multiplicity to be proportional to the target thickness, i.e., $\alpha \simeq 1/3$, in the target fragmentation region ($\eta \sim 0$). However, as WA80 finds, $\alpha(\eta \simeq 0) \simeq 0.8$, which shows that a considerably larger part of the target nucleus is involved in the collision. As the rapidity grows, α falls, eventually to $\simeq 0$ in the projectile fragmentation region ($\eta \gtrsim 4$), where only fast particles from the projectile are found.

The number of pions of given transverse momentum, $p_\perp$, vs. $p_\perp$, falls exponentially in pp collisions. Intriguingly, the spectrum in heavy-ion collisions, as well as pA, appears to have a rise at small $p_\perp$ [18]. The excess of soft pions, approximately as many as under the exponential, appears likely to be a consequence of rescattering as well [19].

Rescattering of secondaries in the target increases the transverse energy and multiplicity primarily through generation of tertiary (and further) particles; two-body elastic scatterings tend to be forward-peaked and lead to little gain of transverse energy. Through rescattering, fast-forward particles become more isotropic. High-energy hadrons typically have an inelastic cross section $\sim$ 20-30 mb, and thus a mean free path in nuclear matter $\sim$ few fm, small compared to the size of the target. A secondary produced near the center of a large target will thus be able to rescatter a few times.

Formation of a secondary takes on average a finite time, τ_0, in the frame in which its rapidity vanishes. Thus, from Lorentz time dilation, excitations at large rapidity are formed primarily outside the target, and their rescattering from target nucleons will be suppressed, while those at smaller rapidity undergo rescattering after formation. Unfortunately, owing to theoretical uncertainties in the nature of the hadronization process, the formation time is not precisely defined. (See reviews in Refs. [20,21].) The simplest estimate is based on the uncertainty principle: in the frame in which the excitation moves solely transverse to the beam, the time for the excitation to appear as an entity distinct from the source is $\tau_0 \sim 1/m_\perp$, where $m_\perp = (E^2 - p_z^2)^{1/2}$ is the transverse mass of the excitation. For $m_\perp \sim 400$ MeV, one has $\tau_0 \sim 0.5 fm/c$. In the lab frame, in which the excitation has rapidity y, one would estimate in this way a formation time $\tau_{form} = \tau_0 \cosh y \sim e^y/2m_\perp$. The Lund string model, which takes into account the internal structure of the hadrons in their formation, has in fact two relevant formation times, the "constituent" time at which strings fragment through quark-antiquark formation, and the "yo-yo" time, when quarks and antiquarks combine to form a physical hadrons [21]; both formation times are larger, for slower excitations, than that above.

To estimate the rapidity range for formation outside the target, consider an excitation resulting from a collision at the center of the target; using (6) we see that if $y \gtrsim y_c = \ln(2R_A m_T) \sim 3.3$ (for $R_A \sim 7$ fm, and $m_\perp \sim 400$ MeV) the secondary will be formed outside, and with $y \lesssim y_c$, inside the nucleus. At CERN, the calorimetry of the NA34 experiment covers the range $y < y_c$, and is thus sensitive to tertiary excitations produced by rescattering of secondary particles formed in the target fragmentation region; the calorimeters of the NA35 experiment, measuring at more forward angles, detect fewer such tertiary particles. At AGS energies, rescattering is always important.

4.1a. Fluctuations in central collisions

The wide Gaussian tails in the measured E_T distributions arise from fluctuations in E_T production in central collisions. Past the "peak" of the Gaussian the observed distribution includes only contributions from central collisions; other impact parameters contribute significantly only to the left of the peak. These fluctuations are a useful probe of the internal dynamics of heavy-ion collisions. The widths of the tails are considerably larger than expected on the basis of independent nucleon-nucleon (NN) subcollisions. What physics is revealed by these fluctuations?

The widths of the Gaussian tails can be written in terms of a scaled variance, $\omega \equiv \langle N \rangle (\langle E_T^2 \rangle - \langle E_T \rangle^2)/\langle E_T \rangle^2$, where $\langle N \rangle$ is the mean number of independent NN subcollisions in a central heavy-ion collision, ~ 100 for O on a heavy nucleus. In an independent NN collision picture, ω is effectively the width of the E_T distribution for each NN collision. NA34 [8] finds that ω, in collisions of ^{32}S on nuclear targets, from Al to U, in the pseudorapidity interval $-0.1 < \eta < 2.9$, ranges from ~ 1.5 - 5.5. The principal contributions to ω are: i) ω_0, the width of the single source E_T distribution, which, from pp-collisions in the comparable energy and rapidity range, is ≈ 0.5; ii) $\omega_{def} = 4\langle N \rangle \delta^2/45$, the fluctuations from the deformation of the target, where δ is the nuclear deformation parameter, particularly large for W and U targets ($\omega_{def}(U) \simeq 1.75$); and iii) $\omega_N = (\langle N^2 \rangle - \langle N \rangle^2)/\langle N \rangle$, the fluctuations in the number of sources, equal to unity for Poisson statistics. These three terms together are not capable of explaining the experimental widths, but fall short by up to 40%. Rescattering can, with proper choice of formation time for secondary excitations, significantly broaden the tails of the E_T distribution and thus increase ω; the results are quite sensitive, however, to the formation time.

One further contribution to the widths arises from fluctuations in the cross sections for interaction of any given nucleon in one nucleus with those in the other [22]. Hadrons have an internal structure arising from their internal quark structure, and thus their cross section, σ, for interacting with other hadrons is a dynamical quantity which depends on their instantaneous internal configurations. The internal configuration of a nucleon is "frozen" in its rest frame on times $\sim 1/2$ fm/c, a characteristic internal dynamic time. Because of Lorentz time dilation, the configuration of a fast nucleon, of energy at least 40 GeV, will be frozen during the time it takes the nucleon to traverse a large nucleus. The cross sections for successive interactions for nucleon i in the projectile with target nucleons j and k (and vice versa) will be correlated:

$\langle \sigma_{ij}\sigma_{ik}\rangle = \langle \sigma_{ij}\rangle^2(1+\omega_\sigma)$, with $\omega_\sigma \simeq 0.2 - 0.3$ for present beam energies. The quantity $\langle \sigma_{ij}\rangle$ is just the mean cross section, quoted in the tables.

As shown in [22], cross-section fluctuations contribute to ω in an AB collision an amount approximately $(N_{pA} + N_{pB})\omega_\sigma$, where N_{pA} and N_{pB} are the mean numbers of collisions a proton makes in traversing the center of A or B. This additional contribution to the fluctuations is just the right size to account for the expermentally large widths.

4.2. Strangeness production

Measurements by NA35 [23], E802 [6,24], and NA34 [25] of production of strangeness-containing particles, e.g., K mesons and Λ's, which probe the quark content of the matter produced in a heavy-ion collision, have yielded good evidence that nuclear collisions produce strangeness beyond that expected in independent nucleon-nucleon interactions. For example, E802 has carried out time-of-flight measurements on the produced particles, sorting them into pions, kaons, protons and anti-protons, as well as heavier particles, and find a surprisingly high ratio of K^+ to π^+ mesons. In Si on Au at 14.5 GeV/A, the ratio K^+/π^+ is $\simeq 19.3\%$ in central collisions, compared with the ratio $K^-/\pi^- \simeq 4.1\%$. By comparison, in pp or pBe collisions at comparable energies, $K^+/\pi^+ \sim 4-8\%$, while $K^-/\pi^- \simeq 2.4\%$; the latter, within error bars is consistent with the heavy ion result. K^+/π^+ in Si+Au shows a strong falloff with rapidity, starting at close to 25% at y=1; the enhancement is also strongest at large transverse momenta, $p_\perp$. The numbers of π^+ and π^- are approximately equal, as one expects in this nearly isospin symmetric situation.

One indeed expects to produce more K^+ than K^-; the K^+ is made of u+$\bar{s}$ quarks, while K^- is made of $\bar{u}$+s, and the nuclear system contains more u than $\bar{u}$. However, before beginning to use the ratio as a measure the quark content of the initial state and to discover new physics, it is necessary to understand the nuclear "background" processes in detail; one must take into account various final state interactions, such as $\pi^+ n \to K^+\Lambda$, which increases K^+/π^+. [There is no analogous reaction that changes a π^- into a K^-.] The Frankfurt group, on the basis of the RQMD simulation, argues that the final state interactions can in fact account for the enhancement [26,13].

A K^+ is made in a strong interaction by first producing an $\bar{s}s$ pair; the $\bar{s}$ is bound in the K^+. For an event with ~ 50 π^+, some 10 s quarks are also produced, which should primarily be in Λ. In a system this rich in s quarks, conditions are very favorable for forming nuclear-like objects with rather large strangeness. Such objects, not observed so far, have been predicted, see e.g., [27], on the basis that in qcd it is easier to bind systems with a large rather than a small strangeness fraction. Indeed, detection of such objects would bring to the fore the question, first raised by Witten [28], of whether the actual ground state of nuclear matter contains a large strange quark fraction, and ordinary matter is only metastable.

4.3 Lepton pairs and charm production

Electromagnetically produced particles, such as e^+e^- and $\mu^+\mu^-$ pairs and direct photons, interact negligibly with the matter after they are made and thus are particularly well-suited to probe conditions in the interior of the interaction volume, in the way that solar neutrinos can provide direct information about the deep interior of the sun. The strongly interacting particles produced in a collision, by contrast undergo considerable interactions before freezeout, and thus do not directly probe conditions at early times in the collision.

Study of $\mu^+\mu^-$ pairs has revealed interesting new physics in heavy-ion collisions. In plotting number of pairs vs. the invariant pair mass, $\sqrt{(p_1+p_2)^2}$, one expects to see a background of pairs formed via the Drell-Yan process, $\bar{q}q \to \mu^+\mu^-$, in the collision volume, together with spikes at the masses of the vector mesons, ρ, ω, ϕ, J/ψ, Υ, etc., which when formed in the plasma can decay into $\mu^+\mu^-$. The region of the J/ψ, with mass 3.1 GeV, has been extensively studied by the NA38 experiment [29] in S on U at 200 GeV/A, Fig. 3. They find the surprising result that the number of J/ψ produced, measured with respect to the Drell-Yan background, in collisions with large E_T is about 50% of the number produced in collisions with small energy deposition. The suppression effect is greatest for J/ψ with lowest $p_\perp$.

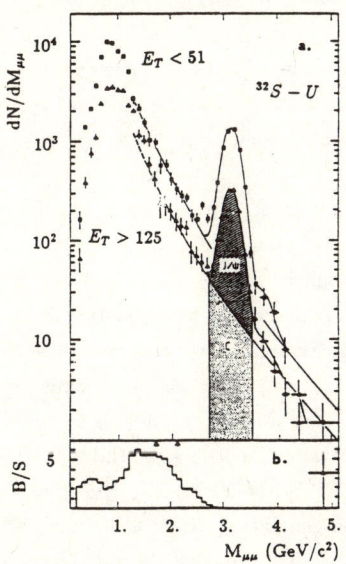

Figure 3. Dimuon mass spectrum in S on U at 200 GeV/A, showing the suppression of the J/ψ. The small shoulder is the ψ'. From [29].

The suppression of J/ψ was proposed, prior to the NA38 experiments, by Matsui and Satz [30] as a possible signal for formation of a quark-gluon plasma. The mechanism was very simple. The J/ψ is a bound state of a $\bar{c}c$ quark pair, arising from an effective potential, string-like at large separation and Coulomb-like at small. In a plasma, the force becomes screened, the binding reduced, and at sufficient plasma density, the bound state should disappear altogether. Thus one would expect less possibility of binding a $\bar{c}c$ pair in a hot dense plasma, corresponding to the events with large E_T. However, before this explanation could be accepted it is again necessary to understand other processes that can suppress J/ψ. The experiment inspired considerable thought to understand the final state interactions of the nascent J/ψ with the matter in the collision volume, as well as the initial interactions prior to J/ψ formation (see, e. g., [31]). Subsequent measurements of J/ψ production in pA collisions at 800 GeV also show suppression with increasing A [32], indicating that some combination of initial and final state interactions is the cause of the suppression, rather than the onset of a plasma.

4.4 Pion and kaon interferometry

Let me now briefly describe the use of pion and kaon pair correlations to probe the geometry of the collision. The basic tool is the Hanbury-Brown Twiss effect, developed for measuring sizes of radio sources, and stars: if two nearby *incoherent* sources of photons of frequency k, at $\vec{r}_1$ and $\vec{r}_2$, are observed at two detectors, a and b, then the correlated intensity $\langle n_a n_b \rangle$ of the number n_a at a and n_b at b is given by

$$C_2 \equiv \frac{\langle n_a n_b \rangle}{\langle n_a \rangle \langle n_b \rangle} = 1 + \cos(\vec{q} \cdot (\vec{r}_1 - \vec{r}_2)),$$

where $\langle n_a \rangle$ and $\langle n_b \rangle$ are the average intensities, and $\vec{q} = \vec{k}_a - \vec{k}_b$ is the difference of the momenta of the photons detected at a and b [33]. Thus the intensity correlations directly reveal the source separation. The same results apply to correlations of any identical bosons, including π and K mesons.

Generally one must average over a distribution of sources, e.g., if the distribution is Gaussian, $\rho(r) \sim \exp(-r^2/R^2)$, then $C_2 = 1 + \exp(-q^2 R^2/2)$. As we see from this simple form, and generally, $C_2 = 2$ at $q = 0$, while at large q, $C_2 \to 1$. The falloff of C_2 with q tells one the source size, R. One normally describes the correlation of intensities as arising from the fact that the emitted particles are bosons, and thus "prefer" to be emitted in the same momentum state. Because of the finite size of the emitting region this implies a correlation over a range in q of order R^{-1}. [This effect is in fact classical physics, a consequence of the superposition principle for field amplitudes.]

Many groups, including NA35 [34], WA80 [35] and E802 [36], have analyzed pion correlations heavy ion collisions. In the second generation experiments, e.g., E859 and NA44, kaon correlations are being studied. Note that one must extract the correlation, C_2, between pairs in individual events, and then average over events. Analysis of the data is complicated by the fact that in a heavy-ion collision the source evolves rapidly on strong interaction time scales, a situation quite unlike the analysis of essentially static astrophysical sources. The pairs carry information about their locations when they become non-interacting, as in photons from surface of sun, since interactions constantly change their momenta. Thus one is faced with the difficult challenge of building up a consistent picture of the collision dynamics and detailed kinetics simultaneously with the geometry. The interpretation of the interferometry experiments are thus very model dependent.

The measured pair correlation functions have been generally analyzed with independent Gaussian distributions for the transverse and longitudinal sizes, varying with rapidity window. The results are consistent with the overall pictures of the dynamics described above, e.g., a fireball of order the projectile radius at AGS energies, and a geometry similar to that in Fig. 1, at SPS energies. Generally the measured C_2 does not rise to 2 at small q, for a number of reasons. Without careful particle identification, pairs of π and K or electrons, between which there are no correlations, are inadvertently included in the data sample and dilute the effect. Also, pions from decay of long-lived meson resonances tend to bloat the apparent geometry. For example, an ω made in the collision will contribute pions a mean distance ~ 23 fm, its mean decay length, λ, from the production point. Resonances contribute to the enhancement of C_2 only at small $q \sim 1/\lambda$. Finally, the assumption of incoherent sources is not accurate; rather, sources are correlated at least over distances ~ 1 fm in the local rest frame. As we have learned from the first experiments, interferometry is a powerful tool for probing the geometry, but much work remains to be done on how correctly to interprete the data.

4.5 Transverse momentum spectra

A probe that may signal reaching the deconfinement phase transition in a heavy-ion collision is the correlation of the mean transverse momentum, $\bar{p}_\perp$, of the produced particles with their mean multiplicity, m. The simple thermodynamic analogy is the boiling of water. As one heats liquid water, putting in entropy, its temperature rises. Then when the boiling temperature is reached, further entropy input converts water to vapor, but T remains fixed. Finally when all the water is in the vapor phase, T again rises with increase in entropy. In a heavy ion collision, $\bar{p}_\perp$ is a measure of the temperature, and m a measure of the entropy (recall, e.g., in black body radiation, that the entropy per photon is 3.6). Thus, one would expect a plot of $\bar{p}_\perp$ vs. m to rise at small m, then at nuclear excitation energies sufficiently high to begin to make a plasma, to flatten (although not completely because of the expansion of the system), and finally to rise again at higher m, corresponding to higher excitation energies. Heavy-ion collisions do not yet provide evidence of such behavior (see, e.g, [18]). Early data from the Fermilab $\bar{p}p$ experiment E735 at $\sqrt{s} = 1.8$ TeV [37] did suggest a flattening and rise, although statistics are not adequate; the onset of minijet production can also mimic the effect.

Conclusion

The present heavy-ion experiments have been very successful. They provide evidence that collisions can deposit a substantial fraction of the beam energy in the collision volume, producing matter at high energy densities. They indicate that with heavier beams at higher energies conditions will be very favorable for discovering new physics, including new states of matter, in collisions. While there is no compelling evidence that the matter in present collisions is anywhere near being in local thermal equilibrium, we can expect that as one goes to heavier beams – increasing the system size – and higher energies – increasing the number of excitations produced and particle mean free paths – conditions will become considerably more favorable to achieve local thermalization. Finally, the present experiments have been very important in showing how to probe the matter produced in ultrarelativistic collisions.

Research described here has been supported in part by U. S. National Science Foundation grant PHY89-21025.

References

1. G. Baym, P. Braun-Munzinger, and S. Nagamiya (eds.), *Quark Matter '88, Proc. 7th Intl. Conf. on Ultra-relativistic Nucleus-Nucleus Collisions, Nucl. Phys.* **A498** (1989).
2. J.-P. Blaizot, C. Gerschel, and A. Romana (eds.), *Quark Matter '90, Proc. 8th Intl. Conf. on Ultra-relativistic Nucleus-Nucleus Collisions, Nucl. Phys.* **A525** (1991).
3. A. Ukawa, *Nucl. Phys.* **B** (Proc. Suppl.) **17** 118 (1990).
4. S. Gottlieb, *Nucl. Phys.* **B20** (Proc. Suppl.), 247 (1991).
5. G. Baym and S. A. Chin, *Phys. Lett.* **62B**, 241 (1976).
6. T. Abbott et al. (E802), *Nucl. Phys.* **A525**, 231c (1991).
7. T. Åkesson et al. (Helios Collaboration), *Z. Phys.* **C46**, 361 (1990).
8. T. Åkesson et al. (Helios Collaboration), *Nucl. Phys.* B353, 1 (1991).
9. G. Baym, H. Monien, C. J. Pethick, and D. G. Ravenhall, Phys. Rev. Lett. 64 (1990) 1867; *Nucl. Phys.* **A525**, 415c (1991).
10. G. Baym, H. Heiselberg, H. Monien, C. J. Pethick, and J. Popp, to be published.
11. V. D. Toneev, N. S. Amelin, K. K. Gudima, and S. Yu. Sivoklokov, *Nucl. Phys.* **A519**, 463c (1990).
12. K. Werner, *Nucl. Phys.* **A525**, 501c (1991).
13. H. Sorge et al., *Nucl. Phys.* **A525**, 95c (1991).
14. T. Csörgö, J. Zimányi, J. Bondorf, and H. Heiselberg, *Phys. Lett.* **222B**, 115 (1989).

15. B. Andersson, G. Gustafson, and B. Nilsson-Almqvist, Nucl. Phys. B281, 289 (1987); T. Sjöstrand and M. Bengtsson, *Comput. Phys. Comm.* 43, 367 (1987).
16. W. Busza and R. Ledoux, *Ann. Rev. Nucl. Part. Sci.* 38, 119 (1988); T. Csörgö, J. Zimányi, J. Bondorf, and H. Heiselberg, *Z. Phys.* C46, 507 (1990).
17. R. Albrecht et al. (WA80), *Phys. Lett.* 202B, 596 (1988).
18. B. Jacak, *Nucl. Phys.* A525, 77c (1991).
19. M. Neubert, G. Baym, G. Friedman, B. Jacak, and Y. Yariv, to be published.
20. N. N. Nikolaev, *Sov. J. Part. Nucl.* 12, 63 (1981).
21. A. Białas and M. Gyulassy, *Nucl. Phys.* B291, 793 (1987).
22. H. Heiselberg, G. Baym, B. Blättel, L. L. Frankfurt, and M. Strikman, *Phys. Rev. Lett.*, in press (1991).
23. R. Stock et al., *Nucl. Phys.* A525, 221c (1991).
24. T. Abbott et al. (E802), *Phys. Rev. Lett.* 64, 847 (1990).
25. H. van Hecke (Helios Collaboration), *Nucl. Phys.* A525, 227c (1991).
26. R. Mattiello et al., *Phys. Rev. Lett.* 63, 1459 (1989).
27. C. Alcock and A. Olinto, *Ann. Rev. Nucl. Part. Sci.* 38, 161 (1988).
28. E. Witten, *Phys. Rev.* D30, 272 (1984).
29. J. Varela, *Nucl. Phys.* A525, 275c (1991).
30. T. Matsui and H. Satz, *Phys. Lett.* 178B, 416 (1986).
31. S. Gavin and M. Gyulassy, *Phys. Lett.* 214B, 241 (1988); J. P. Blaizot and J. Y. Ollitrault, *Phys. Lett.* 217B, 392 (1989).
32. J. M. Moss (E772), *Nucl. Phys.* A525, 285c (1991).
33. G. Baym, *Lectures on Quantum Mechanics*, (W. A. Benjamin, Reading, Mass., 1969), pp. 431-434.
34. J. Baechler et al. (NA35), *Nucl. Phys.* A525, 327c (1991).
35. K. H. Kampert et al. (WA80), *Nucl. Phys.* A525, 327c (1991).
36. T. Abbott et al. (E802), *Nucl. Phys.* A525, 327c (1991).
37. T. Alexopolous et al. (E735), *Phys. Rev. Lett.* 64, 991 (1990); *Nucl. Phys.* A525, 165c. (1991).

Electron Scattering

Ingo Sick

Department of Physics, University of Basel
CH-4056 Basel, Switzerland

Abstract : Investigations of hadronic systems with the electromagnetic probe are illustrated by discussing in detail two special cases: the study of nuclear charge distributions in the lead-region, and the determination of the form factors of the neutron.

1 Introduction

The field of electron scattering is an extremely wide one, and has an impact on many different areas of nuclear- and particle physics. It is obviously impractical to give, in a number of lectures, a complete review. Rather than giving a superficial description of many topics, we below concentrate on a more detailed discussion of two selected examples. We describe in detail the determination of nuclear charge densities, and the impact of these measurements on our understanding of nuclear structure. We then discuss extensively the measurement of the charge- and magnetic form factors of the neutron. While the discussion of the lead-region concerns work that we may classify as accomplishments of the past, the area of the neutron form factors largely concerns the future, as many of the experiments discussed are only in the stage of preparation.

2 Motivation

Investigations of hadronic systems with the electromagnetic probe are singled out by a number of important advantages, and a number of serious drawbacks. When using the electromagnetic probe, we should be aware of both. Among the advantages we should list:

1. The interaction of the electron with the nucleus is well known, and described by an exact theory, QED.

2. The interaction of the electron with the nucleus is weak (of order $\alpha = 1/137$). As a consequence, the scattering process can quantitatively be described using available

theoretical techniques, and the information on the hadronic system under investigation can quantitatively be extracted.

3. Due to the weakness of the electromagnetic interaction scattering at large momentum transfer q indeed corresponds to good spatial resolution, of order $1.5/q$. For strongly interaction probes, scattering at large q results dominantly from multiple probe-target interactions. For n-fold scattering, the spatial resolution then is of order $n \cdot 1.5/q$. If we want to investigate short-distance phenomena — which is imperative for the study of short range properties of nuclear wave functions, for the study of subnucleonic degrees of freedom and for the understanding of quarks in nuclei — we depend on achieving this highest spatial resolution only available in (e,e) at large q.

These advantages of the electromagnetic probe are connected to a number of drawbacks:

1. Due to the weakness of the interaction, cross sections are small; relative to the cross section for strongly interacting probes they are down by a factor $\alpha^2 \sim 10^4$. In order to compensate this effect, and to reach large q, electron scattering experiments depend on high beam intensities $(10 - 100 \mu A)$ and detectors with large solid angle.

2. Due to the smallness of the electron mass, large momentum transfer can only be achieved by high electron energy. At large electron energy, very good *relative* energy resolution $\Delta E/E$ is needed to separate final states of the nucleus. High energy in the past could only be obtained with *pulsed* electron accelerators; duty cycles of $< 10^{-2}$ led to severe limitations when performing coincidence experiments.

3. For highly relativistic projectiles helicity is a conserved quantum number. Polarization experiments of the type $(\vec{e}, e)$ therefore are not providing new information (except for studies of parity-violating processes). To exploit the polarization observables that provide valuable additional information one in general needs to observe the polarization of one additional reaction partner (target nucleus or reaction product). This greatly complicates the exploitation of polarization observables.

As technology advances, the impact of a number of the above drawbacks can be minimized. Superconductive RF cavities allow to maintain *continuously* the high accelerating gradients needed to accelerate intense beams of electrons to GeV-energies. New acceleration schemes such as recirculation and race-track microtrons allow to reach high energies with comparatively small effort. The development of novel polarized electron sources (emission of $\vec{e}$ from crystals of GaAs irradiated with polarized light, for instance) makes polarized electrons available on a routine basis. Polarized targets of light nuclei $(\vec{p}, \vec{d},^3 \vec{H}e)$ are being developed to stand much higher beam intensities than acceptable up to now.

3 Charge Densities in Lead Region

3.1 Determination of $\rho(r)$

One of the classical topics of electron-nucleus scattering is the determination of nuclear charge densities. The interest in densities results from the fact that, apart from a trivial folding with the proton intrinsic charge distribution and a small correction for the contribution of the neutrons, the charge density reflects the spatial distribution of protons in nuclei. Precise measurements of the charge density give us the most accurate information on the geometrical properties of nuclei, both in the nuclear surface *and* the interior.

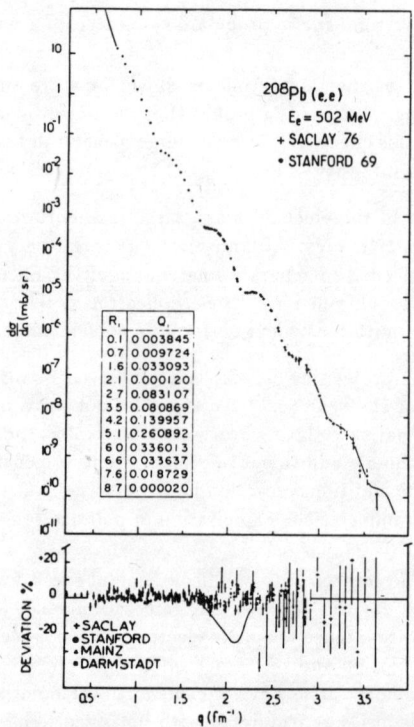

Figure 1: Cross section for $^{208}Pb(e,e)$, and deviations of data relative to fit.

As pointed out in the introduction, we here will concentrate on the discussion of one particular example, the nuclei near the doubly-magic ^{208}Pb. Fig.1 shows the data for elastic electron-^{208}Pb scattering presently available. Much of the data come from 2 experiments we performed at Stanford [1] and Saclay [2]. With these experiments the data reach a maximum momentum transfer of $4 fm^{-1}$, and cover a dynamical range of 12 orders of magnitude in cross section! The accuracy of the data at the lower momentum transfers is of order 2%.

In plane wave Born approximation (PWIA) the relation between the cross section and the charge density is simple

$$\frac{d\sigma}{d\Omega}(E,\theta) = \sigma_{Mott}(E,\theta) \cdot F^2(q) \tag{1}$$

with the momentum transfer (ignoring recoil correction) given by

$$q \simeq 2 \cdot E \cdot sin\theta/2 \tag{2}$$

and the form factor

$$F(q) = \frac{1}{Z} \cdot \int_o^\infty \rho(r) \cdot \frac{sin(qr)}{qr} \cdot 4\pi r^2 dr. \tag{3}$$

The Mott cross section (again ignoring recoil) corresponds to scattering of the electron by a pointlike charge Ze, and is given by

$$\sigma_{Mott} = \left(\frac{Z\alpha}{4E}\right)^2 \cdot \frac{cos^2\theta/2}{sin^4\theta/2} \tag{4}$$

In PWIA it seems easy to derive $\rho(r)$ from measurements of $F^2(q)$. The density is simply obtained by inverting the Fourier transform

$$\rho(r) = \frac{Z}{2\pi^2} \int_o^\infty F(q) \cdot \frac{sin(qr)}{qr} q^2 dq. \tag{5}$$

Equation (3) shows the importance played by the momentum transfer q. Measurements at a given q_o sample the Fourier component of $\rho(r)$ of wave length $\lambda_o = 2\pi/q_o$. As the FWHM of the peaks of $\frac{sin(q_o r)}{q_o r}$ amounts to $\simeq \lambda_o/4$ the spatial resolution of an electron scattering experiment is of order $1.5/q$. If we want to study the nucleus with good spatial resolution, we need large maximum momentum transfer q_{max}.

Two features prevent us from exploiting eq.(5) for a determination of $\rho(r)$:

- For heavy nuclei PWIA is not valid, as the long-ranged Coulomb potential of the nucleus distorts the electron waves.

- Experiments are limited to a certain q_{max}, so that the integral in eq.(5) cannot be extended to $q_{max} = \infty$. Our experimental information is incomplete.

These problems can be dealt with by assuming a model density for $\rho(r)$, depending on a number of parameters. For a given $\rho(r)$ one can numerically solve the Dirac equation for the electron moving in the potential defined by $\rho(r)$, and the cross sections can be calculated. Adjustment of the parameters to fit the data yields the desired density.

This approach, although standard in many areas of physics, has important drawbacks. One cannot specify the degree to which the resulting $\rho(r)$ depends on the model assumed, i.e. one cannot give a sensible error bar $\delta\rho(r)$. This very much limits the usefulness of $\rho(r)$, as an experimentally determined quantity is only as good as specified by its error bar.

Basically two approaches have been developed to derive "modelindependent" densities [3, 4]. To do so, one has to replace the information we lack from experiment ($q > q_{max}$)

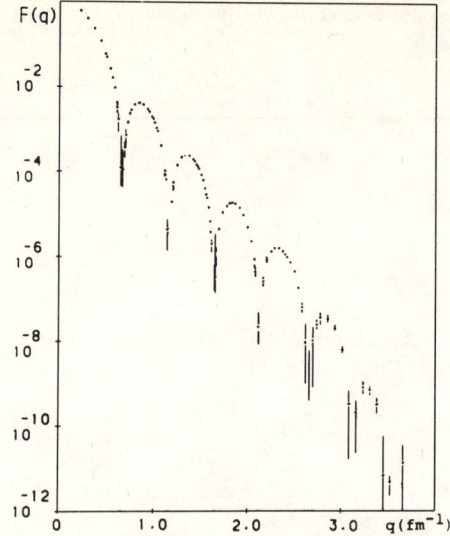

Figure 2: PWIA form factors corresponding to cross sections of figure 1.

by assumptions we can derive from *physical* considerations. In the shell model, densities are sums of squares of radial wave function $R_i^2(r)$, summed over all occupied shells i. These radial wave functions are slowly changing quantities as they result from solutions of a Schrödinger equation. In $\rho(r)$ we do not expect to find structures that are narrower than the narrowest peak in the $R^2(r)$ of all occupied shells. Imposing this condition on $\rho(r)$ allows to estimate the maximum amplitude of the higher, unmeasured Fourier components of $\rho(r)$, and to derive model independent densities with a reliable error bar. Alternatively, one can make sensible assumptions on the upper limit of the form factor in the not measured region $F(q > q_{max})$.

With those widely used approaches one is capable to derive reasonable error bars $\delta\rho(r)$. The procedure, however, still suffers from a non transparent relation between $F(q)$ and $\rho(r)$, as this relation occurs via a numerical solution of the Dirac equation.

One can overcome these limitations if one uses the Direct Fourier Transform (DFT) approach I developed a number of years ago [5]. It is based on the fact that in electron scattering the difference between PWIA and the exact cross section is not very important. One therefore can convert experimental cross sections σ^{exp} into experimental PWIA form factors F_{PWIA}^{exp}. This is achieved by fitting the data (via a solution of the Dirac equation) with any convenient model density $\rho_{Mod}(r)$, and computing for $\rho_{Mod}(r)$ both σ^{Mod} (via the Dirac equation) and

$$F_{PWIA}^{Mod} = \frac{1}{Z} \int \rho_{Mod}(r) \frac{sin(qr)}{qr} 4\pi r^2 dr \qquad (6)$$

The experimental form factors given by

$$F_{PWIA}^{exp} = \left(\frac{\sigma^{exp}}{\sigma^{Mod}}\right)^{1/2} \cdot F_{PWIA}^{Mod} \tag{7}$$

contain the same information as σ^{exp}, but can be interpreted in PWIA. It can be shown that eq.(7) is virtually independent of the model-density ρ_{Mod} employed, a property that results from the fact that the (long-range) Coulomb distortion of the electron waves essentially depends on the charge of the nucleus. In figure 2 we show the data for ^{208}Pb, converted to PWIA form factors. The data now show the diffraction zeroes (with sign change of F(q)) expected in PWIA.

Once one has experimental form factors interpretable in PWIA, one can cope with the second problem, the incompleteness of the experimental information ($q < q_{max}$). This can be done in a very transparent way by studying

$$\rho(r, q_o) = \frac{Z}{2\pi^2} \int_o^{q_o} F_{PWIA}^{exp}(q) \cdot \frac{sin(qr)}{qr} \cdot q^2 dq \tag{8}$$

as a function of the upper integration limit q_o. $\rho(r, q_o)$ is a damped oscillatory function of q_o, with a maximum or minimum at every sign-change of $F(q)$ or $sin(qr)$.

In figure 3 we show $\rho(r, q_o)$ for the most difficult case, $r = 0$, where the $sin(qr)/qr$-term does not help the convergence. At $r = 0$

$$\rho(0, q_o) = \frac{Z}{2\pi^2} \int_o^{q_o} F_{PWIA}^{exp}(q) \cdot q^2 dq, \tag{9}$$

Figure 3 shows that $\rho(0, q_o)$ is a damped oscillatory function of q_o. We may take the difference between the last measured maximum and minimum a an honest estimate of the completeness error of $\rho(0)$. All other errors in $\rho(r)$ can be calculated via the usual error propagation in eq.(8).

We learn from fig.3 that electron scattering can determine the central density of ^{208}Pb to $\pm 1\%$ (and to higher accuracy for $r > 0$). This is true *provided* that the experiment reaches a momentum transfer large enough such that $F(q_{max}) < 10^{-5}$. For an experiment that stops at $q = 2.5 fm^{-1}$, for instance, the difference between the last measured maximum and minimum of $\rho(0, q_o)$ would amount to 14%, leading to a large uncertainty of $\rho(0)$.

The DFT method provides a really model independent analysis of the data, and a transparent relationship between F and ρ. It can serve as a benchmark test for all the other more involved methods.

3.2 Density of lead

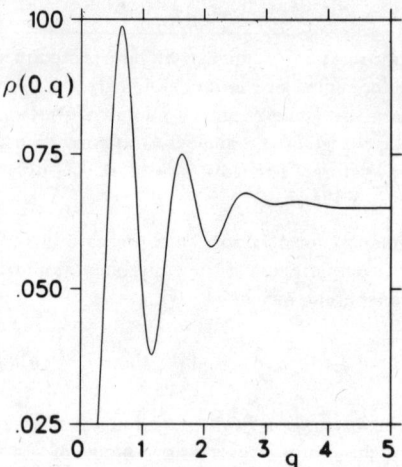

Figure 3: $\rho(0, q)$ as a function of the upper integration limit.

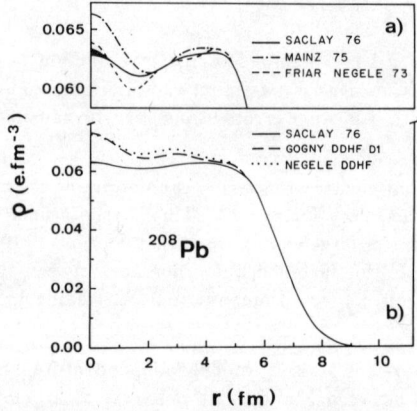

Figure 4: Charge density of ^{208}Pb together with DDHF predictions.

In fig.4 we show the density of ^{208}Pb extracted from the world data for this nucleus. The density exhibits the usual Woods-Saxon type distribution, with superimposed oscillatory structure resulting from the shell structure.

The experimental density is compared to two typical results from Hartree-Fock calculations using density-dependent effective nucleon-nucleon forces [6, 7]. These calculations are based on the assumption that nucleons move, independently from each other, in the mean field calculated from the wave function of all other nucleons and the effective N-N interaction. These DDHF calculations explain the density well in the surface region. In the nuclear interior, the average density is too large, and the calculations predict too much oscillatory structure of $\rho(r)$.

The calculations shown in fig.4 are representative for a large number of other theoretical results that have been obtained with similar approaches. They do, however, not truly represent the degree of our understanding of densities *from first principles*; the density dependence of the effective forces employed has been fitted to nuclear properties (the radius is particular). In figure 5 we show the result of a renormalized Bruckner Hartree Fock (RBHF) calculation [8] that does not make such phenomenological adjustments. In the RBHF approach the properties of infinite nuclear matter at various densities are calculated starting from the true nucleon-nucleon interaction as determined by N-N scattering. In this RBHF calculation the short-range NN correlations, resulting from the repulsive core of the NN interaction, and the occupation of states above the Fermi momentum K_F are accounted for. The resulting wave function is then used to deduce an *effective* density dependent NN interaction that is then used in the independent-particle calculation of the wave function of finite nuclei. Fig.5 shows that, without the phenomenological adjustment inherent in the DDHF approach, an understanding of nuclear densities is still relatively poor.

To explain the discrepancy between experiment and RBHF, a number of options exist a priori. The Independent Particle Shell Model (IPSM), even with density dependent forces, could fail given the high density in the nuclear interior; this seems likely in particular if one realizes that the average distance between nucleons in the interior is $\sim 2fm$, while the nucleon diameter is 1.6 fm. One also may question the neglect of non-nucleonic degrees of freedom, which could be responsible for the fact that the RBHF calculation does not correctly reproduce the nuclear matter saturation properties. Other effects, such as the neglect of relativity or changes due to configuration mixing *etc.* appear to be small on the scale of the discrepancy of fig.5.

When looking at contributions from individual shells to the total density (fig.6), one may in particular suspect the validity of the prediction for the s-shells which give the largest contribution in the nuclear interior. Can we really expect to find, in the high-density environment of the nuclear interior, IPSM wave functions such as they are given by the shell model? Is the IPSM at all relevant in the nuclear interior?

3.3 Density distribution of 3s-shell

In order to learn more on the applicability of the shell model, we can study the density difference between the nuclei Pb and Tl. The last shell filled in ^{208}Pb is the 3s-shell. A precise measurement of the density difference between Pb and Tl gives, modulo corrections discussed below, the density distribution of nucleons in the 3s shell.

The 3s-shell actually would be expected to have a very nice signature in the Pb/Tl cross section ratio. As shown by fig.6 the 3s radial wave function, with 3 peaks and 2 nodes, has a $sin(2\pi r/\lambda_o)/r$-behaviour (all it takes to see it is to move upward a bit the

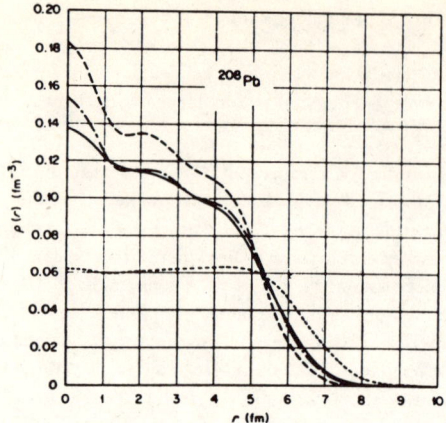

Figure 5: RBHF density (solid line) compared to experimental density (dashed). density.

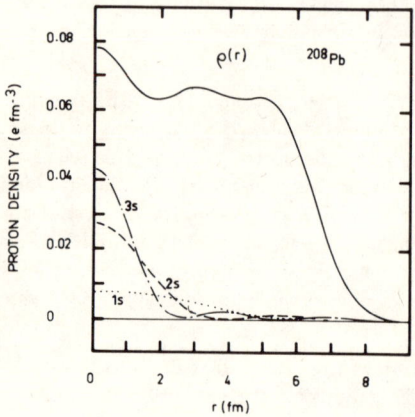

Figure 6: Contribution of s-shells to density of lead.

zero line). The Fourier-transform of such a density distribution is expected to be close to a $\delta(q_o = 2\pi/\lambda_o)$ function. This is borne out by Fig.7, which shows the Tl/Pb cross section ratios obtained when removing from the Pb density the distribution corresponding to one 3s proton as calculated by DDHF.

Of course, when going from Pb to Tl (as ^{207}Tl is unstable we have to compare ^{206}Pb and ^{205}Tl) additional effects need to be considered. As the occupation numbers of shells in Pb,Tl are smaller than the ones given by the extreme single-particle model, due to configuration mixing, the occupation of the other valence shell must be expected to change somewhat as well. As is shown by fig.7, their contribution, which will be $\sim 20\%$ of what

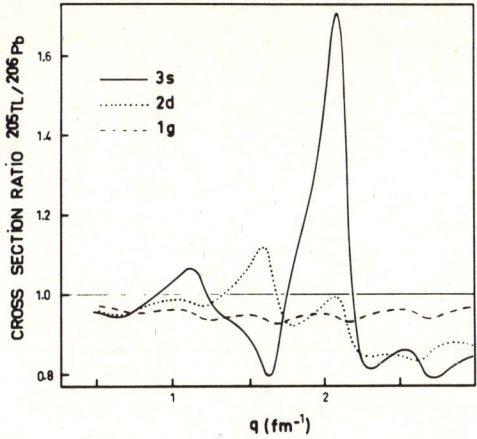

Figure 7: Contribution of shells near Fermi edge to σ^{Tl}/σ^{Pb} for $\Delta n \equiv 1$.

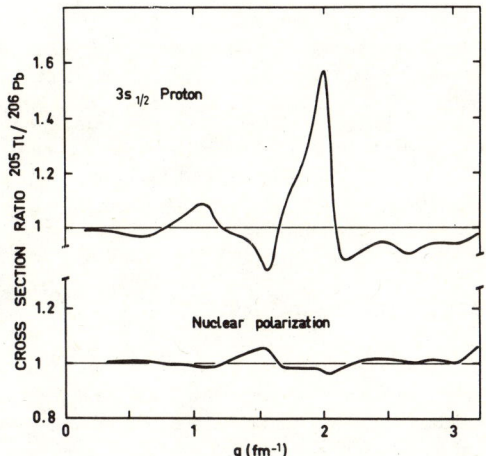

Figure 8: Contribution of core-polarization to σ^{Tl}/σ^{Pb}.

is calculated in fig.7 for a change of *one* in the occupation number in the corresponding shell, will not destroy the 3s-signal at $q \sim 2 fm^{-1}$.

Another complication could result from core polarization. When removing a nucleon from ^{208}Pb, the ^{205}Tl core will adjust a bit; mainly it will change somewhat its radius. Fig.8 shows the contribution of core polarization a calculated by DDHF using a finite-range effective force [9]. In the q-region of interest, the effect of core polarization upon the Pb/Tl cross section ratio is small enough to be easily corrected for.

The experimental results for the $^{205}Tl/^{206}Pb$ cross section ratio [10] is shown in Fig.9. The signal for the 3s-shell contribution, the peak of $2fm^{-1}$, stands out clearly! The curve

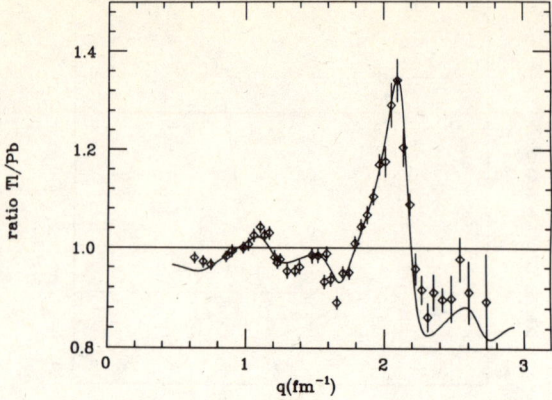

Figure 9: Experimental results for Tl/Pb cross section ratio, together with DDHF prediction with adjusted 3s occupation change.

is obtained from a DDHF calculation by assuming a change of the 3s occupation number of 0.64, and a change of 1 − 0.64 of the other shell expected to be open, the 2d shell. This calculation [9], with only one free parameter (3s occupation) explains the data extremely well, over the entire q-range.

The density difference extracted from the cross sections of ^{205}Tl and ^{206}Pb via DFT is shown in fig.10. This difference, with its 3 maxima and 2 minima, provides the "textbook example" of a 3s orbital! From figs.9 and 10 we must conclude that the IPSM is valid to a surprising degree, even in the nuclear interior.

3.4 Absolute occupation numbers

Spectroscopic factors of nuclear states and occupation numbers of shell model orbitals describe the degree to which real nuclei overlap with a simple model, the IPSM. These quantities have been measured for 3 decades in transfer reactions of type (d,p), *etc.* and contain a large fraction of our knowledge on nuclei.

Despite a longstanding effort, *absolute* spectroscopic factors and occupation numbers still are not available, for basically two reasons:

- Transfer reactions measure the strength at large radii only, where the density has fallen to $< 10^{-2}$ of its central value. From such a measurement one cannot extrapolate reliably to the integral over *all* radii needed to derive a spectroscopic factor.

- Transfer reactions measure the spectroscopic factors for excitation energies $\varepsilon^* < 5 MeV$, typically, of the residual nucleus. For such a limited range in ε^* one cannot deduce the integral over *all* ε^* needed to go from spectroscopic factors to occupation numbers.

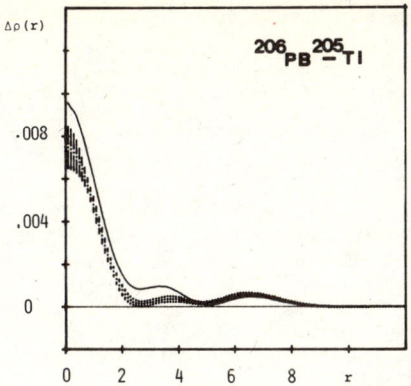

Figure 10: $Pb - Tl$ density difference, together with DDHF prediction.

For these reasons we still do not know, to which degree the shell model, with partial occupation of orbitals below and immediately above the Fermi energy, is valid.

The measurement of the Tl/Pb difference, in combination with a new sumrule analysis of (e,e'p) data, now provides a way to deduce, for the doubly magic ^{208}Pb nucleus, absolute spectroscopic factors and occupation numbers [9].

The approach [11, 12, 11] is based on a Combined Evaluation of Relative spectroscopic factors and Electron Scattering (CERES). The charge density difference between Pb and Tl is assumed to be given by

$$\Delta\rho(r) = \sum_{\alpha} \Delta n_\alpha \cdot \rho_\alpha(r) + \Delta\rho_{cp}(r) \qquad (10)$$

where n_α and ρ_α are the occupation number and single-particle radial density of state α, Δn_α the change $n^{(206)} - n^{(205)}$, and $\Delta\rho_{cp}$ is the change of the charge distribution of the core due to the polarization by the added nucleon. From electron scattering, $z = n^{(206)} - n^{(205)}$ of the 3s-shell has been extracted [10].

Introducing the spectroscopic factor for 3s protons in ^{208}Pb, $s^{(208)}$, with $n^{(208)} = \sum s^{(208)}$ one obtains

$$n^{(208)} = z / \left(\frac{\sum s^{(206)}}{\sum s^{(208)}} - \frac{\sum s^{(205)}}{\sum s^{(208)}} \right) \qquad (11)$$

In this equation, only relative spectroscopic factors occur, thus making the right hand side appropriate for an application to transfer data. The sums, however, in practice must be truncated as the analyses of transfer data are typically limited to excitation energies of $5 MeV$ in the (A-1) system. A significant fraction of the strength may lie at excitation energy $\varepsilon^* \geq 5 MeV$ due to short-range N-N correlations and coupling to giant-resonance excitations. For an approximate solution of this difficulty one assumes equal percentage depletions in neighbouring nuclei, a plausible assumption borne out by theory of short-range correlations of the Jastrow type [13, 14], and by the weak A-dependence of giant resonances. With this assumption one obtains

$$n^{(208)} = z / \left(\frac{\sum' s^{(206)}}{\sum' s^{(208)}} - \frac{\sum' s^{(205)}}{\sum' s^{(208)}} \right) \qquad (12)$$

where the primed sums are taken over a range of excitation energies of typically less than $1\hbar\omega$ (*i.e.* 8MeV for ^{208}Pb).

The above equation now on the right-hand side only contains quantities that are experimentally measurable in transfer reactions, *i.e. relative* spectroscopic strength in a *limited* energy interval. To calculate an absolute occupation number, these quantities are normalized with the z-value.

Employing the (e,e'p) data measured at NIKHEF and our Tl/Pb cross sections, we find [9] an absolute occupation number of the 3s shell of ^{208}Pb of 0.76 ± 0.07. The error bar reflects both the uncertainties of the data, and an estimate for the systematic uncertainty that enters the analysis and the corrections that have to be made.

We can compare this experimental result to results from recent theoretical calculations:

- Mahaux and collaborators [15, 16] exploit the analytical properties of the optical potential (or integral quantities thereof) to extrapolate from the positive energies accessible by proton-nucleus scattering to negative energies (bound states). From the potential they derive spectroscopic factors and occupation numbers of single-particle states. For the 3s state in ^{208}Pb they find $n_{3s} = 0.77$ and $s_{3s} = 0.66$, respectively.

- Pandharipande *et al* [14] use a variational calculation of nuclear matter, which allows to get a realistic estimate for the effect of short-range correlations. The longer-range correlations near the Fermi-edge are calculated in correlated basis perturbation theory, and finite-nucleus effects are added via an RPA estimate. The occupation number for the 3s state in ^{208}Pb is again found to be rather low, $\simeq 0.63 \pm 0.1$, where the error reflects an estimate of the model dependence.

Within the error bars, these results agree with the experimental number of 0.76 ± 0.07. Much of the remaining strength is moved to highlying orbitals by the short-range N-N interaction that leads to the short-range correlations neglected in the shell model.

In many respects the shell model appears to be working much better than one would naively expect for occupation numbers of valence shells of $\sim 75\%$. The success of the shell model can probably be assigned to the fact that, in many observables, the 'background' due to excitation of states far above the Fermi surface does not play an important role. The partial occupation of states near the Fermi surface can be accounted for by the phenomenological renormalizations inherent to many practical applications of the shell model.

3.5 Densities: Summary

From the above discussion of nuclei in the Pb-region we have learned a number of things:

- Electron scattering at large q allows us to determine *precise* densities even in the central high-density region of heavy nuclei. In addition, in combination with a sum-rule analysis, we for the first time can deduce *absolute* occupation numbers.

- Wave functions of nucleons calculated in the IPSM have a surprising degree of validity. Even in magic nuclei, however, the occupation number of these shells is much lower than conventionally assumed. About 25% of the strength is moved to highlying orbits that usually are ignored.

- Hartree-Fock calculations with phenomenological density-dependent NN-forces do quite well in reproducing the experimental densities. The remaining discrepancies — the occurrence of too much oscillatory structure in $\rho(r)$ — can primarily be assigned to the neglect of configuration mixing and occupation of the highlying orbits. It also must be expected that the mocking up of short-range correlations by an IPSM with a density-dependent force leads to an overestimate of oscillatory structure.

- Calculations in the IPSM frame work with *no* phenomenological adjustments of the effective NN force, such as RBHF, still show a serious disagreement with experiments. This disagreement partly results from the neglect of non-nucleonic degrees of freedom (*e.g.* 3-body force) in the treatment of nuclear matter, and partly from the short comings due to the neglect of short-range correlation, *i.e.* the fact that they are accounted for only via a use of an *effective* force in an IPSM frame-work.

4 Neutron form factors

The results for nuclei in the Pb-region may be classified as achievements of the past. The measurement of neutron form factors given in this section largely belongs to the category of projects of the future. It is included here as this topic — besides the obvious interest in the observable — allows to address a number of other areas of electron scattering not touched above, such as polarization, parity violation and coincidence experiments.

The knowledge of the neutron charge and magnetic form factors is important for an understanding of the internal structure of the neutron. These quantities not only are of fundamental interest, they also are needed to calculate *nuclear* form factors that result from a convolution of nuclear body and nucleon intrinsic form factors. Of particular interest is the neutron charge form factor: The positive and negative pieces of the neutron charge density cancel to produce $G_{en}(q = 0) = 0$, with the result that, at non-zero q, the charge form factor $G_{en}(q)$ is very sensitive to the neutron internal structure. Theoretical predictions for G_{en} are very different.

4.1 Deuteron charge form factor and G_{en}

Basically two approaches to determine the neutron charge form factor G_{en} have been used in the past: quasielastic and elastic electron-deuteron scattering. In the former case, G_{en} is obtained after subtracting from the inclusive, quasielastic data the (dominant) contribution from the proton, and after that removing the (dominant) magnetic contribution of the neutron; the resulting error bars are large, both due to the successive subtractions, and the uncertainties due to FSI and MEC corrections. In the latter case, G_{en} is obtained after removing the deuteron structure from the experimental charge form factor, and subtracting the proton contribution; as both operations involve uncertainties and the final quantity G_{en} is very small, again large systematic error bars are unavoidable. Up till recently, the comparatively least uncertain determination of G_{en} came from elastic e-d scattering. A careful study of systematical errors (deuteron structure, MEC, relativistic effects) shows that the resulting values of G_{en} were known only with a (systematical) error bar of nearly ± 100%. Only at very low q^2 our knowledge is better, due to the precise information on the slope of G_{en} at $q^2 = 0$ from thermal neutron-electron scattering.

Recently new data became available from an experiment [17] on elastic electron-deuteron scattering performed by the Saclay-Basel collaboration. These data were measured using a liquid deuterium target, and the 900 MeV/c spectrometer. A special effort was made to reduce the systematical errors on integrated charge, target thickness, spectrometer solid angle and detection efficiency. The overall systematical uncertainty resulting was ±2%.

In order to determine G_{en}, the experimental deuteron structure function $A(q)$ first is corrected for the contributions due to non-nucleonic degrees of freedom, and relativistic effects.

$$A(q)_{IA} = A(q)_{exp} - \Delta A_{rel} - \Delta A_{MEC} \qquad (13)$$

The latter ones seem to be under good control; the relative changes of A(q) predicted by 3 rather different approaches [18, 19, 20] agree within ±3%. The former ones are more uncertain. The $\pi\rho\gamma$-term in particular [21] is still somewhat ambiguous, and introduces a sizeable uncertainty at the highest momentum transfer.

The values of $A_{IA}(q)$ can be interpreted in impulse approximation in terms of nonrelativistic deuteron structure and nucleonic form factors.

$$A(q)_{IA} = (G_{ep}(q) + G_{en}(q))^2 \cdot \left(C_E^2(q) + C_q^2(q) \cdot \frac{8}{9}\tau^2 \right) \qquad (14)$$

Dividing out the appropriate combination of integrals over the deuteron-S- and D-state wave function, calculated with various NN potentials, yields the nucleon isoscalar charge form factor. Subtracting the (dominant) proton piece G_{ep} yields G_{en}. In Fig.11 we show the resulting values of G_{en} for one particular choice of corrections and NN potential.

The error on G_{en} is dominated by the systematical errors stemming from the theoretical ingredients necessary to go from the (large) $A(q)$ to the (small) G_{en}. A careful discussion of these uncertainties is given in ref. [17]. The largest uncertainty comes from the choice of the NN potential used to calculate the deuteron wave functions. This is true even after those potentials (Nijmegen, Argonne V14) that do not give the correct slope of G_{en} at $q^2 = 0$ are discarded. Much of the remaining systematical error comes from the $\rho\pi\gamma$-term. Overall, we estimate that the systematical error of the G_{en} extracted is of order ±30% at intermediate q^2. This is not a great accuracy, but it is a factor of 2-3 better than previous determinations, both due to more accurate data, and due to a better theoretical understanding.

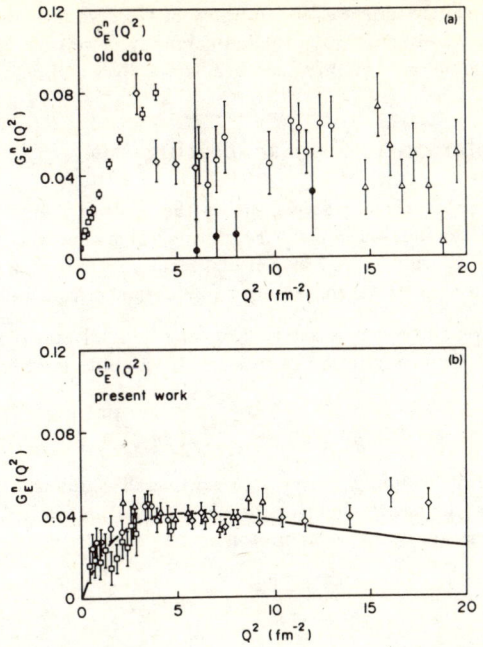

Figure 11: Neutron charge form factor G_{en} extracted from the data of Platchkov et al [17] using the Paris NN potential (bottom) and previous data (top)

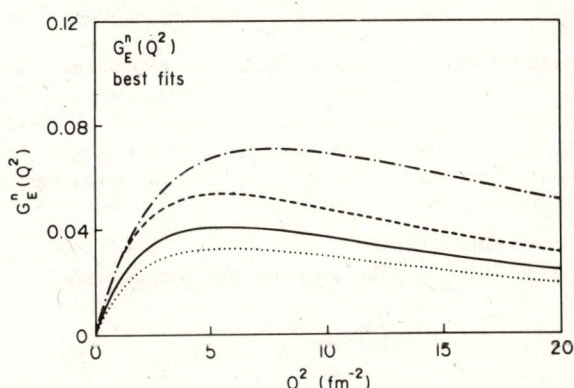

Figure 12: Two-parameter fits to data G_{en} deduced from $A(q^2)$ by unfolding the Paris (solid), RSC (dotted), Argonne V14 (dashed) or Nijmegen (dash-dotted) potentials. For clarity the corresponding sets of data points are not shown.

In order to significantly improve the accuracy of G_{en}, we need to consider new approaches that are much less sensitive to theoretical input. Two such approaches that will be practical with high-energy CW polarized beams are discussed below.

4.2 Parity violation in $\vec{e} - p$ scattering and G_{en}

A radically different approach to get a handle on the neutron charge form factor results from the study of parity violation in $\vec{e} - N$ scattering. This approach, studied in a recent publication of Donnelly, Dubach and myself [22] results in the (at first sight very odd) finding that the neutron electric form factor can be obtained from $\vec{e}$-proton scattering!

The asymmetry for single-arm scattering of longitudinally polarized electrons from an unpolarized target, resulting from the interference amplitude of one-photon and Z^o-exchange, is given by

$$A = \frac{\sigma^+ - \sigma^-}{\sigma^+ + \sigma^-} = \frac{G \cdot \kappa}{2\pi\alpha\sqrt{2}} \cdot q^2 \cdot \frac{W}{F^2} \qquad (15)$$

A depends on the Fermi coupling constant G, an overall coupling constant κ for weak neutral current effects, the parity-violating response W and the familiar parity-conserving electromagnetic form factor F^2. For the proton

$$F^2 = \frac{1}{(1+\tau)\varepsilon}(\varepsilon \cdot G_{ep}^2 + \tau G_{mp}^2) \qquad (16)$$

with

$$\tau = -q^2/4m_p^2, \ \varepsilon = (1 + 2(1+\tau)tg^2\frac{\theta}{2})^{-1} \qquad (17)$$

and

$$W = [\alpha_A \cdot \{\varepsilon \cdot G_{ep}\tilde{G}_{ep} + \tau \cdot G_{mp}\tilde{G}_{mp}\} + \alpha_V \{\sqrt{1-\varepsilon^2}\sqrt{\tau(1+\tau)}G_{mp}\tilde{G}_{Ap}\}]/\varepsilon(1+\tau) \qquad (18)$$

where the α_V, α_A are the leptonic vector and axial vector couplings, $\tilde{G}_{ep}$ and $\tilde{G}_{mp}$ are the form factors of the hadronic weak neutral (vector) current and $\tilde{G}_{Ap}$ the corresponding axial vector current. When one assumes isospin invariance, all the G-form factors occurring in eq.(18) can be written in terms of the *same* isospin-projected matrix elements of the hadronic vector current G_e^o and G_e^1, but with *different* weighting factors for G (electromagnetic) and $\tilde{G}$ (weak).

Starting from the Standard Model ($\kappa = 1, \alpha_r = -(1 - 4sin^2\theta_w), \alpha_A = -1$), and neglecting (for the sake of clarity only) the small term due to axial-vector hadronic currents, yields

$$A = 6.334 \cdot 10^{-4} \cdot \tau \cdot \frac{1}{4}\{(1 - 4sin^2\frac{\theta_W}{2}) - \frac{\varepsilon G_{ep}G_{en} + \tau \cdot G_{mp}G_{mn}}{\varepsilon G_{ep}^2 + \tau G_{mp}^2}\} \qquad (19)$$

Above equation highlights the fact that parity-violating electron scattering is mainly sensitive to the (T=1)–(T=0) term, in contrast to parity-conserving scattering that sees the (T=1) +(T=0) term. Would the Weinberg angle have been $sin^2\theta_W \equiv \frac{1}{4}$ instead of the experimental 0.230 ± 0.005, the weak interaction would be the ideal probe for neutrons, i.e. ((T=1)-(T=0))!

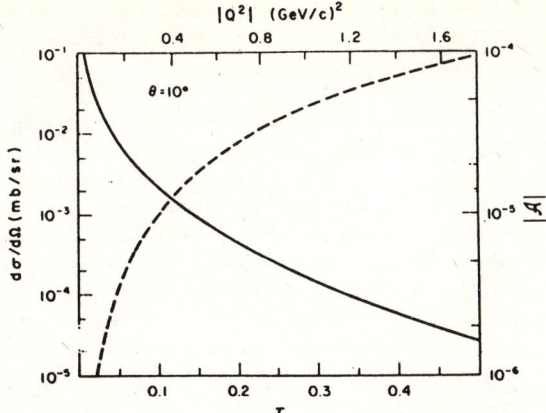

Figure 13: Cross section and asymmetry [22] for parity-violating $p(\vec{e}, e)$

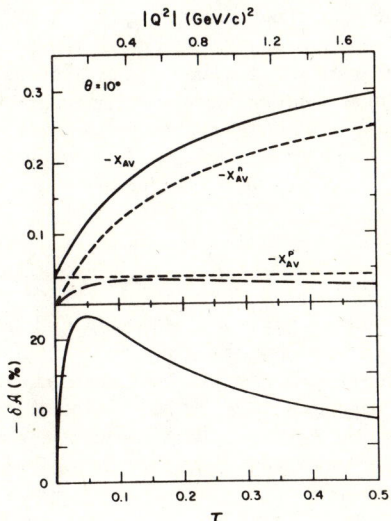

Figure 14: Change of A obtained when setting $G_{en} \equiv 0$

In Fig.13 we show the calculated cross section and asymmetry for a particular kinematical setting ($\theta_e = 10°$) and a standard neutron form factor parametrization [23]. Fig.14 indicates the relative change of A upon setting G_{en} to zero. Equation (18) shows that the asymmetry for a proton target depends on all 4 nucleon form factors. Information on any given one can only be extracted if the others are sufficiently well known. A study has been performed [22] for two typical momentum transfers, $q^2 = 0.4 GeV^2/c^2$ which corresponds to the "maximum" of G_{en}, and $q^2 = 1$ which corresponds to the region of fall-off of G_{en}. At these momentum transfers the kinematical conditions were optimized for a determination of G_{en}, and the error bars on the other form factors were obtained from an analysis of the world data. The conclusions show that at $q^2 = 0.4$ the nucleon form factors are well enough known to determine G_{en} on the 15%-level, while at $q^2 = 1$ todays uncertainty on G_{mn} limits the accuracy to $\pm$ 30%. This latter error bar can be reduced, however, by the type of experiments that are presently under way.

It is a major question whether such parity-violating experiments are practical. For realistic experimental conditions ($P_e = 0.5, 100\mu A, 50 cm LH_2$) and a dedicated detector with azimuthal symmetry ($\Omega = 20 msr$) it would take 350h of beam time for a 10% measurement of G_{en} at $q^2 = 0.4$. For $q^2 = 1$, a number of trade-offs can be made. In about 500h of beamtime a useful result could be obtained.

One caveat should be added. The above analysis [22] is based on the assumption of isospin conservation in the n,p-system. Introducing, e.g., a large component of strange quarks in the nucleon wave function would complicate things. At present, there is no reason to believe in the existence of such a large strange component, although it has been evoked in some attempts to account for the data on the nucleon spin structure functions from deep inelastic scattering [24, 25]. Ultimately, the parity experiment discussed above therefore may be performed not to measure G_{en}, but to determine the $s\bar{s}$-component, particularly if, as discussed below, G_{en} can be accurately determined from $\vec{d}(\vec{e}, e'n)$.

4.3 $\vec{d}(\vec{e}, e'n)$ and G_{en}

The large uncertainties in determinations of G_{en} basically result from two subtractions. In processes not identifying e-n scattering, the dominant contribution comes from e-p scattering, and must be removed. In processes not identifying charge scattering, the dominant contribution comes from magnetic scattering (G_{mn}) and must be removed. To decrease the resulting uncertainties, the electron-neutron scattering must be identified, and the charge form factor should be measured without the need for Rosenbluth separations that produce large error bars for the small (i.e. charge) form factor.

When employing the d(e,e'n) reaction, e-n scattering is identified via the recoiling neutron in the direction of $\vec{q}$; the charge form factor contribution can be isolated if polarization observables are measured.

This type of experiment becomes clear if one discusses the case of scattering of a longitudinally polarized $\vec{e}$ from a free, polarized nucleon [26]. For a nucleon with spin oriented in the scattering plane perpendicular to $\vec{q}$, the cross section is given by

$$\sigma = \sigma_o(1 + P_e \cdot P_t \cdot A) \qquad (20)$$

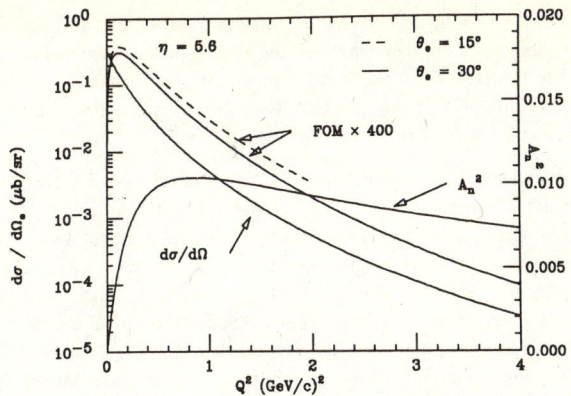

Figure 15: Cross section, asymmetry and figure of merit for $\vec{d}(\vec{e},e'n)$ [28]

where σ_o is the cross section for unpolarized electron and nucleon, P_e and P_t are the polarization of electron and target, and A is the asymmetry

$$A = -2\sqrt{\tau(1+\tau)}G_m G_e tg^2\theta/2 / (G_e^2 + \tau G_m^2(1 + 2(1+\tau)tg^2\theta/2)) \quad (21)$$

with

$$A = \frac{\sigma^+ - \sigma^-}{\sigma^+ + \sigma^-} \quad (22)$$

where the ± refers to the helicity of the incoming electron. Equation (21) demonstrates that the asymmetry depends on the product $G_e \cdot G_m$. Knowing G_m allows to extract G_e without the usual Rosenbluth separation that is inefficient for extracting a very small longitudinal contribution in the presence of a large transverse one.

For the neutron charge form factor, one has to deal with the added complication of a composite target. Polarized deuterium provides the best polarized neutron target. The added complication of the reaction mechanism for quasielastic scattering can be dealt with quantitatively for this 2N-system. Arenhövel has studied the $\vec{d}(\vec{e},en)$ reaction in detail [27] and finds that both the neutron-proton final state interaction, and the poorly known features of the initial state deuteron wave function, do not introduce significant uncertainties. The effects of meson exchange currents also turn out [27] to be manageably small. The $\vec{d}(\vec{e},eN)$ reaction has another outstanding advantage: a check of both experimental procedure and reaction mechanism is possible by doing the corresponding $\vec{d}(\vec{e},e'p)$ reaction for a determination of the (at the q^2 of interest) known proton electric form factor.

Such a $\vec{d}(\vec{e},e'n)$ experiment will be performed by a University of Virginia/Basel collaboration at CEBAF. The asymmetries expected for reasonable assumptions on $G_{en}(q^2)$ are quite large, of order 0.1 (see fig.15). Even for non-maximal beam- and target-polarizations (both will be of order 0.5) such asymmetries are easily measurable insofar as systematical errors go. Only the electron spin has to be flipped (at the $\vec{e}$-source) to measure A. A flip of the target spin (effected by changing the RF-frequency, but not the holding magnetic field) only provides an additional check.

In such a $\vec{d}(\vec{e},e'n)$ experiment one pays for the cleanliness of interpretation with an increased complexity of the apparatus. While longitudinally polarized electrons appear to be a standard item at the new generation of electron accelerators, and while the neutron detection does not impose very stringent conditions (i.e. energy resolution of ~ 50MeV only, no need to know the detector efficiency) the polarized target will require a major effort. The present generation of ND_3-targets accepts only ~ 1 nA of beam current, a value that one can hope to push to 10nA by a brute-force approach of better cooling. Within reasonable counting times, one can hope to reach momentum transfers of 1.5 GeV^2/c^2.

References

[1] J. Heisenberg, R. Hofstadter, J.S. McCarthy, I. Sick, B.C. Clark, R. Herman, and D.G. Ravenhall. *Phys.Rev.Lett.*, 23:1402, 1969.

[2] B. Frois, J.B. Bellicard, J.M. Cavedon, M. Huet, P. Leconte, P. Ludeau, A. Nakada, Phan Xuan Hô, and I. Sick. *Phys. Rev. Lett.*, 38:152, 1977.

[3] I. Sick. Model-independent nuclear charge densities from elastic electron scattering. *Nucl.Phys.*, A218:509, 1974.

[4] B. Dreher, J. Friedrich, K. Merle, H. Rothhaas, and G. Luehrs. *Nucl.Phys.*, A235:219, 1974.

[5] I. Sick. *Lecture Notes in Physics*, 236:137, 1985.

[6] D. Gogny. *in "Nuclear Self-Consistent Fields",*, ed.G. Ripka and M. Porneuf (North-Holland, Amsterdam), 1975.

[7] J.W. Negele. *Phys. Rev. C*, 1:1260, 1970.

[8] K.T.R. Davies, R.J. McCarthy, and P.U. Sauer. *Phys.Rev.*, C6:1461, 1972.

[9] I. Sick and P. de Witt. *Comm.Nucl.Part.Phys.*, to be publ., 1991.

[10] J.M. Cavedon, B. Frois, D. Goutte, M. Huet, Ph. Leconte, C.N. Papanicolas, X.-H. Phan, S.K. Platchkov, S. Williamson, W. Boeglin, and I. Sick. *Phys.Rev.Lett.*, 49:978, 1982.

[11] H.Clement, P. Grabmayr, H. Roehm, and G.J. Wagner. *Phys.Lett.B*, 183:127, 1987.

[12] G.J. Wagner. *Progr.Part.Nucl.Phys.*, 24, 1990.

[13] M.C. Birse and C.F. Clement. *Nucl.Phys.*, A351:112, 1981.

[14] V.R. Pandharipande, C.N. Papanicolas, and J. Wambach. *Phys. Rev. Lett.*, 53:1133, 1984.

[15] C.Mahaux and R. Sartor. *Nucl.Phys.*, A481:381, 1988.

[16] C.Mahaux and R.Sartor. This volume.

[17] S. Platchkov, A. Amroun, S. Auffret, J.M. Cavedon, P. Dreux, J. Duclos, B. Frois, D. Goutte, H. Hachemi, J. Martino, X.H. Phan, and I. Sick. *Nucl.Phys.*, A510:740, 1990.

[18] R.G.Arnold, C.E. Carlson, and F.Gross. *Phys. Rev. C*, 21:1426, 1980.

[19] M.J. Zuilhof and J.A. Tjon. *Phys. Rev. C*, 24:736, 1981.

[20] I. Grach and L. Kondratyuk. *Sov.J.Nucl.Phys.*, 39:198, 1984.

[21] P. Sarriguren, J. Martorell, and D.W.L. Sprung. *McMaster preprint, to be publ.*

[22] T.W. Donnelly, J. Dubach, and I. Sick. *Phys. Rev.C*, 37:2320, 1988.

[23] S. Galster, H. Klein, J. Moritz, K.H. Schmidt, D. Wegener, and J. Bleckwenn. *Nucl. Phys.*, B32:221, 1971.

[24] D.H. Beck. *Phys.Rev.D*, 39:3248, 1989.

[25] D.B. Kaplan and A. Manohar. *Nucl.Phys. B*, 310:527, 1988.

[26] R.G. Arnold, C.E. Carlson, and F. Gross. *Phys. Rev. C*, 23:363, 1981.

[27] H. Arenhoevel. *Prog. Th.Phys.Suppl.*, 91:1, 1987.

[28] D.Day, M. Farkondeh, K. Giovanetti, J. Lichtenstadt, R. Lindgreen, J.S. McCarthy, R.Minehart, B. Norum, D. Pocencic, R. Sealock, J.Jourdan, G. Masson, and I. Sick. *CEBAF proposal*, 1989.

The Role of Coloured Quarks & Gluons in Hadrons & Nuclei

F.E. Close

Rutherford Appleton Laboratory
Chilton, Didcot, Oxon, OX11 OQX, England

Abstract : These lectures describe the role of gluons and colour in hadrons and nuclei at low energies. They examine the particle-nuclear interface including a general introduction to the ideas and applications of colour, the Pauli principle and spin-flavour correlations. They show how the magnetic moments of hadrons relate to the underlying colour degree of freedom and to the polarised deep inelastic lepton scattering.

1 Nucleons and nuclei

Everyone knows that quarks and QCD underwrite nuclear physics at a fundamental level but noone would seriously undertake a programme to "derive" nuclear structure from QCD. Quarks are confined in nucleons and as a first step we need to understand how the proton and neutron are built up and that is a yet unsolved problem.

At a qualitative level one can draw some analogies between QED (electrical charges, atoms and molecules) on the one hand and, on the other, QCD (colour, hadrons and nuclei).

The similarity is such that one could rewrite Bjorken and Drell's QED text by inserting a traceless 3x3 matrix (λ of SU(3)) at the fermion gauge boson vertices and let QED become QCD (with α replaced by $\alpha_s \simeq 1/10$). However, the gluons themselves have colour and so mutually interact via the colour forces (contrast the photon of QED which transmits but does not directly "feel" the electromagnetic force). These new intergluon interactions give rise to vertices involving three of four gluons at a point, and so a text on QCD requires more than just a colouring of Bjorken and Drell.

QED	QCD
electric charge	3 colours
attraction of opposites	attraction of unlike colours
electrically zero atoms	colourless hadrons
radiation photon	radiation gluons
magnetic effects	chromomagnetic effects
hyperfine splitting $^3S_1 - {}^1S_0$	colour hyperfine $m_\rho - m_\pi, m_\Delta - m_N$
Fermi-Breit in hydrogen	Fermi-Breit splittings in hadron spectroscopy

	Carry the charge	$e^- Z^+$	Quarks
Feel the force		$Na^+ Cl^-$	Gluons
	Contain the charge	Atoms Molecules	Hadrons Nuclei
Do not feel the force		Neutrinos	Leptons (ν, e)
Do not contain the charge		Photon	

Now let's make a matrix to summarize how systems variously react to the forces.

Notice that the gluon and photon are in different slots. This small difference gives rise to the different long-range phenomena in QCD compared to QED (e.g. confinement versus ionization).

Within atoms and hadrons one finds analogues. The Coulomb potential of hydrogen has an analogue in quark systems: as the quarks' relative separations $r \to 0$, $V(r) \sim 1/r$, but at large r, $V(r) \sim r$, presumably due to the detailed self-interactions among the gluons that are transmitting the force. In the ground state of hydrogen the magnetic interaction ("one photon exchange") splits the $^3S_1$ and $^1S_0$ levels. In the ground-state quark conglomerates, the high-J conbinations have enhanced masses relative to their low-J counterparts due to "one gluon exchange"; thus the $\Delta(1230)$ resonance, with $J = 3/2$, has greater mass than the $J = 1/2$ nucleon.

Now move up a layer in complexity to the world of molecules (QED) and nuclei (QCD).

At the risk of being accused of oversimplification by the atomic experts, I will divide the interatomic forces into three broad classes, then make analogy in the QCD world with interhadronic forces at the level of the quarks and gluons. If this was the whole story, then nuclear physics from QCD would be a rerun of molecules from QED. However, confinement of colour breaks the simple analogy. The quark exchange at large distances ($\geq 1 fm$) is contained within the confined packages, dominantly pions.

	Covalent	van der Waals	Ionic
Atoms Molecules	e^- exchange	"two photon"	Na^+Cl^-
Hadrons Nuclei	quark exchange	"two gluon"	no analogue if colour confined in neutral clusters

The confinement of gluons in glueballs also breaks the analogy with van der Waals' forces. The hope that QCD would predict observable colour van der Waals' forces in nuclei is most probably flawed, as the gluons will be confined within colourless glueballs. Computer simulations of QCD suggest that the lightest glueballs have masses in excess of 1 GeV and so transmit forces over much less than a nucleon radius. Thus their presence is hidden in nuclear physics.

It is an open question whether analogues of ionic forces occur in dense of hot nuclear systems; whether multiquark clusters occur within nuclei; whether quark-gluon plasma may form in hot- dense systems.

If colour attractions among quarks are the source of internucleon forces, then there could exist analogous clusters of mesons - "meson molecules". The instability of most mesons prevents formation of these systems, but π, K, η are stable on the time scales of the strong interactions and may have the chance to bind. Indeed Weinstein and Isgur find that such attractions occur in S- wave. The $\pi\pi$ system has a strong enhancement above $2m_\pi$ which may be manifested in the $\psi \to \omega\pi\pi$ dipion spectrum. The KK system binds forming nearly degenerate $I = 0,1$ systems 10 MeV below $2m_K$. The S^* (975 MeV) and δ (980 MeV), scalar "mesons", thus appear to be meson molecules; meson analogues of the $I = 0$ deuteron (whose $I = 1$ partner is above $2m_N$).

The colour attractions among quarks and gluons lead to the prediction of glueballs and hybrid hadrons - the latter where gluons play a dynamical role, attracted to quarks to form hybrid mesons and baryons.

The problem in predicting the masses of these states is that we have to simulate the effects of confinement. Perhaps the simplest way of doing this is to suppose that the constituent quarks or gluons are free until they hit an infinitely high wall. This is the essence of cavity or bag models. Confine a massless $J = 1/2$ quark in a radius, R, and it gains an energy that scales as $1/R$. This energy becomes of the order of 350 MeV if R is of the order of the proton radius, hence the proton mass may be modelled. For gluons, one solves the eigenvalue equations for $J = 1$ rather than $J = 1/2$ confined fields. There are electric or magnetic modes (actually TE and TM in the language of classical electrodynamics) with different eigenvalues. If R is the same as for quark systems, the typical confined-mass-scale is some 500 MeV per

TE mode and 750 MeV per TM mode. Thus follows the prediction that the lightest systems consisting of at least two confined gluons weigh in at about (1 GeV) and that the lightest hybrid baryons weigh in at about (1.5 GeV). A problem is that as soon as the hyperfine shifts in energy are taken into account (this involves one first calculating the propagators of confined quarks and gluons), the lowest spin-J systems are pulled down significantly in mass. The lightest hybrid baryon might thus appear to have a mass near that of the proton which suggests either a profound rethink of baryon spectroscopy or that we have unearthed a naievity.

I suspect it is the latter. No one yet has convincingly set up a study of loop effects with renormalization within a cavity. These loop diagrams enter at the same order in perturbation theory to which the hyperfine shifts have been calculated and may alter the naive "effective" energies per confined gluon. In the case of quarks, their effects were subsumed in the MIT bag by an input mass parameter for the quark; this mass fitted to the overall mass scale of the spectroscopy. In the gluonic sector we have no mass scale to set the scale, and until we make sense of the (infinite!) self-energy diagrams, we cannot predict the absolute scale. So the mass separations among the various states may be reliable, but the absolute mass scale is beyond present analysis. To predict the masses of glueballs and hybrid hadrons, we have to resort to computer simulations - lattice QCD. This has proved to be a harder task than was originally thought.

The eventual discovery of the gluonic spectroscopy may give important insights into the nature of confinement of gluons. If lattice calculations, including quarks *and* gluons (to date, people work in the "quenched" approximation, which roughly translated means "ignore the quarks") merely print out masses of states that correspond to the particle data tables, we will confirm QCD but may still require much study to elucidate the analytic dynamics of confinement. The main outcome of such a success may be the advances that will have come in the art of computation and design of machines. Thus the significant questions posed by hadron physics are having a spinoff in the intellectual stimulation they provide to computational science and, in turn, the subsequent ability to encode problems in field theory, condensed matter, and other areas of science. I am reminded of the title of Tony Hey's talk at a recent meeting of the British Association for the Advancement of Science, and it provides an apt one-line summary of the multi-disciplinary efforts flowing from computation at the nuclear-particle interface. It was " Quarks, Supercomputers, and Oil Prospecting".

2 Colour, the Pauli principle and spin-flavour correlations

2.1 Colour

If quarks possess a property called colour, any quark being able to carry any one of three colours (say red, yellow, blue), then the Ω^- (and any baryon) can be built from distinguishable quarks:

$$\Omega^-(S_R^\uparrow S_Y^\uparrow S_B^\uparrow).$$

If quarks carry colour but leptons do not, then it is natural to speculate that colour may be the property that is the source of the strong interquark forces - absent for leptons.

Electric charges obey the rule "like repel, unlike attract" and cluster to net uncharged systems. Colours obey a similar rule: "like colours repel, unlike (can) attract". If the three colours form the basis of an SU(3) group, then they cluster to form "white" systems - viz. the singlets of SU(3). Given a random soup of coloured quarks, the attractions gather them into white clusters, at which point the colour forces are saturated. The residual forces among these clusters are the nuclear forces whose origin will be mentioned later.

If quark (Q) and antiquark $(\bar{Q})$ are the $\underline{3}$ and $\underline{\bar{3}}$ of colour SU(3), then combining up to three together gives SU(3) multiplets of dimensions as follows (see Ref. 3):

$$QQ = \underline{3} \, x \, \underline{3} = \underline{6} + \underline{\bar{3}}$$
$$Q\bar{Q} = \underline{3} \, x \, \underline{\bar{3}} = \underline{8} + \underline{1}$$

The $Q\bar{Q}$ contains a singlet - the physical mesons

$$QQ\bar{Q} = \underline{15} + \underline{6} + \underline{3} + \underline{3}$$
$$QQQ = \underline{10} + \underline{8} + \underline{8} + \underline{1}.$$

Note the singlet in QQQ - the physical baryons.

For clusters of three or less, only $Q\bar{Q}$ and QQQ contain colour singlets and, moreover, these are the only states realized physically. Thus are we led to hypothesize that only colour singlets can exist free in the laboratory; in particular, the quarks will not exist as free particles.

2.2 Symmetries and correlations in baryons

To have three quarks in colour singlet:

$$1 \equiv \frac{1}{\sqrt{6}}[(RB - BR)Y + (YR - RY)B + (BY - YB)R] \tag{2.1}$$

any pair is in the $\underline{\bar{3}}$ and is antisymmetric. Note that $\underline{3} \, x \, \underline{3} = \underline{6} + \underline{\bar{3}}$. These are explicitly

$\underline{\bar{3}}_{anti}$	$\underline{6}_{sym}$
$RB - BR$	$RB + BR$
$RY - YR$	$RY + YR$
$BY - YB$	$BY + YB$

$$RR$$
$$BB$$
$$YY \qquad (2.2)$$

Note well: <u>Any Pair is Colour Antisymmetric</u>

The Pauli principle requires total antisymmetry and therefore any pair must be: <u>Symmetric in all else</u>

("else" means "apart from colour").

This is an important difference from nuclear clusters where the nucleons have no colour (hence are trivially <u>symmetric</u> in colour!). Hence for nucleons Pauli says

$$\underline{\text{Nucleons are Antisymmetric in Pairs}} \qquad (2.3)$$

and for quarks

$$\underline{\text{Quarks are Symmetric in Pairs}} \qquad (2.4)$$

If we forget about colour (colour has taken care of the antisymmetry and won't affect us again), then (i) Two quarks can couple their spins as follows

$$\left\{ \begin{array}{ll} S = 1: & \text{symmetric} \\ S = 0: & \text{antisymmetric} \end{array} \right\} \qquad (2.5)$$

(ii) Two u, d quarks similarly form isospin states

$$\left\{ \begin{array}{ll} I = 1: & \text{symmetric} \\ I = 0: & \text{antisymmetric} \end{array} \right\} \qquad (2.6)$$

(iii) In the ground state $L = 0$ for all quarks; hence the orbital state is trivially symmetric. Thus for pairs in $L = 0$, we have

$$\left\{ \begin{array}{ll} S = 1 \text{ and } I = 1 & \text{correlate} \\ S = 0 \text{ and } I = 0 & \text{correlate} \end{array} \right\} \qquad (2.7)$$

Thus the Σ^0 and Λ^0 which are distinguished by their u, d being $I = 1$ or 0 respectively also have the u, d pair in spin $= 1$ or 0 respectively:

$$\left\{ \begin{array}{ll} \Sigma^0[S(u,d)_{I=1}] & \leftrightarrow \quad S(u,d)_{S=1} \\ \Lambda^0[S(u,d)_{I=0}] & \leftrightarrow \quad S(u,d)_{S=0} \end{array} \right\} \qquad (2.8)$$

Thus, the spin of the Λ^0 is carried entirely by the strange quark.

This is the source of the $\Sigma - \Lambda$ mass difference. The $\vec{S}.\vec{S}$ interaction acts between all possible pairs; thus

$$\Sigma^0[S(u,d)_1] : \quad \langle \vec{S}.\vec{S}\rangle_1 + \langle \vec{S}.\vec{S}\rangle_{s,1}, \tag{2.9}$$

$$\Lambda^0[S(u,d)_0] : \quad \langle \vec{S}.\vec{S}\rangle_0 \tag{2.10}$$

(note $\langle \vec{S}.\vec{S}\rangle$ between a spinless diquark and anything vanishes; hence the absence of $\langle S.S\rangle_{s,0}$).

Now

$$\langle \vec{S}.\vec{S}\rangle_0 = -3\langle \vec{S}.\vec{S}\rangle_1, \tag{2.11}$$

(see p. 91 of Ref. 3). Further, if $m_s = m_{u,d}$, the Σ and Λ become mass degenerate, and so in this limit

$$\langle \vec{S}.\vec{S}\rangle_{s,1} = -4\langle \vec{S}.\vec{S}\rangle_1. \tag{2.12}$$

For unequal masses of u and s, the magnetic interaction scales as the inverse mass. Hence finally

$$\Sigma^0 \sim \langle \vec{S}.\vec{S}\rangle_1 \;\left\{1 - 4\frac{m_u}{m_s}\right\} \tag{2.13}$$

$$\Lambda^0 \sim \langle \vec{S}.\vec{S}\rangle_1 \;\{-3 \quad\}. \tag{2.14}$$

Then with $m_s > m_u$, we find $m_\Sigma > m_\Lambda$ as observed. Increasing m_s/m_u enhances the effect (e.g., for the charmed analogues $\Sigma_c[(u,d)c]$ and $\Lambda_c[(u,d)c]$ the splitting will be larger - again observed).

2.3 Colour, the Pauli principle and magnetic moments

The electrical charge of a baryon is the sum of its constituent quark charges. The magnetic moment is an intimate probe of the correlations between the charges and spins of the constituents. Being wise, today we can say that the neutron magnetic moment was the first clue that the nucleons are not elementary particles. Conversely the facts that quarks appear to have $g \simeq 2$ suggests that they *are* elementary (or that new dynamics is at work if composite).

A very beautiful demonstration of symmetry at work is the magnetic moment of two similar sets of systems of three, viz.

$$\left\{\begin{array}{cc} N; & P \\ ddu; & uud \end{array}\right\} \; \mu_p/\mu_N = -3/2$$

and the nuclei

$$\left\{\begin{array}{cc} H^3; & He^3 \\ NNP; & PPN \end{array}\right\} \; \mu_{He}/\mu_H = -2/3$$

The Pauli principle for nucleons requires He^4 to have *no* magnetic moment:

$$\mu[He^4; P^\uparrow P^\downarrow N^\uparrow N^\downarrow] = 0.$$

Then

$$He^3 \equiv He^4 - N$$
$$H^3 \equiv He^4 - P$$

and so

$$\frac{\mu_{He^3}}{\mu_{H^3}} = \frac{\mu_N}{\mu_p}$$

To get at this result in a way that will bring best comparison with the nucleon three-quark example, let's study the He^3 directly.

$He^3 = ppn$: pp are flavour symmetric; hence, spin antisymmetric; i.e., $S = 0$.

Thus

$$[He^3]^\uparrow \equiv (pp)_0 n^\uparrow \tag{2.15}$$

and so the pp do not contribute to its magnetic moment. The magnetic moment (up to mass scale factors) is

$$\mu_{He^3} = 0 + \mu_n. \tag{2.16}$$

Similarly,

$$\mu_{H^3} = 0 + \mu_p. \tag{2.17}$$

Now let's study the nucleons in an analogous manner.

The proton contains u, u flavour symmetric and *colour antisymmetric*; thus the spin of the "like" pair is symmetric ($S = 1$) in contrast to the nuclear example where this pair had $S = 0$. Thus coupling spin 1 and spin 1/2 together, the Clebsches yield

$$p^\uparrow = \frac{1}{\sqrt{3}}(u,u)_0 d^\uparrow + \frac{2}{\sqrt{3}}(u,u)_1 d^\downarrow \tag{2.18}$$

(contrast Eq. (15)), and (up to mass factors)

$$\mu_p = \frac{1}{3}(0 + d) + \frac{2}{3}(2u - d). \tag{2.19}$$

Suppose that $\mu_{u,d} \propto e_{u,d}$, then

$$\mu_u = -2\mu_d \tag{2.20}$$

so

$$\frac{\mu_p}{\mu_N} = \frac{4u - d}{4d - u} = -\frac{3}{2} \tag{2.21}$$

(the neutron follows from proton by replacing $u \leftrightarrow d$).

I cannot overstress the crucial, hidden role that colour played here in getting the flavour-spin correlation right.

3 Scale invariance

The place where scale invariance was first identified was in $ep \to eX$ (X meaning "anything" i.e. the proton fragments are not observed). It is more straightforward to understand the physical significance by studying e^+e^- annihilation.

In $e^+e^- \to \mu^+\mu^-$ the cross section is

$$\sigma(e^+e^- \to \mu^+\mu^-) = \frac{4\pi\alpha^2}{3Q^2}$$

where Q^2 is the invariant mass squared of the virtual γ in $e^+e^- \to \gamma \to \mu^=\mu^-$ or $Q^2 = 4E_{c.m.}^2$ where $E_{c.m.}$ is the energy of $e^-(e^+)$ in the c.m. frame. For $E >> m_e$ this is the only energy, momentum (or length) scale in the physics (the e and μ being "pointlike" with no discernible structure). So the dimensionless quantity $Q^2\sigma(e^+e^- \to \mu^+\mu^-)$ is *invariant* under change of energy scale: "scale-invariant".

This scale invariance is intimately related to the pointlike nature of the physics. For example, a proton has an intrinsic size and so

$$Q^2\sigma(e^+e^- \to p\bar{p}) = \frac{4\pi\alpha^2}{3}|F(Q^2)|^2$$

where $F(Q^2)$ is the elastic form factor (actually there are two - the electric and magnetic - and F^2 is a combination of these but this need not concern us in this discussion, see ref [3]). $F(Q^2)$ introduces an explicit energy scale dependence; $F(Q^2)$ tends to zero as $Q^2 \to \infty$. In effect $F(Q^2)$ may be considered to be $F(Q^2R^2)$ where R is same length scale associated with the hadron's size. When $Q^2 > R^{-2}$ the hadron's internal structure is probed and the hadron is unlikely to stay whole; many pions are produced viz $e^+e^- \to p\pi\pi\pi\bar{p}\pi\pi\ldots$ rather than the $p\bar{p}$ exclusive channel. So: essential length scale implies that there is a breaking of scale invariance.

However, if you measure the total cross section summed over all final state hadrons, $\sigma(e^+e^- \to$ hadrons$)$, you find scale invariance again:

$$\frac{\sigma(e^+e^- \to hadrons)}{\sigma e^+e^- \to \mu^+\mu^-)} \simeq constant$$

The measure of the hadron size has been lost because the essential dynamics is again pointlike. The e^+e^- produce a point quark-antiquark $(q\bar{q})$ pair; subsequent polarisation of the vacuum generates a shower of hadrons but these can go off in any direction (contrast the exclusive $p\bar{p}$ production where three quarks have to move off in the same direction in order to form the proton and similarly three antiquarks forming the antiproton. The more momentum pumped in, the less chance all three manage "to find" one another, hence the exclusive cross section falls).

The hadron production total cross section is summed over all possible quark flavours (*and* colours) giving

$$\frac{\sigma(e^+e^- \to hadrons)}{\sigma e^+e^- \to \mu^+\mu^-)} \equiv \frac{\sum_i \sigma(e^+e^- \to q_i\bar{q}_i)}{\sigma(e^+e^- \to \mu^+\mu^-)} = 3\sum_i e_i^2$$

Timelike	Spacelike	Scale Invariance?
$e^+e^- \to \mu^+\mu^-$	$e\mu \to e\mu$	Yes
$e^+e^- \to p\bar{p}$	$ep \to ep$	No $F(Q^2)$ suppression
$e^+e^- \to$ hadrons	$ep \to e +$ hadrons	No
$e^+e^- \to q\bar{q}$	$eq \to eq$	Yes

Timelike	Spaceline
$\sigma(e^+e^- \to X) = \Sigma e_i^2 \sigma(e^+e^- \to q\bar{q})$	$\Sigma(ep \to eX) = \sum_i e_i^2 f_i(x) \Sigma(eq(x) \to eq)$

where e_i is the charge of the i-th flavour.

This is confirmed below charm production threshold where the sum over u, d, s gives 2; then you cross charm threshold and the explicit presence of a new mass scale causes scale violation (resonances and thresholds) until a new scale invariance obtains.

For electron scattering we have a parallel development. The comparison is as follows. In the final entry we have shown the elementary subprocess. A direct analogy with $e^+e^- \to$ hadrons would be $e + all \to e + all$; as we need a target proton to begin with the scaling behaviour is slightly more complicated. The elementary quark within the proton carries some fraction x of the proton's momentum with probability $f(x)$. So heuristically the comparisons here are Σ is $Q^4 \frac{d\sigma}{dQ^2}$ because we are measuring a differential cross section for the electron to emerge into a given solid angle, or equivalently with a given energy and momentum transfer (ref. 3 shows this is actually a double differential cross section but we are glossing over this detail in this heuristic description). The energy (ν) and four momentum squared Q^2 are convenient to choose as the two degrees of freedom, or even better Q^2 and $x \equiv Q^2/2M\nu$ (M being the target mass). Note that x is dimensionless whereas Q^2 carries dimensions. So the scale invariance in inelastic electron scattering is essentially that $\Sigma(ep \to eX)/\Sigma(eq \to eq)$ is dependent upon x but not upon Q^2.

The ratio of Σ is known as a "structure function". In fact there are really two of these $F_1(x, Q^2)$ and $F_2(x, Q^2)$ which account for the independent degrees of freedom from transverse and longitudinally polarised virtual photon exchange. The scale invariance phenomenon is that $F_{1,2}$ are functions only of x and not of Q^2.

The model underwriting this is the quark-parton-model (QPM) based on the idea that the proton consists of pointlike quarks. In QCD the quarks have a "structure" due to their radiating gluons and $q\bar{q}$ pairs. This evolving structure (more g and $q\bar{q}$ structure being resolved with increasing Q^2) leads to a logarithmic dependence on Q^2 of the $F_{1,2}(x, Q^2)$.

In most of what follows I shall ignore the Q^2 dependences from QCD. The details of how the structure functions relate to the quark probabilities is standard and can be found in ref. 3.

4 The proton's spin: a quark model perspective

Inelastic lepton scattering from nucleons at high momentum transfer measures the number densities of charged constituents, $q(x), \bar{q}(x)$, as a function of the Bjorken variable x (essentially the ratio of the constituent and target longitudinal momenta in an infinite momentum frame). There is a weak dependence of these distributions on the momentum transfer, Q^2, but I shall suppress this in much of what follows.

If the beam and target are polarized, one can extract the helicity-dependent distributions for quarks or antiquarks polarized parallel $(q^\uparrow(x))$ or antiparallel $(q^\downarrow(x))$ to the target polarization. I shall define $\Delta q(x) \equiv q^\uparrow(x) - q^\downarrow(x); q(x) \equiv q^\uparrow(x) + q^\downarrow(x)$, and similarly for antiquarks, $\bar{q}$.

Data are presented in two ways [1-3]. One is in terms of the polarization asymmetry

$$A(x) = \sum_i e_i^2(\Delta q_i(x) + \Delta \bar{q}_i(x)) / \sum_i e_i^2(x) + \bar{q}_i(x)), \qquad (4.1)$$

(note that $-1 \leq A \leq +1$). The other involves the polarized structure function

$$g_1(x) = \frac{1}{2} \sum_i e_i^2(\Delta q_i(x) + \Delta \bar{q}_i(x)), \qquad (4.2)$$

thus

$$g_1(x) \equiv A(x) F_1(x). \qquad (4.3)$$

In advance of the data, the expectations were that

(i) At $x \geq 0.2$ where valence quarks dominate, $A(x)$ should be large and positive [3,4]. This follows from intuition developed for constituent valence quarks in baryon spectroscopy where the Pauli principle requires $\Delta u > 0, \Delta d < 0$. As the charge-squared weighting of Δu is four times that of Δd in protons, so $A^p(x > 0.2) > 0$, Data confirm this brilliantly. For a neutron target, it is Δd that is weighted 4:1 relative to Δu, hence these tend to cancel and one predicts [4] a small (zero?) asymmetry on the neutron.

(ii) Form $g_1(x)$, which directly shows the charge weighted helicity-dependent distributions and integrate over all x [5,6]. If it were not for the charge weightings, this would measure the net $\Delta q + \Delta \bar{q} (\Delta q \equiv \int_0^1 dx \Delta q(x) =$ net quark polarization). Explicitly, in the quark parton model

$$I^p \equiv \int dx g_1^p(x) = \frac{1}{2} \left\{ \frac{3}{9} \Delta u + \frac{1}{9}(\Delta u + \Delta d + \Delta s) \right\} + (\Delta q_i \leftrightarrow \Delta \bar{q}_i). \qquad (4.4)$$

The surprise [2] is that $I^p(EMC) \simeq 0.12$ and is almost saturated by [7] $\Delta u (\simeq 0.75)$ leaving

$$\sum_i (\Delta q + \Delta \bar{q})_i \simeq 0, \tag{4.5}$$

hence, the much-advertised claim that maybe "none of the proton's spin polarization is carried by quarks". This is a misinterpretation of Eq. (37). The valence quarks are highly polarized (point (i) above); thus, the interpretation of Eq. (37) is that something cancels or hides it. Candidates include a highly polarized sea spinning opposite to the valence quarks, orbital angular momentum, or gluon polarization [8-10].

One can cancel out some charge weighting effects by looking at the difference of proton and neutron for which

$$I^p - I^n = \frac{1}{6}(\Delta u - \Delta d) \equiv \frac{1}{6}\left|\frac{g_A}{g_V}\right|, \tag{4.6}$$

which is Bjorken's sum rule [5]. The various g_A in the baryon octet give information on the *differences* of Δu, Δd, and Δs which are summarized by a measured parameter known as F/D. To extract the *sum*, Δq, we need the proton integral (Eq. (36)) or information on neutral current form factors

$$\tilde{g}_A(\nu p \to \nu p) = \Delta u - \Delta d - \Delta s. \tag{4.7}$$

I shall discuss this at the end of the talk. Preceding that, I shall discuss the question of Δs, since the measured F/D and the measured I^p can be combined to extract a value for Δs. This appears to be substantial; EMC claiming that

$$\Delta s = -0.23 \pm 0.08. \tag{4.8}$$

Implications and criticisms of this startling result will occupy the latter half of this talk. First, I will discuss what we know about the (constituent) quark polarization from static properties of the nucleon (magnetic moments, g_A/g_V) and review the extent to which the new insights do or do not require revision of this simple picture.

4.1 Spin polarization of valence (constituent) quarks

In the constituent quark model where $L_Z = 0$ the charges and the magnetic moments of neutron and proton place the following constraints on the probabilities for finding the flavours and spin correlations of "valence" quarks

$$u_v = 2d_v \ ; \ \frac{\mu_n}{\mu_p} = -\frac{2}{3} \to \Delta u_v = -4\Delta d_v. \tag{4.9}$$

The 56, $L_Z = 0$ wave function of the nonrelativistic quark model (NRQM) satisfies (41) but it is by no means unique. A hybrid state, where a gluon ($J_Z = \pm 1$) is partnered by qqq in 70 (required by the Pauli principle for qqq in colour 8) satisfies

Eq. (41) for the coherent combination [11] $g(^2 8 +^4 8)$ where the superscripts refer to the $2S + 1$ of the net spin of the qqq system. The "valence quarks" here are significantly depolarized relative to 56. One can also have a significant polarized sea without destroying the magnetic moment relations. This is because

$$\frac{\mu_n}{\mu_p} = \frac{2\Delta d - \Delta u + (-2\Delta \bar{u} + \Delta \bar{d} + R\Delta \bar{s})}{2\Delta u - \Delta d + (-2\Delta \bar{u} + \Delta \bar{d} + R\Delta \bar{s})}, \quad (4.10)$$

where $R = m_d/m_s \simeq 3/5$. The electrical neutrality of the sea tends to shield its contribution. A detailed fit is made in Ref. 12.

The (g_A/g_V) for the octet of baryons also relate to the spin polarized probabilities such as

$$\left(\frac{g_A}{g_V}\right)_{np} = \Delta u_v - \Delta d_v \rightarrow -5\Delta d_v, \quad (4.11)$$

where we used Eq. (41). Thus immediately

$$\Delta d_v = -0.25; \Delta u_v = 1. \quad (4.12)$$

In the 56 NRQM one would have [3]

$$\Delta d_v = -1/3; \Delta u_v = 4/3; \Delta u_v + \Delta d_v = 1, \quad (4.13)$$

and the entire spin polarization comes from the quarks. However, from Eq. (44), we see that

$$\Delta u_v + \Delta d_v \simeq 3/4, \quad (4.14)$$

and so, in advance of the EMC data, only naive "quarkists" would have expected 100% for Δq_v. Anyone who worked with four-component spinors, of which the MIT bag is a specific model example, knew that the "orbital dilution" in the lower components played an essential role [13]. In fact, the Δq_v expectation is even less than Eq. (46). When one makes a best fit to all of the baryon octet g_A/g_V, one finds [14]

$$\Delta q_v (\equiv 3F - D) = 0.55 \pm 0.10. \quad (4.15)$$

Note the appearance of F and D which summarizes the g_A/g_V. This parameter will appear later. Note that many analyses of the polarization data use [2,6,10,15] $F/D = 0.63$ (Ref.16). However, this value fitted a value of the neutron lifetime that we now know to have been incorrect [17,18]. The correct current value [19-25] is lower than 0.63 and is dependent upon assumptions about SU(3) flavour breaking.

The earliest predictions for the deep inelastic polarization asymmetry in the valence-dominated region assumed that all Δq and q (valence) have the same x dependence. Thus (see Refs. 3 and 4 for origins of these formulae)

$$A^n(x) \simeq 4\Delta d + \Delta u \rightarrow 0,$$

(the zero following immediately from Eq. (41)) and

$$A^p(x) = \frac{5}{3}(-\Delta d) \rightarrow \frac{1}{3}(g_A/g_V).$$

The prediction that $A^p > 0$ is non-trivial as *a priori* it could be anywhere in the range $-1 \leq A \leq +1$. The presence of a $q\bar{q}$ sea as $x \to 0$ was expected to cause $A(x \to 0) \to 0$. The other qualitative expectation [26,27] was that $A(x \to 1) \to 1$ as follows.

The valence picture above implicitly assumed that $u_v(x) = 2d_v(x)$ for all x. However, unpolarized data show this to be untrue in that it would require that

$$\frac{F_1^n(x)}{F_1^p(x)} = 2/3.$$

In practice, this ratio drops as $x \to 1$, suggesting that the $u(x \to 1) >> d(x \to 1)$, a phenomenon which follows from spin dependence via single gluon exchange. Chromomagnetic hyperfine energy shifts split the $\Delta - N$ masses and elevate $u(x \to 1)$ over $d(x \to 1)$. They also cause $u^\uparrow(x \to 1)$ to dominate over $u^\downarrow(x \to 1)$, which the consequence that $A^{p,n}(x \to 1) \to 1$. Thus, a qualitative expectation for A^p emerged:

$$A^p(x \to 0) \to 0; A^p(x \simeq 1/3) \simeq 1/3 \left|\frac{g_A}{g_V}\right|; A^p(x \to 1) \to 1.$$

These predictions turned out to be remarkably well verified and even agree with the latest EMC data.

Recently Close and Thomas [28] showed that, within the framework of the MIT bag model, one could relate the x-dependent distortion of the valence distributions to the measured chromomagnetic energy shift in the $\Delta - N$ masses. All of this suggests that the valence quark polarizations measured in polarized deep inelastic scattering are similar to the polarizations of the constituent quarks manifested in low-energy spectroscopy. This is an important constraint on model builders. The memory of the constituent quark spins is not lost as one proceeds to the deep inelastic: the *valence quarks are highly polarized*.

If, as is being claimed, the quarks and antiquarks contribute (within errors) nothing to the net spin polarization of the proton, then we must conclude that something is cancelling the contribution of the valence quarks. Candidates include orbital angular momentum polarized gluons or a negatively polarized sea.

4.2 Polarized strange quarks?

One exciting possibility is that the EMC data imply a large polarization of strange quarks and/or antiquarks within the proton. If true, this could have significant consequences. In particular, it could modify earlier analyses of electroweak parity violation in deuterium where Campbell et al argue [15], the polarized strange quarks could give contributions that dominate over electroweak radiative corrections. An extreme claim has appeared in the literature that the large value for Δs is in conflict with perturbative QCD. If true, this would be devastating. This claim comes about, in part, because an incorrect value of F/D has been used in the analyses. It is this parameter, and its implications for Δs, that I will now discuss.

Given the integral, I_p, of the polarized structure function $g_1^p(x,Q^2)$, one extracts Δs (including new QCD corrections)

$$I_p \equiv \int dx g_1^p(x,Q^2) = \frac{1}{18}\left(\frac{g_A}{g_V}\right)\left[\frac{9f-1}{f+1} - \frac{\alpha_s(Q^2)}{\pi}\frac{3f+1}{f+1}\right] + \frac{\delta s}{3}, \qquad (4.16)$$

where $f \equiv F/D$ with $\alpha_s(Q^2) = 0.27, g_A/g_V = 1.254 \pm 0.006$ and $I_p = 0.126 \pm 0.022$. A feeling for the sensitivity of Δs to f can be gauged from the approximate relation

$$-\Delta s \simeq (f - 0.40) \pm 0.07. \qquad (4.17)$$

The widely used value, following the much-quoted fit of Ref. 16 has been

$$F/D = 0.63 \pm 0.02 \longrightarrow \Delta s = -0.23 \pm 0.09. \qquad (4.18)$$

If the sea is flavor-independent, then Eq. (50) summarizes the widely accepted interpretation for the EMC polarized structure function data where a significant negative polarization of the sea cancels out the positive polarization of the valence quarks.

This value was based on the original value for I_p quoted by EMC [2], namely $I_p = 0.116 \pm 0.022$. However, the revised value [29], $I_p = 0.126 \pm 0.022$, reduces the magnitude of Δs by 0.03, and so $\Delta s = -0.20 \pm 0.09$ should replace Eq. (50).

However, it does not seem to be widely appreciated that the F/D of Ref. 16 was much constrained by an outdated value of the neutron lifetime, and that Ref. 16 chose "to omit from (their) fit the neutron decay correlation (which yields) $g_A = 1.258 \pm 0.009$, which differs significantly from the result 1.239 ± 0.009 required by the neutron lifetime measurements". The value accepted as correct today [18] differs by some 3σ from the old value, and this, together with other data on hyperon beta decays [16,18,19], shows that F/D is much smaller than the old value. Flavour symmetry breaking causes a spread in values of F/D, depending on which partial set of data one uses; indeed, the symmetry breaking even calls into question the utility of the F/D parameter [20], and so Refs. 17 and 21 set up their analyses without direct reference to F/D. Translating their work into F/D, one finds that the value subsumed in Ref. 17 is $F/D = 0.56$ consistent with that implicit in Ref. 21 and, within errors, with the fitted value in Ref. 22. Reference 23 obtained an even smaller value of $F/D = 0.545 \pm 0.02$. Recent improvements in the Σn beta decay data, in particular, may raise F/D to 0.58 (Ref. 24), but nowhere as high as the 0.63 used previously.

The magnitudes for Δs implied by these values for F/D are

$$F/D = 0.548 \pm 0.01 \longrightarrow \Delta s = -0.12 \pm 0.06 (\text{Ref. 23}). \qquad (4.19)$$

$$F/D = 0.58 \pm 0.01 \longrightarrow \Delta s = -0.15 \pm 0.08 (\text{Ref. 24}). \qquad (4.20)$$

Thus we see that the magnitude of the (negative) strange polarization may be only half as big as that previously assumed.

g_A	F, D	$\Delta q^{(p)}$	Data
np	$F+D$	$\Delta u - \Delta d$	1.26 ± 0.005
Λp	$F+\frac{1}{3}D$	$\frac{1}{3}(2\Delta u - \Delta d - \Delta s)$	0.72 ± 0.02
$\Xi\Lambda$	$F-\frac{1}{3}D$	$\frac{1}{3}(\Delta u + \Delta d - 2\Delta s)$	0.25 ± 0.05
Σn	$F-D$	$\Delta d - \Delta s$	-0.33 ± 0.02

4.3 Polarized gluons?

It has recently been realized [8] that the perturbative QCD correction to the singlet part of $g_1^p(x)$ effectively scales (to $0(\alpha_s^2)$) and may be important. This may be incorporated by replacing the Δq in Section 1 by $\tilde{\Delta} q \equiv \Delta q - \alpha_s/2\pi \Delta G$., where $\Delta G \equiv \int_0^1 dx \Delta g(x)$ and $\Delta G(x) = g_\uparrow(x) - G_\downarrow(x)$ is the polarized gluon distribution. This modifies the polarized lepton analysis, but cancels out in the expressions for (g_A/g_v) and does not enter the magnetic moment (Section 2) analysis.

One consequence is that there may be a continuity between the low-energy polarization revealed in constituent quarks (magnetic moments and spin dependence of resonance excitation) and the deep inelastic polarization.

First of all, we summarize the data on the Δq (or equivalently $\tilde{\Delta} q$) from the various (g_A/g_V).

If we assume $SU(3)_F$ symmetry in the sense that $s(\Sigma^+) \equiv d(P)$, then we may write the various g_A in terms of F, D, or Δq as follows: Thus $F/D \equiv (\Delta u - \Delta s)/(\Delta u + \Delta s - 2\Delta d)$. Extracting the individual contributions involves a correlated fit. The EMC values, corrected for F/D, become $\tilde{\Delta} u = 0.80 \pm 0.06, \tilde{\Delta} d = -0.45 \pm 0.06$, and $\tilde{\Delta} s = -0.15 \pm 0.06$. One possibility is that $\Delta s = 0$, so that $\tilde{\Delta} s = -\alpha/2\pi \Delta G$. In this case, we obtain for

$$\Delta u \equiv \tilde{\Delta} u - \tilde{\Delta} s = 0.95 \pm 0.06 \quad (4.21)$$

$$\Delta d \equiv \tilde{\Delta} d - \tilde{\Delta} s = -0.30 \pm 0.06 \quad (4.22)$$

It is interesting to note that these values are consistent with those extracted from the magnetic moments (Eq. (4)) viz

$$\Delta u_v = 1, \Delta d_v = -0.25$$

When the news of the EMC polarisation data first broke, some people were suggesting that the quark model was dead. However, we see now that if polarised protons contain polarised gluons then the quark structure of both low energy and deep inelastic polarised protons may be rather similar. Why gluons should feel significant polarisation is a question for the future.

References

[1] G. Baum et al., Phys. Rev. Lett. **51**, 1135 (1983); V.W. Hughes and J. Kuti, Ann. Rev. Nucl. Part. Sci. **33**, 611 (1983).

[2] J. Ashman et al., (EMC), Phys. Lett. **B206**, 364 (1983); V. W. Hughes et al., Phys. Lett. **B212**, 511 (1988).

[3] F.E. Close, "Introduction to Quarks and Partons," Academic, New York (1979) Chap. 13.

[4] J. Kuti and V. Weisskopf, Phys. Rev. **D4**, 3418 (1971).

[5] J.D. Bjorken, Phys. Rev. **D1**, 1976 (1970).

[6] J. Ellis and R.L. Jaffe, Phys. Rev. **D9**, 1444 (1984).

[7] J. Ellis, R.A. Flores and S. Ritz, Phys. Lett. **B194**, 493 (1987).

[8] G. Altarelli and G.G. Ross, Phys. Lett. **B212**, 391 (1988); R. Carlitz, J. Collins and A. Mueller, Phys. Lett. **B214**, 229 (1988).

[9] L. M. Sehgal, Phys. Rev. **D10**, 1663 (1974).

[10] S.J. Brodsky, J. Ellis and M. Karliner, Phys. Lett. **B206**, 309 (1988).

[11] T. Barnes and F.E. Close, Phys. Lett. **128B**, 277 (1983); F. Wagner, Proc. XVI Rencontre de Moriond (1982), ed. J. Tranthanhvan.

[12] C. Carlson and J. Milano, College of William & Mary report, WM-89-101. The role of gluon exchange is discussed by H. Hogassen and F. Myhrer, Phys. Rev. **D37**, 1950 (1988).

[13] For example, p. 117 in Ref. 3.

[14] This comes from combinations of g_A/g_V for np with that for $\Lambda p, \Sigma n$, and $\Xi\Lambda$; see Eq. (11) in Ref. 17.

[15] B.A. Campbell, J. Ellis and R.A. Flores, CERN-TH- 5342/89.

[16] M. Bourquin et al., Z. Phys. **C21**, 27 (1983).

[17] F.E. Close and R.G. Roberts, Phys. Rev. Lett. **60**, 1471 (1988).

[18] M. Aguilar-Benitez et al., (Particle Data Group) Phys. Lett. **B204**, 1 (1988).

[19] S. Hsueh et al., (E715 collaboration) Phys. Rev. **D38**, 2056 (1988).

[20] H.J. Lipkin, Phys. Lett. **214B**, 429 (1988).

[21] M. Anselmino, B. Ioffe and E. Leader, Santa Barbara ITP report (1988) unpublished.

[22] D. Kaplan and A. Manohar, Nucl. Phys. **B310**, 527 (1988).

[23] J. Donoghue, B. Holstein and S. Klint, Phys. Rev. **D35**, 934 (1987).

[24] A. Beretvas, private communication; Z. Dziembowski and J. Franklin, Temple University, Philadelphia report TUHE-89- 11 (1989).

[25] R. Jaffe and A. Manohar, MIT report, MIT-CTP-1706 (1989) (but note that this inputs outdated neutron lifetime, which artificially increases the errors).

[26] F.E. Close, Phys. Lett. **43B**, 422 (1973).

[27] G. Farrar and D. Jackson, Phys. Rev. Lett. **35**, 1416 (1975).

[28] F.E. Close and A.W. Thomas, Phys. Lett. **B212**, 227 (1988).

[29] EMC collaboration, Nucl. Phys. B328, 1, (1989)

Quarks in nuclei

A. Ferrando[1], P. González and V. Vento [2]

[1] Departamento de Física Teórica, Universidad Autónoma de Madrid
E-28049 Canto Blanco (Madrid), Spain
[2] Departament de Física Teòrica and I.F.I.C., Universitat de València – C.S.I.C.
E-46100 Burjassot (València), Spain

Abstract: We review some properties of Quantum Chromodynamics, the theory of the hadronic interactions, which serve as guidelines to introduce low energy models of hadron structure. Among these we shall center our attention in the non relativistic quark model and the topological bag model. We present some of their applications to actual problems in experimental and theoretical nuclear physics. In particular we discuss exotic nuclei, quark matter, deep inelastic scattering, proton spin,... and their relation to such phenomena as quark Pauli blocking, strangeness enhancement, nuclear structure functions, bosonization,...

1 Introduction

Quantum Chromodynamics (QCD) is generally accepted as the fundamental theory of the hadronic interactions. It is a local gauge invariant field theory whose dynamical degree of freedom is color and whose gauge group is SU(3) [1]. It is described in terms of quark ($q_i^f(x)$) and gluon ($A_\mu^a(x)$) fields, i.e.,

$$\mathcal{L}_{QCD} = \bar{q}_i^f(x) i\gamma^\mu (D_\mu)_{ij} q_j^f(x) - \frac{1}{4} F_{\mu\nu}^a F^{a\mu\nu}, \tag{1}$$

where

$$(D_\mu)_{ij} = \delta_{ij}\partial_\mu - ig(t^a)_{ij} A_\mu^a(x), \tag{2}$$

and

$$F_{\mu\nu}^a(x) = \partial_\mu A_\nu^a(x) - \partial_\nu A_\mu^a(x) + g f^{abc} A_\mu^b(x) A_\nu^c(x). \tag{3}$$

Here g is the bare coupling constant and t^a are the generators of the $SU_c(3)$ Lie algebra satisfying

$$[t^a, t^b] = i f^{abc} t^c, (a, b, c = 1, 2, \cdots, 8), \tag{4}$$

where f^{abc} are the structure constants of the corresponding color group. The last term in Eq.(1) includes gluon self-interactions, which are characteristic of a non-abelian theory. The index f denotes the flavors. Equations (1) through (4) describe the QCD lagrangian for massless quarks.

In order to account for non-zero quark masses, we have to add a mass term,

$$\bar{q}_i^f(x) m_f q_i^f(x). \tag{5}$$

These masses are the so-called current quark masses, and for the flavors we are interested, $m_u \approx m_d \approx 0$ and $m_s \approx 175$MeV. Note that this term contains the only flavor dependence of the theory.

QCD has some features which make it very attractive and appealing as a theory for the strong interactions. They are

i. Renormalizability : all the ultraviolet infinities of the theory can be reabsorbed into the constants of the theory.

ii. Universality : gauge invariance implies a unique coupling constant for all hadronic interactions.

iii. Asymptotic freedom : the effective coupling vanishes at short distances.

iv. Confinement : contrarily to what happens at short distances, the effective coupling constant, when perturbatively computed, increases as the distance between the quarks increases. This fact has been taken as a strong indication that QCD is a confining theory which does not allow free quarks and gluons to exist[2].

Besides the dynamical local symmetry, the QCD lagrangian possesses other symmetries of global nature. In particular one has $U(1)$ baryon number symmetry, and if one assumes N_f massless flavors, QCD is invariant under chiral $SU(N_f) \otimes SU(N_f)$ symmetry.

The coupling constant g appearing in the lagrangian, Eqs.(1) to (4), is the bare coupling constant. All physical processes however are described in terms of the running coupling constant, $\alpha_S(Q^2) = \frac{g^2(Q^2)}{4\pi}$, characterizing the interaction at momenta Q^2. It is well known [3] that

$$\alpha_S(Q^2) = \frac{2\pi}{b \ln Q/\Lambda} \qquad (6)$$

where b is the first coefficient of the beta function and Λ is the scale parameter of QCD. The logarithmic fall of this effective coupling constant leads to a perturbative behavior of the theory at short distances, which is named asymptotic freedom. This feature allows for the interpretation of *Bjorken scaling* and deviations from it. The process in which short-distance contributions are important are called hard. In any reaction, whatever the characteristic distances are, experimentalists do not detect quarks and gluons but pions, nucleons, etc... The fact that physicists have learned how to calculate observable quantities in terms of the short-distance dynamics (asymptotic QCD), eliminating or parametrizing appropriately the process of hadronization, is the greatest achievement of the theory of hard processes. Within the limits of applicability all the results are understood [3].

The genuine hadronic theory starts at distances of the order of 0.5 fm and larger. At such distances the coupling constant becomes large and non perturbative effects play the key role. In spite of more than a decade of intensive efforts, color confinement has still to be accepted as a starting postulate. It is therefore assumed that in any non-abelian gauge theory the spectrum of the observed states consists of color singlet composite objects (hadrons), while isolated colored objects have infinite energy [4].

Accepting color confinement as a starting hypothesis we are now able to explain qualitatively the main regularities in the hadronic world. Let us now review some of the most crucial results obtained from the theory, which give indications about its behavior in the non-perturbative regime.

1.1 The 't Hooft self-consistency condition

The principle which guides the bound-state picture of any theory of composites is based on chiral invariance [5]: *A composite particle must reproduce the axial anomaly due to its fermionic constituents.*

The main assumption that goes into the proof is that the quarks are permanently confined. In the chiral limit of the theory (massless quarks) the existence of the anomaly [6] implies that the absorptive part of the three current correlation function is non zero only when the four-momentum squared associated with the axial probe vanishes. Confined quarks do not contribute to the correlator since they are not physical states and therefore there must be massless physical states of one of the following two types:

i. spin zero particles, the Goldstone bosons of the spontaneously broken of chiral symmetry

ii. spin $\frac{1}{2}$ fermions if chiral symmetry remains unbroken

Particles of higher spin cannot occur, because they couple to external sources with derivative couplings whose contribution to the absorptive part vanishes at threshold.

1.2 Exact inequalities

There has been a certain degree of progress in proving rigorous results about QCD. In particular it has been shown that of all color singlet channels with non-zero isospin, the lowest threshold is in the pseudoscalar channel [7]. This together with 't Hooft's principle asserts that QCD with zero mass quarks must have a massless pseudoscalar of isospin 1, presumably a Goldstone boson. Moreover another exact result states that in vector-like gauge theories (QCD is an example) vector-like symmetries, like isospin or baryon number, cannot be spontaneously broken [8]. Putting this two results together one may convene that there is a strong indication that in nature the following additional principle is realized: the axial symmetries are spontaneously broken and the vector symmetries are unbroken.

1.3 The large N limit

In our world the number of colors N is equal to three. 't Hooft noticed that it is extremely useful to treat N as a free parameter and to consider N large [9]. The hope being that one may be able to solve the theory in the large N limit and that the N= 3 theory may be qualitatively and quantitatively close to the large N limit.

QCD simplifies notably in this limit and there exists a systematic expansion in powers of $\frac{1}{N}$. This simplification has provided qualitative insight into the behavior of the theory, but as of yet none of the quantitative results that are available seem to be quite sufficient for encouragement. The main results of the large N expansion are [9, 10]

i. Non planar graphs are suppressed by factors of $\frac{1}{N^2}$

ii. Internal quark loops are suppressed by factors of $\frac{1}{N}$

Thus in the large N limit only planar graphs with the minimal number of quark loops survive. The theory in this approximation leads to a infinite stable spectrum of almost free mesons and glueballs, with couplings which are at least $\frac{1}{N}$. Thus for large N, QCD

behaves as a field theory of weakly coupled bosons with effective local interactions of order $\frac{1}{N}$.

Weakly coupled field theories sometimes possess states whose masses diverge, for weak coupling, like the inverse of the coupling. These states are usually called solitons, kinks, monopoles,... The result of Witten's work [10] strongly suggests that the baryons are such states. Indeed the baryon mass turns out to be of order N, which is analogous to the fact that a soliton has mass $\frac{1}{g}$.

Large N arguments applied to QCD provide partial confirmation of Skyrme's old idea that baryons are solitons in a meson theory [11]. Thus in this limit QCD becomes equivalent to an effective field theory of mesons describing the hadronic interactions, from which the baryons arise as classical topologically non trivial solutions.

QCD in two dimensions has served as a testing grounds for many of the ideas just exposed. In this exactly solvable model one is able to study all the wishful properties of the theory, although their realization is intimately related to the dimensionality of space time and therefore no generalization to four dimensions has been obtained [12].

The fact that QCD is, in the low energy regime, highly non-perturbative and unsolved, has motivated the construction of models, whose aim is to substitute for the unknown solution of the theory[13]. The properties just described serve as guiding principles for most of the models. In these lecture we shall describe their aspects which are most related to current research. Our aim is not to provide technical details, but to give an overview and to address to the adequate bibliography.

2 The non-relativistic quark model

The naive quark model appeared as an immediate consequence after the quarks were accepted as the elementary constituents of hadronic matter. Many years later, soon after the proposal of QCD as the theory of the strong interactions, De Rújula, Giorgi and Glashow [14] proceeded to interpret the quark model within the quark dynamics provided by this theory. Their basic assumptions, which now constitute the foundations of most of the potential models, were:

i. The quark structure of hadrons is described by a non-relativistic $SU(2 \times N_f) \supset SU(N_f) \otimes SU(2)_{spin}$ scheme.

ii. Long-range flavor and spin independent scalar confining forces.

iii. Asymptotic freedom implemented by requiring that the residual interaction arises from the non–relativistic reduction of the one gluon exchange (OGE) between quarks.

This paper led to a reanalysis of the hadron spectrum with considerable success [15].

The elementary constituents of the model are not the current quarks of the QCD Lagrangian but rather a sort of *dressed* quarks. These have masses which are not well known, but the values

$$m_u \approx m_d \approx 340 MeV, m_s \approx 550 MeV, m_c \approx 1600 MeV, m_b \approx 4900 MeV$$

are the most commonly used. These masses are just effective quantities which parametrize our ignorance of the confinement mechanism, the relativistic corrections, etc...

In this model a simple type of confining force is used, namely

$$V_{qq}(r_{12}) = -V(r_{12})\vec{t_1} \cdot \vec{t_2} \qquad (7)$$

where $\vec{t}_i$ represent the color generators of particle i in vector notation and $V(r_{12})$ goes to infinity when $r_{12} \to \infty$. Some insight about its radial dependence can be obtained by looking at the spectra of the heavy $q\bar{q}$ systems, such as charmonium ($c\bar{c}$) and bottonium ($b\bar{b}$) for which a non-relativistic model is certainly applicable [3].

With only the confinement forces just described, we would have among others, the nucleon (N) and the isobar (Δ) degenerate. In order to remedy this situation a non-relativistic one gluon exchange interaction between quarks was introduced [14]. With the assumption of non-relativistic quark dynamics, the interaction between two massive quarks which exchange a massless (vector) gluon coincides, except for the color part, with the exchange of one photon between two electrons and leads to the familiar Coulomb potential plus the Fermi–Breit relativistic correction [13].

Chiral symmetry realized in a spontaneously broken manner is a major ingredient of QCD. The model lacks this very crucial ingredient, which appears very distressing when one attempts to study long distance properties, in particular, the behavior of the interaction between two nucleons. The need for a new degree of freedom, a pion, is inevitable [16]. We shall not describe the implementation of this feature in non–relativistic models, but shall discuss about it in some detail when describing bag models.

Quarks move with a considerable velocity inside the potential and therefore a relativistic treatment would be more appropriate. Relativistic equations have though a very complex center of mass motion, which makes their treatment quite approximate. In particular the large discrepancy in g_A may be solved in a relativistic treatment due to the small components [3, 17]. These generalizations lead to enormous calculational difficulties without a major improvement in the predictions.

This model has been extended to study systems of several baryons. In particular one major attempt has been to study the nucleon–nucleon interaction. For a review on some early attempts see R.F. Alvarez-Estrada et al. [13]. Recently, the possibility of detecting quark effects when studying properties of nuclei has been analyzed by several authors [18].

2.1 Quarks in light nuclei

Nucleons, deltas and other baryonic excitations appearing in nuclear physics are large objects (rms radius $\simeq 0.8$ fm). Light nuclei, as for example ^{3}He and ^{3}H are not much larger (rms radius $\simeq 1.9$ fm). Thus some overlap between the internal structures of the hadrons involved in physical processes of light nuclei seems inevitable. In order to determine the magnitude of the effect the knowledge of hadron structure is necessary. We shall assume in what follows that the hadron structure is well represented by a naive quark model and shall analyze how this compositeness contributes to some observed phenomena of nuclear systems. Our point of view, together with that of other authors, has not been to interpret nuclear physics fully in terms of subnucleonic degrees of freedom or effective QCD based theories, but to search for new physics in the low and intermediate energy regime, hopefully related to those subnucleonic degrees of freedom [18].

We have proposed the use of many body effects as an alternative tool for unveiling the quark structure in nuclei. This effort should lead to complementary information to that obtained by the conventional well developed use of short distance probes. Since the dynamics of quarks at low energies is not well determined we have looked for effects associated with fundamental principles. In this respect characteristic many body phenomena arise due to the Symmetrization Principle and the Spin Statistics Connection Theorem. These phenomena are qualitatively independent of the dynamics involved. Moreover if we look at overlapping systems (or non local operators) we obtain observable effects due to the Pauli principle whose magnitude depends on the degree of overlap (or non locality).

Our aim has been to look for observables and physical systems where the extreme circumstance named maximal Pauli blocking takes place. We have for this purpose studied observables related to one body (at the quark level) operators, in particular charge densities, for both conventional light nuclei and non conventional or exotic nuclear systems.

2.2 One body observables and maximal Pauli effects

In order to study quark effects in nuclear systems a convenient mathematical tool is the Hadronic Quark Cluster Decomposition (HQCD) [19]. The natural implementation of the dynamics in this scheme is by considering that the center of the cluster is subject to a certain force, the remnants of the nuclear force, and the motion of the quarks is governed by some intracluster potential whose origin is confinement. For simplicity we have taken everywhere harmonic forces.

If 3N quarks occupy the available levels of an harmonic oscillator potential of parameter A (quark shell model limit), there is no contribution from antisymmetrization at the quark level to the expectation value of one body observables. But, if we have N clusters of 3 quarks each, the dynamics of the clusters governed by an harmonic oscillator parameter B and that of the quarks within each cluster by one of parameter A, in general, expectation values of one body operators do show effects due to the Pauli principle at the quark level. This last statement represents the scenario we have chosen to develop the contributions due to quark exchange effects. There exist a relation between A and B for which one reaches the quark shell model limit where the effects due to antisymmetrization vanish.

We have performed the study of observables in a simplified model of nature. Never mind the dramatic simplifications involved in our toy model, all the different possible scenarios arise. Two types of nuclear systems, with very different behaviour, appear in the analysis. One, in which all the quarks have different internal quantum numbers. For this kind of system, even for large radial overlaps, there is almost no effect associated with the exchange of quarks. Another, in which many quarks have the same internal quantum numbers. For this one, even for small radial overlaps, spectacular observable effects appear due to the Pauli principle at the quark level. These latter type of systems are the so called maximal Pauli blocked nuclei. Unluckily conventional light nuclear systems, the ones which have been most closely scrutinized, are of the former type [20].

Several quasi realistic calculations for light nuclei have been performed. In this case one body observables do not show large non dynamical effects at low momenta ($q < 3 fm^{-1}$) associated with the exchange of quarks. The effects become larger at higher momenta. In particular for $3 fm^{-1} < q < 6 fm^{-1}$ they are comparable to meson exchange current contributions. However one would need a more fundamental approach to disentangle quark Pauli contributions from these more conventional ones.

The search for maximal Pauli blocked nuclear systems, in order to avoid the kind of problems just mentioned, has been launched. Several non traditional nuclear systems have been proposed: dibaryons, light hypernuclei and delta-nuclei. In this last proposal, we have analyzed delta production mechanisms on light nuclei. Quark Pauli effects are sizeable only for large momenta ($q > 4 fm^{-1}$). But several features appear which might lead to a clear experimental determination of these effects at lower q's. One very suggestive result is that electric type transitions are allowed in nuclei due to the exchange terms and forbidden for free baryons. Thus the comparison of data obtained from nuclei and free baryons may lead to disentangle the quark substructure. Moreover different delta channels have different Pauli blocking behavior. Therefore comparison of various channels might also shed some light into the discovery of substructure effects at low energies [21].

2.3 The two nucleon Coulomb energy

We have developed the formalism to calculate the contribution to the expectation value of two body operators from the exchange of quarks. In particular we have performed a quasi realistic calculation of the Coulomb energy for a two nucleon system with deuteron quantum numbers. The dynamical dependence in this calculation enters in the chosen radial forms of the wave functions. Different models imply different radial functions and therefore the exchange contribution changes from one model to another. However we can try to extract some dynamical dependence by changing arbitrarily the parameters of our gaussian wave functions, which implies in turn modifying the overlap between the nucleon. We have shown that the exchange contribution to the Coulomb energy is strongly dependent on the overlap between the nucleons. If the overlap is small, like in the deuteron (rms radius $\simeq$ 2.116 fm) the exchange effect is small ($\simeq$ 50 KeV). In denser systems (rms radius $\simeq$ 1 fm) it can be as high as 600 KeV. The order of magnitude is within the experimental result ($\simeq$ 300 KeV for the difference between the Coulomb energy of ^{3}He and that of ^{3}H). Many indeterminacies remain still in the calculation of this quantity which have to be removed before a more quantitative statement can be advanced [20].

3 The Bag Model

Initially the bag model was developed by a group of physicist of the Massachussets Institute of Technology[22]-[24]. The crucial idea behind the model is the implementation of confinement and asymptotic freedom. A hadron is described as a hypertube (the bag) which divides space-time in two very distinct regions. The interior one contains quarks and gluons whose dynamical behavior is described by perturbative QCD. Inside the tube one considers a constant energy density, which allows the bag of perturbative vacuum to stabilize. The exterior region contains no color degrees of freedom and represents the complicated non-perturbative vacuum. The model may be formally described in terms of a cavity field theory[25]-[28]. A complete pedagogical account can be found in refs.[29]-[31].

3.1 The static spherical cavity approximation

We proceed with a brief presentation of the model omitting gluons. Let $q_r^f(x)$ be a quark field inside the bag, where f is the flavor index and r the color index (they will not appear unless required for clarity). The QCD lagrangian (Eqs.(1) to (5)) imply that inside the bag

$$i\gamma^\mu \partial_\mu q^f = m_f q^f \qquad (8)$$

Outside the bag the quark fields must vanish as demanded by confinement. A possible way of realizing that there is no color flux away from the bag is

$$i n_\mu \gamma^\mu q_r = q_r, \text{ on the surface} \qquad (9)$$

where n_μ is a space-like unit four vector normal to the surface. This is the so called Linear Boundary Condition (LBC).

Moreover no energy–momentum flux is to leave the bag, this leads to

$$B = \frac{1}{2}n_\mu \partial^\mu (\sum_r \bar{q}_r q_r), \text{ at the surface} \tag{10}$$

where B is a constant energy density playing the role of a pressure on the bag surface, which prevents it from collapsing. This is the so called Non Linear Boundary Condition (NLBC).

In dealing with the MIT bag model, the first practical problem one encounters is to find the exact solutions to the bag equations. Indeed, no realistic solutions exist thus far. The conventional treatment is based on the so called static spherical cavity approximation. It simply means that one takes the bag as a static sphere of radius R. This violates causality and translational invariance, but is certainly the easiest calculation one may attempt.

Let us perform an exercise to present the simplicity and beauty of the bag model approach. We assume N quarks inside the bag and no residual quark–gluon interaction ($g = 0$). The lowest energy single particle wave functions are given by

$$q = \mathcal{N} \begin{pmatrix} ij_0(\omega r)\xi \\ -j_1(\omega r)\vec{\sigma} \cdot \hat{r}\xi \end{pmatrix} \exp(-i\omega t). \tag{11}$$

Here $\mathcal{N}$ is a normalization constant, the j's are spherical Bessel functions and ξ a Pauli spinor. Within the static spherical cavity approximation the LBC is given by

$$j_0(\omega R) = j_1(\omega R), \tag{12}$$

leading to a first eigenfrequency with value $\varepsilon_0 = \omega_0 R = 2.04$. (Complete details of the various modes may be looked after in refs.[13, 31].) The mass of the corresponding hadronic groundstate is

$$m_H = N\frac{2.04}{R} + \frac{4}{3}\pi R^3 B. \tag{13}$$

The NLBC implies

$$R = \left(\frac{2.04N}{4\pi B}\right)^{1/4}, \tag{14}$$

and thus

$$m_H = \frac{4}{3}N\omega_0. \tag{15}$$

By taking $N = 3$ and the nucleon mass as input we obtain $R_{nucleon} \approx 1.7\text{fm}$, certainly a rather extended object.

In this model the ratio of baryon to meson mass is simply $(\frac{2}{3})^{3/4} \approx 0.74$, which is rather accurate for m_ρ/m_N, but quite wrong for m_π/m_N. This is the first indication that pions play a special role in low energy phenomenology.

Several corrections have to be applied to this simple scheme : quark masses, gluon exchange mechanism, etc.. Let us emphasize that the resulting overall picture of hadron structure one obtains from the description just developed is quite appealing [3, 13, 24, 29]. Let us describe briefly the quark-gluon phase transition as an exercise with the bag model in its most simple form

3.2 Quark matter

Speculations about the structure of matter at very high density and/or temperature have led to the conjecture that a phase transition from nuclear to quark takes place [32]. Such a phenomena might occur in the case of a heavy neutron star [33] or in a heavy ion collision [34]. At the phase transition region, the extended structure of hadrons becomes important, which requires the use of models. In particular the MIT bag model is easy to implement in that regime [35].

Let us study the region of the phase diagram characterized by T=0 and $\mu \neq 0$. The plasma will be described as a relativistic gas of quarks with no global color change. The dynamics of confinement is introduced in a bag like scenario:

i. quarks are separated from vacuum by a phase boundary.

ii. the interior region is endowed with a universal energy density B.

We shall neglect in this presentation corrections due to gluon exchanges, i.e., we work in the $\alpha_c = 0$ approximation (see [35] for $\alpha_c \neq 0$ effects).

For thermodynamic systems boundary conditions are unimportant since they fall off like $V^{-1/3}$. Therefore the change in the thermodynamic potential due to the confinement dynamics will be

$$\Omega \to \Omega + B \tag{16}$$

and consequently the energy density and pressure are modified as

$$\begin{aligned} \varepsilon &\to \varepsilon + B \\ p &\to p - B \end{aligned} \tag{17}$$

Thus B has the interpretation of the thermodynamic potential of the vacuum.

The model of quark matter we will use is based on the production of an equilibrium distribution of quarks and electrons by beta decay

$$d \rightleftharpoons \mu + e^- + \bar{\nu} \tag{18}$$

$$s \rightleftharpoons \mu + e^- + \bar{\nu} \tag{19}$$

If we neglect the contribution from the $\bar{\nu}$, the thermodynamic potential in the T=0 limit becomes.

$$\Omega = \Omega_u + \Omega_d + \Omega_s + \Omega_e + B \tag{20}$$

where

$$\Omega_{u,d} = -\frac{\mu_{u,d}^4}{4\pi^2} \tag{21}$$

$$\Omega_s = -\frac{1}{4\pi^2}\left[\mu_s(\mu_s^2 - m_s^2)^{1/2}(\mu_s^2 - \frac{5}{2}m_s^2) + \frac{3}{2}m_s^4 \ln\left(\frac{\mu_s + (\mu_s^2 - m_s^2)^{1/2}}{m_s}\right)\right] \tag{22}$$

and

$$\Omega_e = -\frac{1}{3}\frac{1}{4\pi^2}\mu_e^4 \tag{23}$$

We have considered $m_u = m_d = m_e = 0$, $m_s \neq 0$ and $\mu_u, \mu_d, \mu_s, \mu_e$ are the corresponding chemical potentials. If quark matter is to be electrically neutral, the following relation holds

$$e\left(\frac{2}{3}n_u - \frac{1}{3}n_d - \frac{1}{3}n_s - n_e\right) = 0 \tag{24}$$

where n_i is the number density of particle i.

Moreover since
$$n_i = -\frac{\partial \Omega}{\partial \mu_i} \qquad (25)$$
all chemical potentials can be eliminated in favor of the baryon number density
$$n_B = \frac{1}{3}(n_u + n_d + n_s) \qquad (26)$$

Let us proceed to impose the restrictions due to the equilibrium condition. In first place β decay implies that
$$\varepsilon = \Omega + \sum_i \mu_i n_i \qquad (27)$$
is a minimum with respect to the density of strange quarks and electrons, i.e.,
$$\mu_e = \mu_d - \mu_u \qquad (28)$$
$$\mu_s = \mu_d \qquad (29)$$
For practical purposes electrons may be ignored at all densities and thus
$$n_B = n_u - n_e \simeq n_u \qquad (30)$$

Thus all thermodynamics quantities may be expressed in terms of the baryon number density.

Let us study the high density limit, which is characterized by
$$\mu_u, \mu_d, \mu_s \gg m_s \qquad (31)$$
and therefore $n_B \gg n_o$ ($n_o \simeq .17 fm^{-3}$ is nuclear matter density). If we use the $m_s \simeq 0$ approximation we arrive at the following flavor symmetric scenario
$$\mu_u = \mu_d = \mu_s \quad ; \quad n_u = n_d = n_s \qquad (32)$$
and therefore
$$p = \frac{1}{3}\varepsilon - B \qquad (33)$$
$$\varepsilon = \frac{9}{4}\pi^{2/3} n_u^{4/3} + B \qquad (34)$$
$$n_i = \frac{\mu_i^3}{\pi^2} \qquad (35)$$
If we choose $B \approx 50 MeV fm^{-3}$, then
$$0.5 fm^{-3} < n_u < 1.0 fm^{-3} \to 75 MeV fm^{-3} < \varepsilon < 1200 MeV fm^{-3} \qquad (36)$$

Therefore the exact value of B still plays a significant role for $n_u < 1 fm^{-3}$, but not for higher values of the baryon density.

Let us now proceed to the low density limit. In this case the strange quark may be ignored and
$$\mu_u = \mu_d \qquad (37)$$
thus
$$p = \frac{1}{3}\varepsilon - \frac{4}{3}B \qquad (38)$$

$$\varepsilon = \frac{3}{2}\pi^{2/3}n_u^{4/3} + B \tag{39}$$

$$n_{u,d} = \frac{1}{\pi^2}\mu_{u,d}^3 \tag{40}$$

The above equations imply a phase transition from nuclear to quark matter as one moves from low to high densities. The ratio n_s/n_u increases implying a large production of strange quarks in the phase transition to quark matter [35].

A simple explanation of this phenomenon may be given in terms of the Pauli Principle: at high densities the Fermi energy of the u and d quarks become greater than m_s and it pays to invert the β decay process to create strange particles.

In order to describe in full detail this scheme, interactions due to gluonic coupling to quarks ($\alpha_c \neq 0$) have to be incorporated. These features will change quantitatively but not qualitatively our results.

4 Chiral symmetry and the bag model

The MIT bag model violates chiral symmetry even in the case of massless quarks. This is certainly a major disagreement with QCD. The vector and axial vector currents in the bag are obtained by conventional field theoretic procedures leading to

$$\vec{V}^\mu = \bar{q}(x)\gamma^\mu \frac{\vec{\tau}}{2} q(x)\Theta_B, \tag{41}$$

and

$$\vec{A}^\mu = \bar{q}(x)\gamma^\mu \gamma_5 \frac{\vec{\tau}}{2} q(x)\Theta_B, \tag{42}$$

where Θ_B is the Heaviside function for the bag. Because of the LBC we obtain for massless quarks

$$\partial_\mu \vec{V}^\mu(x) = 0, \tag{43}$$

and

$$\partial_\mu \vec{A}^\mu(x) = -\bar{q}(x)\gamma_5 \frac{\vec{\tau}}{2} q(x)\delta_S \neq 0. \tag{44}$$

Here the divergence of the axial current does not vanish on the bag surface and therefore chiral symmetry is violated.

The main idea behind chiral bag models is to incorporate besides asymptotic freedom and confinement, chiral symmetry as a fundamental ingredient. QCD possesses spontaneous chiral flavor symmetry breaking in the medium-strong coupling regime and this has to be incorporated into any phenomenological model of hadrons. The spontaneous breaking of chiral symmetry is immediately connected to the role of the pion as a Goldstone boson. Pionic degrees of freedom are therefore implicit into any mode to be incorporated to restore the chiral symmetry. A two phase picture arises naturally in terms of the realization of chiral symmetry in the two possible modes : Wigner mode inside the confinement region and the Goldstone mode outside it.

The chiral bag model[36] has its origin in this two phase picture. The interior region is well represented by the MIT description. The exterior region reflects the structure of the complicated non-perturbative vacuum. We consider that there exists a pionic mode imposed by conservation of the axial current. In this phase the global $SU(N_f) \times SU(N_f)$ chiral symmetry is realized in the Goldstone mode via a phenomenological pion field and is spontaneously broken to $SU(N_f)$. Finally, the communication between the two phases is implemented by the sole requirement of chiral symmetry conservation.

A main feature of bag models is that the effects associated with gluons are treated perturbatively and we therefore omit them from the presentation[13, 31]. To zeroth order in the color coupling constant the boundary condition for the quark field becomes

$$in_\mu \gamma^\mu q(x) = U_5(x) q(x), \text{ on the surface,} \qquad (45)$$

where U_5 is related to the chiral angle by

$$U_5(x) = \exp(i\vec{\tau} \cdot \hat{\pi} \gamma_5 \theta). \qquad (46)$$

The pion field outside satisfies the equation of motion associated with its lagrangian density, in our case the non-linear sigma model, i.e.,

$$\frac{f_\pi^2}{4} \partial_\mu U^\dagger \partial^\mu U, \qquad (47)$$

where f_π is the pion decay constant, whose value is 93MeV, and U is the unit quarternion

$$U = \exp(i\vec{\tau} \cdot \hat{\pi} \theta). \qquad (48)$$

The boundary condition for the pion field arises from equating the normal components of the axial current

$$n_\mu \vec{A}^\mu = i\bar{q}\gamma_5 \frac{\vec{\tau}}{2} n_\mu \gamma^\mu q. \qquad (49)$$

As we shall see later one might incorporate higher order momentum terms to Eq.(48), although they do not play quantitatively an important role.

There are two regimes in which we may look for different types of solutions of the above set of equations. Let us express them in terms of the pion field. If the pion field is small, this occurs for large bag radii, the dominant feature of chirality is associated with single particle pionic excitations which can be studied in a perturbative manner[13, 31, 37]. On the other hand if the pion field is large, which occurs for small bag radii, the dominant feature associated with the implementation of chirality is due to the so called hedgehog solution, a non-perturbative soliton solution for the pion field carrying baryon number[10, 38].

4.1 Deep inelastic structure functions

As an application of the perturbative scheme we briefly discuss the description of lepton scattering on nucleons in a perturbative formalism.

At the early stages of the development of the bag model theory, Jaffe analyzed in detail how the MIT bag model, which had been constructed to incorporated asymptotic freedom and confinement, realized scaling[40]. Later it was shown that the Bjorken limit of the structure functions did not depend on the bag boundaries and therefore was not affected by the dynamical structure of these[41]. This analysis was performed in the cavity approximation. Following a similar approach we describe here the same calculation in the Chiral Bag Model (CBM)[42].

The cross section for inelastic lepton scattering is given by the imaginary part of the appropriate forward Compton scattering amplitude. Viewing the lepto-production process in this way allows one to use the cavity approximation for the calculation of the correlation function associated with the product of two currents keeping the hadron at rest. The hadronic tensor [3, 43], $W_{\mu\nu}$, in the case of bag models, a situation where the states are not momentum eigenstates, is defined by[40],

$$W_{\mu\nu}(q) = \frac{M}{2\pi} \int dt \int d^3r \exp(iq^0 t - i\vec{q} \cdot (\vec{y} - \vec{x})) \langle B | [J_\mu(\vec{x}, t), J_\nu(\vec{y}, 0)] | B \rangle, \qquad (50)$$

where q^μ is the four momentum of the virtual photon, $J_\mu(x)$ the corresponding electromagnetic current and the target is described as a bag state at rest normalized to unity, i.e., $\langle B|B\rangle = 1$.

In the CBM the electromagnetic current is written as a sum of the various degrees of freedom

$$J_\mu(x) = i\sum_f e_f \bar{q}_f(x)\gamma_\mu q_f(x)\Theta(R-r) + e(\vec{\pi}\times\partial_\mu\vec{\pi})_3\Theta(r-R) = J_\mu^q(x) + \Pi_\mu(x). \quad (51)$$

Here e_f denotes the electromagnetic charge for the quarks of flavor f and e the positron charge. Quarks and pions are coupled through the boundary conditions (Eqs.(45) and (49)).

The CBM used in a perturbative fashion introduces two additional contributions to the structure functions compared with the analogous calculation performed in the MIT bag model. On the one hand, the pion is charged and therefore the photon will couple to it. This corresponds to meson exchange current corrections. On the other hand the pion field modifies the quark wave functions and the baryon mass formula leading to additional contributions.

The pion of the cloud contribute in the Bjorken limit only to the longitudinal structure function. In physical terms this implies that the photon is detecting a point like particle of spin zero in that regime.

The pionic correction to the quark wave function contribute in the Bjorken limit solely to the transverse structure functions. Hence the CBM behaves in the deep inelastic regime as if the pions of the cloud were elementary constituents. In order to eliminate this anomalous behavior we have to incorporate the structure of the pion.

The role of the one pion exchange in deep inelastic lepton scattering was analysed by J. Sullivan[44]. He proved that the processes involving the scattering of leptons from nucleons with vertices γNN give rise to contributions that do not scale but vanish in the Bjorken limit. However in the inclusive case the contribution is non vanishing and scales appropriately. In particular the residue of the pion pole is proportional to the structure function of the scattering of leptons from pions. We have applied this analysis to the pions of the cloud in the CBM. Since the high momentum of the virtual photon only flows through the pion in the meson exchange current term to leading order, it will be only modified to incorporate the structure of the pion. The application of the convolution formula leads to

$$F_2^N(x) = \frac{15}{32\pi^2}\left(\frac{\varepsilon}{\varepsilon-1}\right)^2 \frac{M}{f_\pi^2 R}\int_x^1 dy j_0^2(MRx\left(\frac{1-x}{y-x}\right))F_2^\pi\left(\frac{x}{y}\right), \quad (52)$$

where

$$F_2^\pi(z) = F_2^\pi(0)(1-z), \quad (53)$$

and x is the Bjorken variable[3]. From this expression it is immediate to obtain the number of pions, which comes out to be

$$N = \frac{15}{32\pi^2}\left(\frac{\varepsilon}{\varepsilon-1}\right)^2 \frac{M}{f_\pi^2 R}. \quad (54)$$

Note that after having included the structure of the pion not only scaling but also Regge behavior is satisfied.

As can bee seen in [42], once the pion structure is incorporated, the exchange diagram dominates the low x behavior of the F_2 structure function. However not agreement with the data is expected, since we are describing the data by means of a low energy model. In order to connect our calculation with the experimental data we have to evolve the calculated structure functions from a low energy scale to some high energy scale [45]. In this way one incorporates soft gluons and additional sea quarks, thus reducing the enormous valence quark contribution present in our calculation. Preliminary results are quite encouraging [46, 47].

Finally the philosophy just stated, namely to use low energy models to describe phenomena such as confinement, hard gluons, chiral symmetry breaking, etc... and use the evolution equations to incorporate soft processes, may be applied to our simplified models of nuclei. This leads to study contributions to the structure functions such as quark overlaps, clustering, six quark states, quark antisymmetrization, etc.. which should become observable for large momentum transfers in the $x > 1$ region [47].

4.2 The skyrmion

Standard current algebra is described by global $SU(N_f) \times SU(N_f)$, presumably as a result of the underlying $SU(N)$ color gauge interaction. The former symmetry is spontaneously broken to $SU(N_f)$. Standard current algebra can be described by a field $U(x)$ governed by an effective action of the form[10]

$$S = \frac{f_\pi^2}{4} \int d^4x Tr(\partial_\mu U^\dagger \partial^\mu U) + N\Gamma, \tag{55}$$

where Γ is the Wess-Zumino-Witten term

$$\Gamma = -\frac{i}{240\pi^2} \int_{V \times S^1 \times [0,1]} d^5x \varepsilon^{\alpha\beta\gamma\delta\epsilon} Tr(U^\dagger \partial_\alpha U \cdots U^\dagger \partial_\epsilon U). \tag{56}$$

Here

$$U = e^{iT \cdot \pi} \tag{57}$$

where T^a, $a = 1, 2, \cdots, N_f^2 - 1$ are the corresponding **generators of the flavor group** in the fundamental representation. V is the three-space volume, S^1 the compactified time and $[0, 1]$ the extension needed to write a local form for the Wess-Zumino-Witten term.

For any finite energy configuration $U(x)$ must approach a constant at spatial infinity, therefore $U \in \pi_3(SU(N_f))$. But since $\pi_3(SU(N_f)) \approx Z$, there are soliton excitations and they obey an additive conservation law. Actually in order to circumvent Derrick's theorem, higher order terms are required in (55) which stabilize the soliton. In the minimal model[11] we have to add

$$\frac{1}{32e^2} Tr \left[U^\dagger \partial_\mu U, U^\dagger \partial_\nu U \right]^2. \tag{58}$$

Skyrme[11] proposed and it has been recently confirmed[10, 38] that the winding number associated with the solutions of current algebra is the same as the baryon number. This has open a new field of research namely the search for a mesonic theory which describes low energy hadron physics [39, 48].

Let us come back to our two-phase picture and assume that our pion field is a topologically non-trivial soliton : the hedgehog[36, 49]. For the case of two flavors it is

$$U_0 = \exp(i\vec{\tau} \cdot \hat{r}\theta(r)), \tag{59}$$

with $\theta(r)$ constrained to

$$\theta(0) = \pi, \quad \theta(\infty) = 0. \tag{60}$$

What is the response of the quarks inside to this external field? The hedgehog ansatz brings great simplification to the boundary conditions[36, 50]. In order to satisfy them we have to construct quark states with new quantum numbers K, K_z defined from

$$\vec{K} = \vec{J} + \vec{I}, \tag{61}$$

in the usual way. Here $\vec{J}$ represents the total angular momentum and $\vec{I}$ the total isospin. If the valence quarks are in the lowest possible angular momentum states $J^P = \frac{1}{2}^+$, then the lowest energy occurs for $K^P = 0^+$. It is a striking feature of the hedgehog solution that the coupling of quarks and pions is reflected in $\varepsilon = \omega R$. Moreover there is a critical value of the radius for which $\varepsilon \longrightarrow 0$, that is a zero mode appears in the spectrum. Zero modes are associated with peculiar properties of the vacuum in a field theory[51]. What has happened is that the coupling of the pion to the quarks has produced a quark spectrum which is non-symmetrical with respect to positive and negative quark states. This symmetry is recovered again at the critical radius when the lowest positive energy states plunge into the negative energy sea. This lack of symmetry of the spectrum leads to a vacuum carrying baryon number[13, 52, 53]. However it leads to a situation where the interior phase diminishes its baryon number. Where has it gone? Certainly we now know the answer. The topologically non-trivial pion field also carries baryon number and therefore the total baryon number is maintained equal to one.

The procedure just described is the so-called charge fractionization. The bag model was created to confine color. An unwanted result was that it confined many more baryonic properties. Once the pion field is introduced we are free from such oppression. Many technical problems remain : quantization of the soliton, Casimir effects,... [54]. We shall not discuss them here.

4.3 The chiral bag and the Cheshire cat principle

The fact that the confinement boundary condition confines only color leads to what is known as the Cheshire Cat phenomenon. Simply stated the physics should not depends upon the bag radius R. This phenomenon can be rigorously established in $1+1$ dimensions [56] but the situation is not clear at all in the relevant $(3+1)$ dimensions. The reason is that in $(1+1)$ dimensions fermionic theories can be bosonized but in $(3+1)$ dimensions there are no known exact bosonization rules. However large N arguments tend to indicate that bosonization in the latter case is feasible [10].

Suppose we invoke the Cheshire Cat and shrink the bag to $R \leq 0.5$fm. Because of the anomaly induced leakage of the "spin" or more precisely the flavor singlet axial current (FSAC) matrix element of the proton lodged in the bag is small [55]. Since the "spin" leaked out into the mesonic sector disappears into an orbital part, the FSAC cannot detect it, so the total "spin" seen by the FSAC is predicted to be small, ≈ 0.14, at the "magic" radius $R(\Theta = \pi/2)$ [57]. This is small enough to be consistent with the EMC and νP data. How does one explain the increasing "spin" as the bag radius becomes bigger? This is a Cheshire Cat principle question and has been analysed in refs.[58].

In order to answer this question, one has to take into account the axial anomaly [59, 60], which says that even if one ignores the quark masses, the FSAC is not conserved

$$\partial_\mu A^\mu = \frac{N_f \alpha_s}{4\pi} Tr(F_{\mu\nu} \tilde{F}^{\mu\nu}), \tag{62}$$

where α_s is the strong fine-structure constant and $\tilde{F}^{\mu\nu} = \varepsilon^{\mu\nu\lambda\rho} F_{\lambda\rho}$. In the CBM, this is just a piece relevant to the interior of the bag. Outside, current algebra arguments tell us that it is the η' that contributes[60]. To incorporate the two sectors correctly, write[58, 61]

$$A^\mu(x) = A^\mu_B \Theta_B + A^\mu_M \Theta_M, \tag{63}$$

where the subscript $B(M)$ denotes the bag(mesonic) component and Θ's are the appropriate Heaviside functions. With this definition A^μ_B satisfies Eq.(62) and A^μ_M satisfies

$$\partial_\mu A^\mu_M = f_{\eta'} m^2_{\eta'} \eta', \tag{64}$$

(We will drop the prime from now on and write simply η meaning η'.) From the expected structure of the divergence of the FSAC, we infer the boundary condition

$$n_\mu A^\mu_B = n_\mu A^\mu_M, \text{ at } r = R. \tag{65}$$

Because of (color) gauge invariance, this condition specifies to

$$\frac{1}{2}\hat{n} \cdot \bar{q}\vec{\gamma}\gamma_5 q = f_\eta \hat{n} \cdot \vec{\nabla}\eta \text{ at } r = R, \tag{66}$$

which implies that

$$\hat{n} \cdot \vec{A}_{B_{gluon}} = 0 \tag{67}$$

for the gluonic contribution inside, what ever it might be. Equations (62) through (66) lead to the desired matrix element.

$$\frac{1}{2}\mathcal{M} = \langle P| \int_V d^3r \vec{A}(\vec{r})|P\rangle$$
$$= \langle P| \int_{V_B} d^3r \vec{A}_{eff}(\vec{r})|P\rangle, \tag{68}$$

where

$$\vec{A}_{eff} = \frac{3(1+y) + y^2}{2(1+y) + y^2}\left(\frac{1}{2}\bar{q}\vec{\gamma}\gamma_5 q - \frac{n_f \alpha_s}{2\pi}\vec{r}(\vec{E}^a \cdot \vec{B}^a)\right) \tag{69}$$

with $y = m_\eta R$, V the total volume of the proton and V_B the bag volume. $\vec{E}^a$ and $\vec{B}^a$ represent the color electric and color magnetic fields respectively. Note that the boundary condition Eq.(66) has allowed us to replace the outside η contribution by an interior quark contribution with, however, the parameters of the η field. With strictly confined color electric and magnetic fields[58], the two terms in (69) compensate each other to a large extent for all bag radii. For small bag radii, the gluonic contribution is small as one expects from asymptotic freedom and the quarkish contribution is also small because of the leakage mentioned above. As $R \to 0$, $\mathcal{M} \to 0$ which is the known skyrmion result[62]. For a finite bag size, we have to work out the Casimir effects[58] for one and two body operators. For the relevant range of R, i.e., $0 \leq R \leq 1.2$fm, the result is

$$0 \leq \mathcal{M} \leq 0.10 \tag{70}$$

which agrees with the observed value.

5 Concluding remarks

The models we have discussed are substitutes for QCD. They incorporate by fiat some of the proven or expected properties of the theory. If and how these models arise from QCD is not even qualitatively understood. Their phenomenological success serves as the most convincing evidence of their validity.

The mechanism for confinement has not as of yet been disentangled [4]. Chirality seems to be on better grounds [5, 63]. This situation has led us to study a toy model, namely two dimensional QCD. This theory has a trivial confining mechanism, namely a linearly rising two body potential. However a full understanding of the mechanism in more complex situations is not yet fully grasped. Bosonization, which is exact in two dimensions, has proven extremely powerful [12].

We have obtained a description of the spectrum of bosonized QCD_2 by establishing, through the bosonization rules, a correspondence between characteristic bosonic Green functions and the well known fermionic amplitudes. The crucial role of the chiral sector has been enlightened, both for mesons and baryons [12].

The chiral sector of the theory consists solely of zero mass mesons. These are intimately related with a *naive* way of realizing chiral symmetry consistent with Coleman's theorem [64], namely à la BKT [65]. In this way the impossibility of spontaneous symmetry breaking in two dimensions is transformed into a long range order for the various correlators. Parity doublets are non-existent due to explicit symmetry breaking dynamically generated by the color axial anomaly in such a beautiful way as to preserve parity [12]. The excited mesons have their origin in the gluonic interactions, and therefore arise in the color sector, not in the chiral sector.

The interpretation of the baryonic spectrum in the bosonized theory relies on the existence of the zero mass baryons. The chiral and color sectors of bosonized QCD_2 are completely decoupled and the former is trivial. However to describe baryon number, boundary conditions have to be admitted, which render the theory topologically non-trivial [66, 67]. Still, although there exist baryons in the chiral sector, they are trivial free quark states confined by the dimensionality of space-time [68]. However their presence is crucial to describe the remaining non-trivial spectrum associated with the color sector, i.e., with the gluonic interactions. These baryonic excitations arise as pseudo-boundstates of mesonic character on top of the massless baryons. The massless states exist to provide the baryon number to the massive baryons. The massless baryons are the supporting soliton of the baryonic spectrum.

It is important to stress, that the chiral sector is completely decoupled from the color sector and therefore no gluonic interactions are present in it. It is the flavor structure of the vacuum and the dimensionality of space-time which provide the scenario which determines the meson (baryon) dynamics. This realization of the chiral sector is a two-dimensional image of the corresponding realization in four dimensions, where one obtains an effective lagrangian in terms of the chiral fields (the pseudo-scalar mesons) only, by omitting explicit gluonic interactions [69]. On the other hand in the interpretation of the excited meson and baryon spectrum the color sector plays a crucial role. In the former, the purpose of this investigation, the gluonic interactions bind the pseudo-mesons giving the baryons their mass. This reflects also an image of the real world, when gluonic effects or higher mesons are considered into the effective schemes [70, 71].

It is worth looking at our results with the perspective of time. In the beautiful work of Amati et. al. [68] it is stated that the picture that arises in the massless theory, both for the spectrum as for chiral symmetry realization is too naive. Their analysis was carried out only in the chiral sector. Once the color sector is incorporated, the richness of the theory is fully grasped, but moreover the chiral sector, although naive, turns out to be crucial. By unveiling this behavior one is capable of establishing a connection with the real world, which leads to an understanding of some of the low energy approximations to the four dimensional theory.

Finally low energy models of four dimensional QCD using chiral lagrangians which describe the baryon spectrum, the Skyrme models [11, 10], correspond in the two-dimensional world just to the chiral sector. Its dynamics turns out to be trivial, beyond the acceptance of topological non-trivial boundary conditions. This correspondence is certainly more transparent if one includes mass [72]. The present analysis is connected with the bosonization beyond the chiral sector. Our conclusion is that higher mesonic and baryonic excitations result fundamentally from color dynamics,i.e., gluons play a crucial dynamical role. However if one integrates out the gluonic degrees of freedom, the Green functions may be described in terms of color singlet states which are the meson excitations and "boundstates" of mesons and solitons [73].

The road leading from QCD_2 to the true theory is very steep. In the meantime models represent the closest we have to QCD in the low energy regime. Although they have proven extremely successful, in my opinion, their power lies not so much there, but in their usefulness to expose new phenomena of interacting hadronic systems.

References

[1] H. Fritsch, M. Gell-Mann and H. Leutwyler, Phys. Lett. **47B** (1973) 365; W. Marciano and H. Pagels, Phys. Rep. **36C**(1978) 137; F. Yndurain, *Quantum Chromodynamics*, Springer Verlag, Heidelberg(1983).

[2] K. Huang, *Quarks, Leptons and Gauge Fields*, World Scientific, Singapore(1982).

[3] F. E. Close, *An Introduction to Quarks and Partons*, Academic Press, London(1979); P. Pascual and R. Tarrach, *Lecture Notes in Physics*, **194**, Springer Verlag, Heidelberg(1984).

[4] J. B. Kogut, Rev. Mod. Phys. **51**(1979) 1802; G. K. Savvidy, Phys. Lett. **71B**(1977) 133; N. K. Nielsen and P. Olesen, Nucl. Phys. **B144** (1978) 376; F. Zachariasen, Mod. Phys. Lett. **1** (1986) 255.

[5] G. 't Hooft, in *Recent developments in Gauge Theories*, G. 't Hooft et al. eds. (Plenum Press, N.Y., 1980).

[6] S.L. Adler, Phys. Rev. **177** (1969) 2426; J.S. Bell and R. Jackiw, Nuovo Cimento **A51** (1967) 47.

[7] D. Weingarten, Phys. Rev. Lett. **51** (1983) 1830.

[8] C. Vafa and E. Witten, Nucl. Phys. **B234** (1984) 173.

[9] G. 't Hooft, Nucl. Phys. **B72** (1974) 461; Nucl. Phys. **B75** (1974) 461

[10] E. Witten, Nucl. Phys. **100** (1979) 57; Nucl. Phys. **B223**(1983) 423; *ibid* 433

[11] T. H. R. Skyrme, Proc. Roy. Soc. London, **A260**(1961) 127; Nucl. Phys. **31**(1962) 556.

[12] A. Ferrando and V. Vento, Phys. Lett. **B256** (1991) 503; Universitat de Valéncia Preprint FTUV91-21 (to be published in Phys. Lett. **B**; A. Ferrando, Ph. D. Thesis, Universitat de Valéncia 1991.

[13] R. F. Alvarez-Estrada, F. Fernández, J. L. Sánchez Gómez and V. Vento, *Models of Hadron Structure based on Quantum Chromodynamics*, Lecture Notes in Physics 259, Springer-Verlag, Heidelberg (1986).

[14] A. De Rújula, H. Giorgi and S.L. Glashow, Phys. Rev. **D12** (1975) 147.

[15] A.J.G. Hey and R.L.Kelly, Phys. Rep. **96** (1983) 72.

[16] J. Navarro and V. Vento, Nucl. Phys. **A440** (1985) 617; M.V.N. Murhty and R.K. Badhuri, Phys. Rev. Lett. **54** (1985) 745; H. Ito and A. Faessler, Nucl. Phys. **A470** (1987) 626.

[17] R.P. Feynman, M. Kislinger and F. Ravndal, Phys. Rev. **D3** (1971) 2706.

[18] H. Toki, Y. Suzuki and K.T. Hecht, Phys. Rev. **C26** (1982) 736; K. Maltman, Nucl. Phys. **A439** (1985) 648; S. Takeuchi, K. Shimizu and K. Yazaki, Nucl. Phys. **A449** (1986) 617 ; P. González, V. Sanjosé and V. Vento, Phys. Lett. **B196** (1987) 1; P. Hoodbhoy, Nucl. Phys. **A465** (1987) 637; T. De Forest and P.J. Mulders, Phys. Rev. **D35** (1987) 2849.

[19] P.González and V. Vento, Few Body Systems **2** (1987) 145.

[20] P. González and V. Vento, Nucl. Phys. **A501** (1989) 710

[21] P. González and V. Vento, Nucl. Phys. **A485** (1988) 413.

[22] A. Chodos, R. L. Jaffe, K. Johnson, C. B. Thorn and V. F. Weisskopf, Phys. Rev. **D9** (1974) 3471.

[23] A. Chodos, R. L. Jaffe, K. Johnson and C. B. Thorn, Phys. Rev. **D10** (1974) 2599.

[24] T. deGrand, R. L. Jaffe, K. Johnson and J. Kiskis, Phys. Rev. **D21**(1975) 2060.

[25] T. D. Lee, Phys. Rev. **D19**(1979) 1802.

[26] J. D. Breit, Nucl. Phys. **B202**(1982) 147.

[27] O. V. Maxwell and V. Vento, Nucl. Phys. **A407**(1983) 366.

[28] T. H. Hansson and R. L. Jaffe, Phys. Rev. **D28**(1983) 882 ; Ann. Phys. **151**(1983) 204.

[29] K. Johnson, Acta Physica Polonica **B6**(1975) 865.

[30] R. L. Jaffe, *Proceedings of the Ettore Majorana School of Subnuclear Physics*, Ed. A. Zichichi, Bologna(1979).

[31] P. González and V. Vento, *Proceedings of the VIIIth Autumn School on Few Body Problems*, Ed. A. Fonseca, Springer Verlag, Heidelberg (1987).

[32] Quark Matter '84, Proceedings of the forth International Conference on Ultrarelativistic Nucleus-Nucleus Collisions, Lecture Notes in Physics **221** (Springer, Heidelberg 1985)

[33] K. Brecher and G. Caperaro, Nature **259** (1976) 377; G. Baym and S.A. Chin, Phys. Lett. **62B** (1976) 241; G. Chapline and M. Nauenberg, Nature **264** (1976) 235.

[34] L.V. Hove, Nucl. Phys. **A447** (1985) 443.

[35] J.C. Collins and M. J. Perry, Phys. Rev. Lett. **34** (1975) 1353; V. Baluni, Phys. Lett **72B** (1978) 381; Phys. Rev **D17** (1978) 2092; B.A. Freedman and L.D. McLerran, Phys. Rev **D16** (1978) 1130; ibid 1147; ibid 1169; Phys. Rev. **17** (1978) 1109.

[36] G. E. Brown and M. Rho, Phys. Lett. **82B**(1979) 177; G. E. Brown, M. Rho and V. Vento, Phys. Lett. **84B**(1980) 383; V. Vento, M. Rho, E. M. Nyman, J. H. Jun and G. E. Brown, Nucl. Phys. **A345**(1980) 413.

[37] P. González and V. Vento, Nucl. Phys. **A415**(1984) 413.

[38] J. Goldstone and F. Wilczek, Phys. Rev. Lett. **47**(1981) 986.

[39] G. S. Adkins, C. P. Nappi and E. Witten, Nucl. Phys. **B228**(1983) 552.

[40] R. L. Jaffe, Phys. Rev. **D11**(1975) 1953.

[41] R. L. Jaffe and A. Patrosciou, Phys Rev. **D12**(1975) 1314.

[42] V. Sanjosé and V. Vento, Phys. Lett. **225B**(1989) 15 and Nucl. Phys. **A501** (1989) 672.

[43] G. B. West, Phys. Rep. **18C**(1975) 264.

[44] J. D. Sullivan, Phys. Rev. **D5**(1972) 1732.

[45] A. González-Arroyo et al., Nucl. Phys. **B153** (179) 161; Nucl. Phys. **159** (1979) 51; Nucl. Phys. **B166** (1980) 429.

[46] A.W. Thomas, Hadrons and Hadronic matter, D. Vautherin et al. eds. (Plenum Press, N.Y. 1990).

[47] P. González and V. Vento (work in preparation)

[48] I. Zahed and G. E. Brown, Phys. Rep. **142**(1986) 1.

[49] A. Chodos and C. B. Thorn, Phys. Rev. **D12**(1975) 2733.

[50] V. Vento and M. Rho, Nucl. Phys. **A412**(1984) 413.

[51] R. Jackiw and C. Rebbi, Phys. Rev. **D13**(1976) 3398.

[52] M. Rho, A. S. Goldhaber and G. E. Brown, Phys. Rev. Lett. **51**(1983) 747.

[53] J. Goldstone and R. L. Jaffe, Phys. Rev. Lett. **51**(1983) 1518.

[54] G. E. Brown, A. D. Jackson, M. Rho and V. Vento, Phys. Lett. **B140**(1984) 285; P. J. Mulders, Phys. Rev. **D30**(1984) 1073.

[55] B.-Y. Park and M. Rho, Z. Phys. **A331**(1988) 151.

[56] S. Nadkarni, H. B. Nielsen and I. Zahed, Nucl. Phys. **B253**(1985) 308; R. Perry and M. Rho, Phys. Rev. **D34**(1986) 1169.

[57] M. Rho, G.E. Brown and B.-Y. Park, Phys. Rev. **C39** (1989) 1173.

[58] B.-Y. Park, V. Vento, M. Rho and G. E. Brown, Nucl. Phys. **A504** (1989) 829; B.-Y. Park and V. Vento, Nucl. Phys. **A513** (1990) 413.

[59] G. Altarelli and G. Ross, Phys. Lett. **B212** (1988) 391; R.D.Carlitz, J.C. Collins and A.H. Mueller, Phys. Lett. **B214** (1988) 229.

[60] C. Rosenzweig, J. Schechter and C. G. Trahern, Phys. Rev. **D21**(1980) 3388; E. Witten, Ann. Phys. **128**(1980); P. di Vechia and G. Veneziano, Nucl. Phys. **B171**(1980) 253.

[61] T. Hatsuda and I. Zahed, Phys. Lett. **B221**(1989) 173; H. Dreiner, J. Ellis and R. A. Flores, Phys. Lett. **B221**(1989) 167.

[62] S. Brodsky, J. Ellis and M. Karliner, Phys. Lett. **B206**(1988) 309.

[63] D.I. Dyakonov and Yu.V. Petrov, Nucl. Phys. **B272** (1986) 457; D. Ebert and H. Reinhardt, Nucl. Phys. **271**(1986) 188.

[64] S. Coleman, Comm. Math. Phys. **31** (1973) 259.

[65] V.L. Berezinskii, Sov. Phys. JETP **32** (1970) 493; ibid **34** (1971) 610; J.M. Kosterlitz and D.J. Thouless, J. Phys. **C6** (1973) 1181.

[66] M.B. Halpern, Phys. Rev. **D12** (1976) 1684

[67] H. Galić, *Alchemy in 1+1 dimension: from bosons to fermions*, SLAC-PUB-5381 (1990).

[68] D. Amati, K.-C. Chou and S. Yankielowicz, Phys. Lett. **110B** (1982) 309.

[69] D. Espriu, E. de Rafael and J. Taron, Nucl. Phys. **B345** (1990) 22.

[70] A. Pich and E. de Rafael, CERN-TH.5906/90 (to be published in Nucl. Phys. **B**); G. Ecker, J. Gasser, H. Leutwyler, A. Pich and E. de Rafael, Phys. Lett. **B223** (1989) 425; G. Ecker, J. Gasser, A. Pich and E. de Rafael, Nucl. Phys. **321B** (1989) 311.

[71] U. Meissner, Phys. Rep. **161** (1988) 213.

[72] I. Affleck, Nucl. Phys. **B265** (1986) 409; ibid 448; Y. Frishman and J. Sonnenschein, Nucl. Phys. **B294** (1987) 801.

[73] C.G. Callan and I. Klebanov, Nucl. Phys. **B262** (1985) 365.

Chiral Models of Low Energy QCD

Georges Ripka

Service de Physique Théorique
Laboratoire de la Direction des Sciences
de la Matière du Commissariat à l'Energie Atomique
Centre d'Etudes de Saclay, F-91191 Gif-sur-Yvette Cedex, France

Abstract: Two processes may be distinguished when a hadron propagates in a dense baryonic medium: the polarization of the medium and the change in the quark structure of the hadron. The polarization of the medium is better described in terms of colorless mesons and nucleons while the intrinsic change of the hadron is better described by quark models. We show how to couple the two processes. We also relate the scaling of effective lagrangians, based on the QCD anomaly, to changes in the quark constituent masses.

1. Ambiguity and compatibility in chiral quark models.

Today, quark models are fashionable and flourishing. They are used to describe *inter alias* nucleons, chiral symmetry breaking and the variation of hadron masses in dense and hot nuclear matter. And, of course, people are excited because it is all so successful. Some fifteen years earlier meson physics was flourishing. It was used *inter alias* to discuss PCAC, $\pi\pi$ scattering and the variation of nucleon masses in dense and hot nuclear matter. And, of course, people were quite excited because it was so successful. Still fifteen years earlier the hamiltonian was $H = T + V$, a sum of kinetic energy and 2-body interactions. It was used to describe *inter alias* nuclear binding, nuclear shapes and effective masses of nucleons propagating in nuclear matter. And, of course, people were very excited because it was so successful.

Are we simply restating the same physics in different terms? The question is more acute today than it was yesterday because the *dramatis personnae* seem to have changed. Although low energy QCD is, for the most part, practised without gluons, quarks are not only present, but they often replace nucleons without anybody really noticing or caring.

The π^0 electromagnetic decay rate.

One famous example is the pion electromagnetic decay rate usually described in your local field theory book [1] by the process:

$$\pi_0 \text{ --- } \langle \text{ loop } \rangle \text{ === } \gamma, \gamma \qquad (1.1)$$

It turns out that, whether you calculate it with quarks or nucleons in the fermion loop, you get the same answer which even fits experiment. Which of the two descriptions is correct? The quark one, obviously, since quarks are the fundamental constituents. Why, you even get an independent estimate of the number of colors (N_c=3 is obtained). The nucleon description is obviously better because the quark description is ambiguous: it is willing to admit quarks in the virtual intermediate states but it forgets that the pion itself is a $q\bar{q}$ excitation. The graph has the anomalous feature of not depending on the fermion (quark or nucleon) mass provided a Goldberger Treiman relation is used. The result is thus insensitive to quark confinement. Furthermore one cannot complain that one is making a perturbative QCD calculation for a low energy process because one never uses the quark-gluon running coupling constant. It is all buried in the π-nucleon coupling constant. In some way, the calculation of the pion decay rate is right in either description.

One might feel happy, if uneasy, about the fact that both descriptions of the pion decay give the same result and claim that the whole thing is a non-problem. But is it always so that one can replace nucleons by quarks and get the same answer? Suprisingly, it is often almost true, but never exactly so and sometimes downright wrong.

Spontaneous chiral symmetry breaking in the vacuum.

As a next example, consider spontaneous chiral symmetry breaking in the vacuum. The idea, originally due to Nambu and Jona-Lasinio [2] is deceivingly simple. Imagine quarks had a mass m usually called the constituent quark mass. The Dirac sea would then consist of quarks in negative energy plane wave states and its energy would be $-\nu \sum_{k<\Lambda} \sqrt{k^2 + m^2}$ where ν is the degeneracy of the plane wave states ($\nu=12$ for u and d quarks) and Λ is a suitable cut-off of unknown origin, introduced to avoid the infinity in the sum. The Dirac sea energy decreases when the constituent mass increases and, were they free to do so, quarks would acquire infinite mass m and lower the vacuum energy indefinitely. Massive quarks in a Dirac sea break the chiral symmetry of the vacuum and this is the essence of the Nambu process. Since the Dirac sea energy decreases at most linearly with m, Nambu added a chiral symmetry restoring term to the energy, which is quadratic in m. The energy per unit volume is then:

$$\frac{E_{vacuum}}{\Omega} = -\frac{\nu}{\Omega} \sum_{k<\Lambda} \sqrt{k^2 + m^2} + \frac{a^2}{2} m^2 \qquad (1.2)$$

and the vacuum constituent quark mass is the value of m which minimizes the energy (1.2) as shown by the full line of Figure 1.

In fact, this is not exactly what Nambu said. The important difference is not the setting. (He used a 4-fermion interaction which generated m in a Hartree approximation but that is only a formal difference.) The difference is that he said all this before quarks were invented and so he filled the Dirac sea with nucleons instead of quarks. Today we use the same model with quarks instead of nucleons [3-7] and we call it low-energy QCD. The only change is that now the degeneracy is $\nu = N_c \times 4 = 12$ whereas Nambu used $\nu=4$ for his nucleons. Accordingly, Nambu assumed that the vacuum value of m was the 938 MeV nucleon mass. But now that ν has increased from 4 to 12 we require the vacuum value of the quark constitutent mass m to be (very roughly) N_c times smaller, that is, 300 MeV, a third of the nucleon mass. Indeed it takes $N_c=3$ quark masses to substitute each nucleon mass.

Antiquarks in constituent quark models.

We consider 300 MeV to be a "reasonable value" because it is close, although higher, to the value 220 MeV used in constituent quark models [8]. The latter confine quarks by making them interact with confining potentials and they are therefore incompatible with the idea of a Dirac sea of quarks. Why should quarks feel free to float around in plane-wave states ignoring confinement when they are in the Dirac sea? Could it be because, with a roughly 1 GeV cut-off Λ, they form a baryon soup about 64 times more dense than normal nuclear matter, which has a Fermi momentum $k_F = 1/4$ GeV ? Attempts to mine the Dirac sea have always produced mesons and anti-protons but never free anti-quarks. In fact the mini-hadronization process required to produce a meson or an anti-proton has not, to my knowledge, been studied much. In constituent quark models a Dirac sea cannot exist as such and anti-quarks are introduced as distinct particles which happen to have the mass (of unspecified origin) and other quantum numbers of Dirac sea quarks. But no Dirac sea is mentioned.

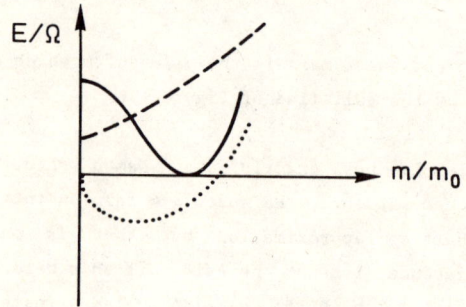

Fig.1: Contributions to the energy per unit volume. The full line is the Dirac sea energy (1.2) given by the Nambu Jona-Lasinio model, the lower dotted line is the vacuum energy (3.8) of the gluonium field which reproduces the QCD scale anomaly, and the upper dashed line is the Fermi sea contribution (1.3).

Partial chiral symmetry restoration at finite baryonic density.

The same quark/nucleon ambiguity is encountered when one considers partial chiral symmetry restoration at finite baryonic density

[9,10,5,11,12]. If a quark Dirac sea is used to describe the chirally broken phase of the vacuum, then certainly a Fermi sea of quarks should be used to describe nuclear matter at finite density. This has often been assumed [5,10]. However, who believes that nuclear matter is a Fermi sea of quarks? Does it matter whether we put nucleons or quarks in the Fermi sea? The process of partial chiral symmetry restoration is as simple as Nambu's spontaneous chiral symmetry breaking in the vacuum. It is the same process, whether you use the linear σ-model, the Nambu Jona-Lasinio model or Walecka's σ-ω model. Nuclear matter is assumed to be a Fermi sea of nucleons, but allow, for the time being, the theorist to choose quarks if he insists. Assume simply a Fermi sea of fermions of mass m. The Fermi sea contribution to the energy is then $\nu \sum_{k<k_F} \sqrt{k^2 + m^2}$. Since k_F is about 1/4 GeV and m usually higher, the Fermi sea energy is roughly a linearly increasing function of m (dashed curve of Fig.1) which should be added to the vacuum energy (1.2). The result is that the minimum energy occurs for a value $m^* < m$ thus partially restoring the chiral symmetry of the vacuum. The vacuum energy (1.2) has different interpretations in various models, but the process of partial chiral symmetry restoration is always the same. Indeed, in the linear σ-model, the Nambu energy (1.1) is represented by the wine bottle shaped potential $\frac{\kappa^2}{8}(\sigma^2 + \pi^2 - f_\pi^2)^2$. In the Walecka σ-ω model, it is represented by the σ-meson mass term which is nothing but the quadratic approximation of the σ-model potential, namely $\frac{1}{2} m_\sigma^2 (\sigma - f_\pi)^2$ with $m_\sigma = \kappa f_\pi$.

Models, like Nambu's, which use a Dirac sea of quarks, may be formulated in such a way as to allow for a Fermi sea of either quarks or nucleons and to avoid any double counting. This way, the partial chiral symmetry restoration produced by quarks or nucleons in the Fermi sea may be compared [7,13]. What is compared, in fact, is the difference between the two sums:

$$E_{Fermi} = \nu_q \sum_{k<k_F} \sqrt{k^2 + m_q^2} \quad \text{(in the case of quarks)} \quad (1.3a)$$

and:

$$E_{Fermi} = \nu_N \sum_{k<k_F} \sqrt{k^2 + m_N^2} \quad \text{(in the case of nucleons)} \quad (1.3b)$$

where the indices q and N refer to quarks and nucleons respectively. Notice that the same Fermi momentum k_F is required for either quarks or nucleons in order to obtain a given baryonic density. Again, the main difference is between the nucleon degeneracy $\nu_N=4$ and the quark degeneracy $\nu_q=12$. Notice also that, to the extent that the momenta k in the sums can be neglected compared to m, the quark and nucleon contributions to the energy become equal when:

$$m_N = N_c m_q \qquad (1.4)$$

as in constituent quark models. Eq.(1.3) would imply constituent quark masses of about 300 MeV. A more precise evaluation shows that the quark and nucleon Fermi sea energies are about the same for constituent quark masses of about 400 MeV [13]. So our crude estimates are not so far off.

Modification of the nuclear medium by a propagating meson.

Mesons are $\bar{q}q$ excitations of the vacuum. If we have a model to build such excitations, we should be able to calculate how their structure is modified in a medium. Such calculations have been pursued in the framework of the Nambu Jona-Lasinio model for example [5,7,13,14]. It has also been speculated that all meson and baryon masses are reduced by the same factor at a given density [15,16]. Relative variation of the Φ and K masses could explain the relative production rate of these mesons as observed in dilepton production experiments [17]. The calculations run into the same nucleon/quark ambiguities we have discussed above. In the Nambu Jona-Lasinio model the meson is represented by propagating and interacting $\bar{q}q$ pairs:

$$\qquad (1.5)$$

These processes are summed by a Bethe-Salpeter equation and, in the Nambu Jona-Lasinio model, they can desribe a *bound* $\bar{q}q$ state, but not a *confined* $\bar{q}q$ pair. In the vacuum, the $\bar{q}q$ pairs are quark-antiquark excitations in which Dirac sea quarks are excited into positive energy orbits:

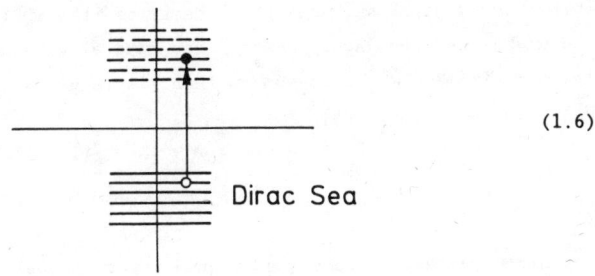

(1.6)

In nuclear matter, there is a Fermi sea so that new excitations can occur as shown in (1.7). The first kind involves the excitations (1.7a) of the Fermi sea. The second kind involves excitations (1.7b) of Dirac sea quarks, but these excitations differ from the vacuum excitations (1.6) by a Pauli blocking which forbids the excitation (1.7c) of Dirac sea quarks into the Fermi sea:

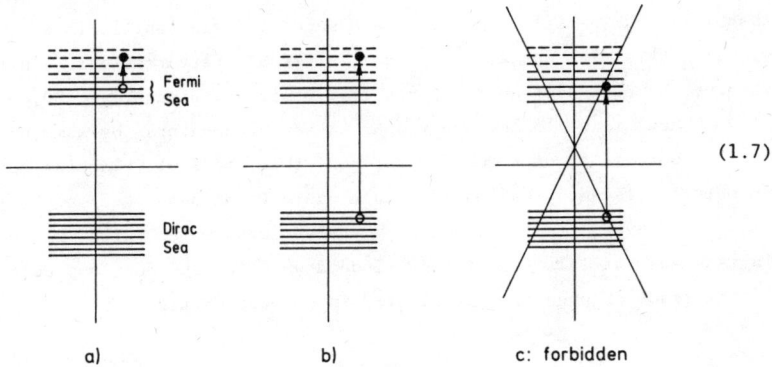

(1.7)

We are immediately faced with more problems and ambiguities. The process (1.7a) simply represents the excitation of the Fermi sea by the propagating meson. These excitations are better described when nucleons fill the Fermi sea (and are allowed to become deltas). The reason for this is that only a small energy gap separates filled and empty orbits in the Fermi sea. As a consequence, intermediate states occur which are degenerate with the energy transfered by the meson and the meson can be absorbed. This leads to an imaginary part in the meson mass operator. The threshold for meson absorbtion depends on the masses of the colorless real hadrons and these would be hopelessly wrong if the Fermi sea were represented by quarks. The process (1.7b) describes excitations of the Dirac sea by the meson. The process (1.7c) corrects for the $q\bar{q}$ excitations of the Dirac sea which normally occur in the vacuum but which are are forbidden in nuclear matter by the Pauli principle. It is referred to as *Pauli blocking*.

Processes such as (1.7) describe the modification of the nuclear medium by the propagating meson. They have already been extensively studied in the framework of meson-nucleon lagrangians [18-25] especially in Walecka's σ-ω model [26]:

$$\mathcal{L} = \bar{N}\left(i\partial_\mu\gamma^\mu - g_\sigma\sigma - g_\omega\omega_\mu\gamma^\mu\right)N + \frac{1}{2}(\partial_\mu\sigma)(\partial\sigma^\mu) - \frac{1}{2}m_\sigma^2\sigma^2 - \frac{1}{4}F_{\mu\nu}F^{\mu\nu} - \frac{1}{2}m_\omega^2\omega_\mu\omega^\mu \tag{1.8}$$

Mesonic excitation modes and liquid-vapor transitions of nuclear matter have been calculated this way. Also, instabilitites have been found, both of the vacuum and of nuclear matter. Vacuum instabilities do not seem to appear in the Nambu Jona-lasinio model but they do occur in the σ-model [27,28] and in meson-nucleon calculations of nuclear matter [18,22]. They have been coined as *tachyon* modes [29] and it is also sometimes said that their occurence makes the theory *acausal*. These are unfortunate terms because the instabilities in no way imply that the model predicts particles travelling faster than light nor that causality is lost. They simply imply that, in a specific model, the assumed translationally invariant state vacuum state is unstable against external perturbations. This occurs whenever a meson self-energy turns from positive to negative [30].

Now if we calculate vacuum meson propagators by summing the $\bar{q}q$ excitations (1.6) of a quark Dirac sea, then the Pauli blocking correction is only consistently evaluated with a quark Fermi sea.

However Pauli blocking evaluations made with meson-nucleon lagrangians are also consistent. Indeed when the nucleon loop contribution to the meson propagator is evaluated in nuclear matter:

$$\tag{1.9}$$

only the *change* in the propagator is actually evaluated, the vacuum contribution is included in the lagrangian mass term and it can be subtracted off by renormalisation.

Modification of the meson structure in dense matter.

Another process contributes to the modification of the meson propagators. It is due to the reduction of the constituent quark mass in dense matter, as seen above. In the Nambu Jona-Lasinio model this mass reduction corresponds to a reduction of the quark condensate $\langle \bar{q}q \rangle$. Such processes are not included in the meson-nucleon lagragians used to calculate the modification of the nuclear medium. Indeed the quark propagators used in summing the $\bar{q}q$ excitations (1.5) are dressed by a scalar field insertion:

(1.10)

which produces the constituent quark mass and this constituent quark mass varies with density. Consider the process in which a meson is exchanged between nucleons:

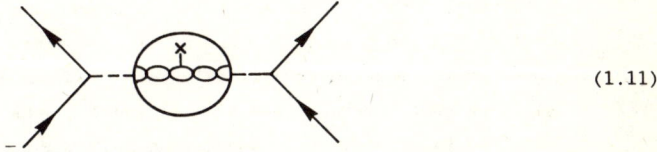
(1.11)

The circle is a blow-up displaying the $\bar{q}q$ nature of the exchanged meson. The quark constituents of the meson feel the medium by interacting with the scalar field as in (1.10). Meson masses vary linearly with the constituent quark mass [13]. So we should modify the mass terms of meson-nucleon lagrangians accordingly. Lagrangians such as (1.8), possibly supplemented with other meson fields, then take the form:

$$\mathcal{L} = \bar{N}\left(i\partial_\mu \gamma^\mu - g_\sigma \sigma - g_\omega \omega_\mu \gamma^\mu - \ldots \right)N + \frac{1}{2}(\partial_\mu \sigma)(\partial \sigma^\mu) - \frac{1}{2}\left(\frac{\varphi}{\varphi_0}\right)^2 m_\sigma^2 \sigma^2$$

$$- \frac{1}{4} F_{\mu\nu}F^{\mu\nu} - \frac{1}{2}m_\omega^2 \left(\frac{\varphi}{\varphi_0}\right)^2 \omega_\mu \omega^\mu + \ldots \quad (1.12)$$

where φ/φ_0 is the relative modification of the constituent quark mass in the medium. The modified meson-nucleon lagrangian (1.12) is similar to the one which is obtained by making the lagrangian scale-invariant. This is discussed in section 3. The process (1.11) is the physical process which is mocked up by the scaling.

In several of the examples discussed above, quarks actually complicate the physics more than they simplify it so is it really essential to use them? The fact that mesons are quark $q\bar{q}$ excitations (rather than $N\bar{N}$ and $\wedge\bar{\wedge}$) has been settled long ago. So we must face the fact that the quark and gluon structure of hadrons can be modified in a dense or hot medium. The problem is to avoid ambiguities and double counting when we take into account both the modification of the internal structure of the meson and the modification of the nuclear medium caused by the propagation of the meson. Perhaps we should endeavor to reconcile the quark structure of hadrons with the meson-nucleon lagrangians. But even there ambiguities remain. For example, any quark model will predict that quark exchange occurs [59] when the hadrons overlap. How are such processes parametrized in effective meson-nucleon lagrangians? And should they?

One sure way of *not* finding answers to these questions is to claim *urbi et orbi* that the model one happens to be using is dictated by QCD and $1/N_c$ expansions and to ignore all the other approaches on the basis that they are heretical. I was once privately told by an eminent theorist from CERN, that I was not even *allowed* to talk about quarks and mesons. I was only allowed to talk about quarks and gluons *or* about colorless hadrons, but not about both at the same time. Suprised, I asked him why and he told me that I should understand this as a frenchman, since in France we have an Academy which also tells us what we are allowed to ...

2.Chiral quark models of baryons in dense media.

Further problems appear when chiral models of baryons are considered. Bag models are discussed in other lectures. I wish to point out however that bag models offer an alternate mechanism to the Nambu process of spontaneous chiral symmetry breaking. One can speculate that the QCD ground state is a condensate of bound colorless gluon 0^{++} states called glueballs [31], in which light quarks can be trapped forming bag-like cavity modes. The light quarks break the chiral symmetry in the vacuum by forming "shallow Dirac seas" in the glueballs [32]. I quote these works to stress that the Nambu process is far from being the only candidate, be it the simplest one. For example, it is not even obvious that the Nambu Jona-Lasinio lagrangian does a better job than, say, a σ-model lagrangian to describe quarks propagating in the QCD vacuum.

Others have speculated that the QCD vacuum is an instanton liquid [33]. Chiral symmetry is broken by the propagation of quarks in this medium. Other approaches have been suggested, based on a parametrization of the quark 2-point functions in the QCD ground state [34,35], or, still, on quarks propagating in a color-dielectric medium [36]. Our lack of knowledge of the nature of the QCD ground state (even in the absence of quarks) forbids us to be too doctrinary about effective theories and about the meaning of their parameters.

We have less trouble in predicting the modifications a nucleon undergoes in a dense medium than in understanding the nature of the medium or the existence of the baryon. This is true to the extent that we can represent the medium by a scalar field σ. It is usually assumed that it is the prevailing average value $\langle\sigma\rangle$ of the scalar field in the medium that determines the modification of baryons in a nuclear medium. It is well to remember that this is a low density approximation. As discussed above most models predict a reduction of $\langle\sigma\rangle$ in dense nuclear matter. The baryon mass and the inverse baryon size scale as $\langle\sigma\rangle$ so that baryons are expected to decrease their mass and to increase their size as the density increases [9]. In fact this modification of baryon mass is the exact analogue of the meson mass modification discussed in section 1. It is entirely determined by the curve (Fig.1) representing the energy per unit volume as a function of $\langle\sigma\rangle$, and, at densities not too close to the critical density at which chiral symmetry is restored, only the curvature of this curve at its minimum counts. A more significant parameter is the constituent mass $g\langle\sigma\rangle$ where g is the coupling constant of the σ-field to the fermions.

Quarks bound by hedgehog shaped chiral fields.

In chiral quark models, the nucleon is usually represented as a *bound* (and *not* confined) state of quarks in a chiral field of hedgehog shape. From the very first calculations [37,38], using the linear σ-model lagrangian [39]:

$$\mathcal{L} = \bar{q}\left(i\partial_\mu\gamma_\mu + g(\sigma+i\gamma_5\vec{\pi}.\vec{\tau})\right)q + \frac{1}{2}(\partial_\mu\sigma)^2 + \frac{1}{2}(\partial_\mu\vec{\pi})^2 - \frac{\kappa^2}{8}\left(\sigma^2+\vec{\pi}^2 - f_\pi^2\right)^2 \quad (2.1)$$

it became clear that a coupling constant $g \gtrsim 3.5$ was required to form a bound state [37,38]. This means that constituent quark masses $gf_\pi \gtrsim 325$ MeV are required to form bound states. The soliton ressembles more a Skyrmion than a Friedberg-Lee soliton [40] in that the chiral field stays close to the chiral circle, making the last term of (2.1) not important. Solitons were also obtained in the presence of the ω,ρ,and a_1 fields [41] using the Lee and Nieh lagrangian [42]. In this case a soliton is obtained with a slightly weaker coupling of the scalar mesons to the quarks because considerable attraction is obtained from the ρ and a_1 fields. The coupling constants of the ρ and ω fields to the quarks is of the order of 4-5, considerably larger than those deduced from the Nambu Jona-Lasinio model [43,44,13].

Problems arise when the Dirac sea orbits are included. These are related to the vacuum instability of the renormalized σ-model against the formation of high gradients [27,28]. When the Dirac sea is included, the renormalized soliton energy has no lower bound. This prevents one from making a variational calculation of the soliton energy and makes it difficult to evaluate small changes in the shape of a soliton embedded in a dense medium.

The high gradient instabilities do not occur in the Nambu Jona-Lasinio model [2], described by the lagrangian:

$$\mathcal{L} = \bar{q}(i\partial_\mu \gamma_\mu - \varphi U_5)q - \frac{a^2 \varphi^2}{2} \qquad (2.2)$$

with $U_5 = \exp(i\gamma_5 \vec{\theta}\cdot\vec{\tau})$. (See the lectures of Ref.[3] for an introduction). However, the model forms only weakly bound states of quarks (solitons) with hedgehog shaped fields [45-47]. The reason is that, in this model, the Dirac sea yields a greater repulsion to the hedgehog field than is suggested by the kinetic term of the σ-model lagrangian (2.1). Soliton sizes are too small for the gradient expansion to be valid [48,49] so that the assimilation of the NJL model to a linear σ-model [14] is dangerous. The weak binding makes it unreliable to evaluate distortions of the soliton embedded in a dense medium so that we are not much better of than in the σ-model with the Dirac sea included.

The main difference between the σ-model (2.1) and the Nambu Jona-Lasinio model (2.2) is the absence of a kinetic term for the meson fields which are pure $\bar{q}q$ excitations. The potential term of the linear σ-model keeps the fields on the chiral circle and does not affect significantly the nucleon properties [50] so that nucleon properties do not change much in the limit κ → ∞ of the non-linear σ-model. This also means that κ is not a parameter on which nucleon properties can be fitted. If the pion decay constant f_π is fixed to its measured value 93 MeV, then the nucleon properties depend only on one parameter, the coupling constant g. It is found [50] that with values g ≈ 5-6 nucleon masses of about 1 GeV are obtained and that the axial coupling constant is g_A ≈ 1.5, closer to the non-relativistic quark model value than to the Skyrme type model prediction of about 0.7-0.9 [51]. But g_A may be adjusted by adding an extra term to the lagrangian [52,46].

The Nambu Jona-Lasinio model has fewer parameters than the linear σ-model because each field has only one associated mass parameter but no coupling constant because the fields have arbitrary normalisation. However, a cut-off Λ needs to be introduced to regularize the Dirac sea constribution and this adds one parameter, making nucleon properties depend on the same number of parameters as in the σ-model. However, fewer parameters occur when vector fields are introduced [4,5,43,13].

We do not yet know whether the inclusion of vector fields in the Nambu Jona-Lasinio lagrangian will yield more binding for the quarks. The only available calculation [53] does show attraction obtained from the ρ and a_1 fields, as expected from previous gauged σ-model calculations [41], but it omits the repulsion of the ω-meson.

Baryons as bound quark-diquark composites.

The weak binding obtained with hedgehog fields in the Nambu Jona-Lasinio model need not be a weakness of the model. Indeed alternative consfigurations may turn out to be more bound. For example, diquarks may form sufficiently bound states [54,55,34,35] for the nucleon to prefer a quark-diquark configuration [55,34]. The original argument in favor of quark-diquark configurations comes from the observation that mesons and baryons have parallel Regge trajectories. Indeed a pair of quarks, coupled to form a color triplet, can be seen from afar by another quark as a color source identical to an anti-quark and, together, they may be bound by gluons as quark-antiquark pairs are in mesons. Nucleons and mesons formed by light quarks would then appear more like color strings [56] and would yield parallel Regge trajectories. The calculation of a nucleon is then achieved in two steps: the calculation of a bound diquark and the subsequent coupling of the diquark to a third quark. Diquarks are usually calculated by solving a Bethe-Salpeter equation which sums the ladders:

 (2.3)

(the same which occur in the study of pairing vibrations) and complex poles of the diquark propagators indicate an instability of the assumed vacuum againt condensation of diquark pairs as in superconductivity [30].

Diquarks have been calculated with Nambu Jona-Lasinio type lagrangians [54] as well as with lagrangians originating from gluon exchange [55,34,44]. Considerable binding of the order of 300 MeV is obtained for a scalar color-triplet diquark in the Nambu Jona-Lasinio model when a 300 MeV constituent mass is used [54]. A diquark bound state can be simply expressed in terms of an auxiliary field which is used to bosonize the Nambu Jona-Lasinio action in the usual way. Auxiliary fields can be used to represent diquark qq pairs as well as quark-antiquark $q\bar{q}$ pairs. Condensation of the $q\bar{q}$ pairs leads to the usual chiral symmetry breaking and to the associated self-energy giving constituent quarks their mass. Condensation of the diquark pairs leads to spontaneous breaking of baryon number and color in the vacuum.

However the method of auxiliary fields suffers from an ambiguity related to the choice of either qq or $q\bar{q}$ pairs with which one decides to linearize the 4-fermion interactions. In Ref.[54] the scalar interactions are bosonized in terms of $q\bar{q}$ pairs and the vector interactions in terms of (diquark) qq pairs. Such ambiguities may however be avoided by a resummation of Feynmann diagrams, in which one is free to choose the self energies [34].

Whether the quark-diquark representation of the nucleon [55,35] is an improvement remains to be seen. Indeed several problems must be faced. The third quark may interact with the diquark by processes such as [35]:

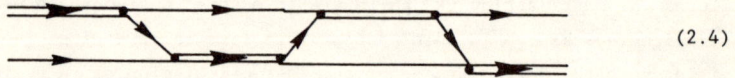 (2.4)

and such interactions may well destroy the correlations which bind the diquark. Furthermore, it is not at all clear that the Nambu Jona-Lasinio model will yield sufficient binding for the quark-diquark system describing the nucleon. The binding could depend on the confining forces which are absent in the lagrangian. Finally it is also possible that, once the quark-diquark interactions are taken into account, the difference between this representation and any other way of coupling three quarks to form a nucleon, will be smeared out in the nucleon ground state.

3. Scale invariance and the nature of scalar mesons.

Because the vacuum is a 0^+ state, scalar fields are usually introduced to represent vacuum expectation values. Quite often, we only use the vacuum expectation value and not the full dynamics of the corresponding scalar meson. For example, it is the vacuum expectation value $\langle\sigma\rangle = f_\pi$ which is used in the σ-model for pion and soliton calculations and it makes little difference whether they are made with the linear or non-linear σ-model. As long as the fields remain on the chiral circle, meaning that $\sigma^2 + \pi^2 = f_\pi^2$, we are not sensitive to the σ-meson mass parameter κ. Similarly, in the standard model, only the average value ($\approx$ 250 GeV) of the Higgs field is used and the Higgs mass is not predicted. However the hunt for the Higgs is open and, similarly, we would like to identify the meson associated to the scalar field occuring in low-energy chiral models. This is important for understanding what process controls partial or complete chiral symetry restoration in dense and hot matter. In the linear σ-model (2.1) the nature of the scalar field σ is not specified. In the Nambu Jona-Lasinio model (2.2) the scalar field is $\varphi = -\langle\bar{q}q\rangle/a^2$ and it represents a quark condensate. It is an order parameter measuring the amount of chiral symmetry breaking. However this may not be the whole story.

Scalar fields need not be order parameters associated to a spontaneously broken symmetry. They may, for example represent quadratic fluctuations of some field in the vacuum. It has been suggested by the Syracuse group [57-59,50] to associate the scalar field appearing in low-energy chiral lagrangians to the QCD scale anomaly [60]. This is done done in by introducing a new scalar field which represents quadratic fluctuations of the gluon field (gluonium) in the vacumm. This idea has recently become rather fashionable [16,61].

The discussion is made easier by rewriting the chiral field in the form:

$$\sigma + i\vec{\pi}\cdot\vec{\tau} \equiv \varphi U \qquad U = e^{i\vec{\theta}\cdot\vec{\tau}}$$

$$\sigma + i\gamma_5\vec{\pi}\cdot\vec{\tau} \equiv \varphi U_5 \qquad U_5 = e^{i\gamma_5\vec{\theta}\cdot\vec{\tau}} \qquad (3.1)$$

Consider the linear σ-model lagrangian (2.1) written with this form:

$$\mathcal{L} = \bar{q}(i\partial_\mu\gamma_\mu - g\varphi U_5)q + \frac{\varphi^2}{4}\mathrm{Tr}(\partial_\mu U)(\partial^\mu U^\dagger)$$

$$+ \frac{1}{2}(\partial_\mu\varphi)(\partial^\mu\varphi) - \frac{\kappa^2}{8}(\varphi^2 - f_\pi^2)^2 \qquad (3.2)$$

In the limit $\kappa \Rightarrow \infty$ the field φ acquires the value f_π at all points x and the lagrangian (3.2) reduces to the the non-linear σ-model represented by the first line in which φ is replaced by f_π.

We now follow Refs.[57,58,50], where a more detailed discussion can be found, and we consider the QCD trace anomaly. In a scale transformation the quark and gluon fields transform as follows:

$$A_{\mu a}(x) \Rightarrow \lambda A_{\mu a}(\lambda x) \qquad q(x) \Rightarrow \lambda^{3/2} q(\lambda x) \qquad (3.3)$$

The scale transformation leaves the QCD action invariant. However the divergence of the current s_μ, associated to the scale transformation (3.3), does not vanish because an anomaly occurs. One finds that the divergence is equal to [60]:

$$\partial_\mu s^\mu = \frac{\beta(g)}{g}\langle F_{\mu\nu}F_{\mu\nu}\rangle \qquad (3.4)$$

where $\langle\ \rangle$ means a vacuum expectation value. We introduce a scalar field χ to represent the right-hand side of (3.4):

$$\chi^4 \equiv \langle F_{\mu\nu}F_{\mu\nu}\rangle \qquad (3.5)$$

The field χ is called the *gluonium field* and it is assumed to have scaling dimension 1 so that it tranforms as:

$$\chi(x) \Rightarrow \lambda\chi(\lambda x) \qquad (3.6)$$

The idea is to write down a lagrangian for χ, such that the divergence of the current associated to the scale transformation (3.6) satisfies the equation:

$$\partial_\mu s^\mu = \chi^4 \qquad (3.7)$$

Such as lagrangian is:

$$\mathcal{L} = \frac{a}{2}(\partial_\mu \chi)(\partial^\mu \chi) - \chi^4 \ln\frac{\chi}{\Lambda} \qquad (3.8)$$

where Λ is a mass parameter.

The lagrangian (3.8) represents pure gluonium and it should describe the vacuum in the absence of quarks. The potential term has a minimum at $\chi_0 = \Lambda/(e)^{1/4}$ as shown by the dotted line of Fig.1. Small amplitude vibrations about this minimum represent a glueball with squared mass $m_\chi^2 = 4\chi_0^2/a$. It has been estimated using QCD sum rule methods [62] that $\chi_0 \approx 350$ MeV. If the scalar glueball is expected to have a mass in the range 800-1600 MeV [61] then the dimensionless constant a lies in the range 0.2-0.75.

Consider the scale invariance of the σ-model action obtained from the lagrangian (3.2). Assume that the field φ has a scaling dimension 1 and that the chiral angle θ, appearing in U, has a scaling dimension zero:

$$\varphi(x) \to \lambda\varphi(\lambda x) \qquad \theta(x) \to \theta(\lambda x) \qquad (3.9)$$

Then all the terms of (3.2) yield a scale invariant action, except for the potential term $(\kappa^2/8)(\varphi^2 - f_\pi^2)^2$. The latter can be made scale invariant if we replace f_π by a term proportional to the gluonium field χ, thus yielding a potential term of the form $(\kappa^2/8)(\varphi^2 - R\chi^2)^2$ where R is a constant. This yields a coupling between the gluonium field χ and the field φ, which we call the quarkonium field. Adding the pure gluonium lagrangian (3.8), we obtain the "scaled" σ-model lagrangian:

$$\mathcal{L} = \bar{q}(i\partial_\mu\gamma_\mu - g\varphi U_5)q + \frac{\varphi^2}{4}\text{Tr}(\partial_\mu U)(\partial^\mu U^\dagger)$$

$$+ \frac{1}{2}(\partial_\mu\varphi)(\partial^\mu\varphi) - \frac{\kappa^2}{8}(\varphi^2 - R\chi^2)^2 + \frac{a}{2}(\partial_\mu\chi)(\partial^\mu\chi) - \chi^4 \ln\frac{\chi}{\Lambda} \qquad (3.10)$$

Note that, in this form, chiral symmetry breaking is driven by the gluonium potential term $\chi^4 \ln(\chi/\Lambda)$. This will also be the case in the examples described below.

In limit $\kappa \to \infty$ of the non-linear σ-model, we have $\chi = \varphi/R$ and $f_\pi = R\chi_0$. Thus $\varphi = f_\pi \chi/\chi_0$ and we obtain a "scaled" non-linear σ-model [50]:

$$\mathcal{L} = \bar{q}\left(i\partial_\mu \gamma_\mu - gf_\pi \frac{\chi}{\chi_0} U_5\right)q + \frac{f_\pi^2}{4}\frac{\chi^2}{\chi_0^2} \text{Tr}(\partial_\mu U)(\partial^\mu U^\dagger) + \frac{a}{2}(\partial_\mu \chi)(\partial^\mu \chi) - \chi^4 \ln\frac{\chi}{\Lambda}$$

(3.11)

This scaled non-linear σ-model is in fact quite similar to the linear σ-model. Indeed, the only difference is the shape of the potential term for χ in (3.11) and for φ in (3.2). They both have a minimum at a non-vanishing value of the respective scalar fields, so that, provided the scalar fields do not deviate too much from their value at the minimum, the dynamics are similar. Actually this is the whole idea of the form (3.11). In the non-linear σ-model, the fields are not allowed to deviate from the chiral circle. The lagrangian (3.11) is a model in which it is assumed that deviations from the chiral circle are controlled by the potential term of the gluonium field. Whether this is strictly true or whether actually both the quarkonium and gluonium fields φ and χ exist is interesting physics and it is discussed in Ref.[50,57,59]. In many instances, the scaled non-linear σ-model, as well as the scaled Skyrme model, discussed below, yield results which are similar to the linear σ-model.

All but the last term of (3.11) may be obtained by modifying the non-linear σ-model terms with a field χ, carrying scale dimension 1, so as to make each term scale invariant. This is exactly how the Skyrme lagrangian or any other lagrangian can be modified, so as to saturate the divergence of the scaling current s_μ with a single gluonium field. For example, the scaled Skyrme lagrangian has the form [58]:

$$\mathcal{L} = \frac{f_\pi^2 \chi^2}{4\chi_0^2} \text{tr}(\partial_\mu U)(\partial^\mu U^\dagger) + \frac{1}{32e^2} \text{tr}\left[(\partial_\mu U)U^\dagger, (\partial_\nu U)U^\dagger\right]^2$$

$$+ \frac{a}{2}(\partial_\mu \chi)(\partial^\mu \chi) - \chi^4 \ln\frac{\chi}{\Lambda} \quad (3.12)$$

Current quark mass terms can of course be added. The form (3.12) is then the one used in the more recent applications where chiral symmetry restoration [61] and the variation of meson masses [16] are studied.

Scaled Skyrme-like lagrangians have been considered as a possible explanations of bag formation. The measure of bag formation is the ratio χ/χ_0 which should vanish or at least decrease inside a bag. The bag formation depends on the vacuum expectation χ_0, which is estimated to be 340 MeV [62]. Shallow bags are formed with this value and deeper bags are formed for smaller values. Solitons and bag formation with scaled Skyrme-like lagrangians with vector fields have also been calculated [63]. Rather shallow bags and a better, although still too small value $g_A \approx 0.9$ are obtained. It should be stressed that this bag formation does not mean confinement. Indeed, the quarks still have a finite mass outside the soliton.

There is another way of understanding the scaled Skyrme lagrangian (3.12). A soliton with winding number B=1 (such as a nucleon), constructed from this lagrangian, has an energy proportional to $(f_\pi \chi)/(e\chi_0)$. This means that the nucleon has a mass which is proportional to the gluonium field χ. This is exactly what is expressed by a scaled meson-nucleon lagrangian such as (1.12). The question raised here is: at which level should we scale the lagrangians? At the quark level as suggested in Refs.[50,57,59] or at the meson-nucleon (or Skyrme) level? If we decide to scale at the meson-nucleon level, then we can effectively reconcile the Nambu process of spontaneous chiral symmetry breaking with the scaling. The scaling of the meson mass terms simply mocks up the change in the meson mass due to the change in the constituent quark mass, as illustrated in (1.11). There remains however one difference. In the Nambu Jona-Lasinio lagrangian, the chiral symmetry is regulated by the Dirac sea quark mass. In the scaling lagrangians it is regulated by the term $-\chi^4 \ln(\chi/\Lambda)$ designed to saturate the scale anomaly with the single field χ. Is this important? Notice also that in the scaled lagrangians, the coupling of the meson fields to the fermions is unaltered and does not change with density. This should be checked.

Conclusion.

Nothing of what I have said is really profound. On the contrary I feel that much progress can be achieved in the study of low energy hadronic processes once it is realized that even the best physicists are more inclined to do what they can than what they ought to do. The reason is that they do not, in spite of so many claims to the contrary, really know what they should be doing. In such a situation it is better for Koestler's *sleepwalkers* to regain consciousness and to face squarely the fact that low energy hadronic physics is difficult and ambiguous. This is also true of experimental hadronic physics. Thank God because, there at least, the source of Query Confusion and Doubt about low energy QCD is not in our mind but is embedded somewhere in the apparatus. Where it should be.

Acknowledgement

Most of the ideas in these lectures stem from discussions and collaboration with Martine Jaminon, Ramon Mendez Galain and Pierre Stassart.

References

[1] C.Itzykson and J.B.Zuber, "Quantum Field Theory", McGraw Hill (1980) p.552.
[2] Y.Nambu and G.Jona-Lasinio, Phys.Rev.122 (1961) 345; 124 (1961) 246.
[3] For introductory lectures to recent applications of the Nambu Jona-Lasinio models see the following lectures:
G.Ripka, M.Jaminon and P.Stassart, in ""Hadrons and Hadronic matter", Ed.D.Vautherin, F.Lenz and J.W.Negele, Plenum Press (1990);
G. Ripka, in "XII workshop on Nuclear Physics, Ed.M.C.Cambaggio, A.J.Kreiner and E.Ventura, World Scientific (1990);
R.Mendez Galain and G.Ripka, in "Hadronic Physics with Multi-GeV Electrons, Ed. B.Desplanques and D.Goutte, Nova Science (1991).
[4] D.Ebert and H.Reinhardt, Nucl.Phys.B271 (1986) 188.
[5] V.Bernard and Ulf-G.Meissner, Nucl.Phys.A489 (1988) 647.
[6] M.Wakamatsu and W.Weise, Zeit.Phys.A331 (1988) 173.
[7] M.Jaminon, G.Ripka and P.Stassart, Nucl.Phys.A504 (1989) 733
[8] S.Godfrey and N.Isgur, Phys.Rev. D32 (1985) 189.
S.Capstick and N.Isgur, Phys.Rev. D34 (1986) 2809.
P.Kokowski and N.Isgur, Phys.Rev. D35 (1987) 907.
[9] G.Ripka in "Windsurfing the Fermi Sea", vol.2, Ed.T.T.S.Kuo and J.Speth, North Holland (1986).
[10] H.Reinhardt and B.V.Dang, Journal Phys.G13 (1987) 1179.
[11] Chr.V.Christov, E.Ruiz Arriola and K.Goecke, Nucl.Phys. A510 (1990) 689.

[12] R.Mendez Galain, G.Ripka, M.Jaminon and P.Stassart, Europhys.Lett.14 (1991) 7.
[13] M.Jaminon, R.Mendez Galain, G.Ripka and P.Stassart, SPhT/91-037 (1991)
[14] Chr.V.Christov, E.Ruiz Arriola and K.Goecke, Nucl.Phys. A510 (1990) 689.
[15] G.E.Brown, V.Koch and M.Rho, "The pion at finite temperature and density" (1991) to be published.
[16] G.E.Brown and M.Rho, "Scaling effective lagrangians in dense medium", Saclay preprint (1990) SPhT/90/191.
[17] J.P.Blaizot and R.Mendez Galain, "Φ and K mesons in hot dense matter", Saclay preprint (1991) SPhT/91-045.
[18] K.Wehrberger, R.Wittman and Brian D.Serot, Phys.Rev C42 (1990) 2680
[19] T.Matsui, Nucl.Phys. A370 (1981) 365.
[20] K.Lim and C.J.Horowitz, Nucl.Phys. A501 (1989) 729.
[21] J.Diaz Alonso and A. Perez Canyellas, Nucl.Phys. A526 (1991) 623.
[22] B.L.Friman and A.Henning, Phys.Lett.B206 (1988) 579.
[23] R.J.Furnstahl and C.J.Horowitz, Nucl.Phys. A485 (1988) 632.
[24] A.R.Bodmer, Nucl.Phys. A526 (1991) 703.
[25] R.J.Furnsdahl, R.J.Perry and B.D.Serot, Phys.Rev.C40 (1989) 321.
[26] B.D.Serot and J.D.Walecka, Adv.Nucl.Phys. 16 (1986) 1
[27] V.Soni, Phys.Lett.B183 (1987) 91.
[28] G.Ripka and S.Kahana, Phys.Rev. D36 (1987) 1233.
[29] R.J.Perry, Phys.Lett. B199 (1987) 489.
[30] J.P.Blaizot and G.Ripka, Quantum theory of finite systems, MIT Press (1986), pp.322 and 536.
[31] T.H.Hansson, K.Johnson and C.Peterson, Phys.Rev. D26 (1982) 2069.
[32] T.H.Hansson and I.Zahed, Phys.Rev. D36 (1987) 221.
  T.H.Hansson, D.Klabucar and I.Zahed, Phys.Rev. D36 (1987) 233.
[33] D.Diakonov and Yu.V.Petrov, Nucl.Phys. B272 (1986) 457.
[34] J.Praschifka, R.T.Cahill and C.D.Roberts, Int.J.Mod.Phys.A4 (1989) 4929.
[35] R.D.Ball, Int.J.Mod.Phys.A5 (1990) 4391; Phys.Lett. B.245 (1990) 213.
[36] I.F.Mathiot, G.Chanfray and H.J.Pirner, Nucl.Phys. A500 (1989) 605.
[37] M.C.Birse and M.K.Banerjee, Phys.Lett. B136 (1984) 284.
[38] S.Kahana, G.Ripka and V.Soni, Nucl.Phys.A415 (1984) 351.
[39] M.Gell-Mann and M.Lévy, Nuovo Cim. 16 (1960) 705.
[40] R.Friedberg and T.D.Lee, Phys.Rev. D16 (1977) 1096.
[41] W.Broniowski and M.K.Banerjee, Phys.Rev. D34 (1986) 849.
[42] B.W.Lee and T.H.Nieh, Phys.Rev.166 (1968) 1507
[43] S.Klimt, M.Lutz, U.Vogel and W.Weise, Nucl.Phys. A516 (1990) 429.

[44] R.D.Ball, Int.J.Mod.Phys.A5 (1990) 4391; Phys.Lett. B.245 (1990) 213.
[45] H.Reinhardt and R.Wünsch, Phys.Lett. B227(1989) 296; B230 (1989) 93.
[46] M.Praszalowicz, Phys.Rev. D42 (1990) 216.
[47] Th.Meissner, F.Grümmer and K.Goecke, Phys.Lett. B227 (1989) 296.
[48] I.J.R.Aitchison, C.M.Fraser, E.Tudor and J.Zuk, "Failure of the derivative expansion for studying the stability of the baryon as a chiral soliton", University of Oxford preprint, 1985. (It seems incredible that such useful papers do not get published.)
[49] I.J.R.Aitchison and C.M.Fraser, Phys.Rev. D31 (1985) 2605.
[50] P.Jain, R.Johnson and J.Schechter, Phys.Rev.D38 (1988) 1571.
[51] Ulf.-G.Meissner, N.Kaiser, A.Wirzba and W.Weise, Phys.Rev.Lett. 57 (1986) 1676.
[52] A.Manohar and H.Georgi, Nucl.phys. B234 (1984) 189.
[53] R.Alkofer and H.Reinhardt, Phys.Lett. B244 (1990) 461.
[54] U.Vogl, Zeit.Phys.A337 (1990) 191.
D.Kahana and U.Vogl, Phys.Lett. B244 (1990) 10.
[55] G.V.Efimov, M.A.Ivanov and V.E.Lyubovitskij, Zeit.Phys.C47 (1990) 583.
[56] A.B.Migdal, Nucl.Phys. A518 (1990) 358.
[57] H.Gomm, P.jain, R.Johnson and J.Schechter, Phys.Rev. D33 (1986) 801.
[58] H.Gomm, P.Jain, R.Johnson and J.Schechter, Phys.Rev.D33 (1986) 3476.
P.Jain, R.Johnson and J.Schechter, Phys.Rev. D35 (1987) 2230.
[59] Ulf.-G.Meissner and N.Kaiser, Phys.Rev. D35 (1987) 2859.
Ulf-G.Meissner, R.Johnson, N.W.Pak and J.Schechter, Phys.Rev.D37 (1988) 1285.
[60] A.Chodos and C.B.Thorn, Phys.Rev. D12 (1975) 2733.
[61] B.A.Campbell, J.Ellis and K.A.Olive, Nucl.Phys.B345 (1990) 57.
[62] M.A.Schifman, A.I.Vainstein and Z.I.Zakharov, Nucl.Phys. B147 (1979) 385 and 448.

Random-Matrix Modelling of Stochastic Nuclear Properties

Hans A. Weidenmüller

Max-Planck-Institut für Kernphysik
Heidelberg, F.R. of Germany

<u>Abstract</u> : This is a short "guided tour" through the three lectures given at La Rábida Summer School, with references to more detailed papers published recently. The topics covered are (i) justification for random matrix models (RMM); (ii) Compound Nucleus (CN) scattering and isospin mixing as simple examples of RMM; (iii) technical issues encountered in evaluating RMM; (iv) tests of fundamental symmetries in CN scattering; and (v) RMM in mesoscopic physics as a natural generalization of RMM used in nuclear physics.

1. Introduction

When asked by the organizers of the Summer School at La Rábida to supply them and the students with a set of lecture notes, I responded by saying: Most of what I am going to discuss has been published recently, both in the form of conference proceedings and in the form of refereed papers, and I do not believe in duplicating these efforts. I was very relieved when the organizers suggested that I prepare a few pages introducing the topic, with references to my previously published work. The present text is the result of such a compromise, and I am grateful to the organizers for both having invited me, and having been so flexible about these notes.

2. First lecture

In this lecture, I covered three topics: The justification for using RMM in nuclei, and CN scattering and isospin mixing as simple examples for such modelling.

2.1 The justification for using RMM in nuclei

This issue is discussed in considerable detail in Ref. [1,2].
The solid empirical evidence in favour of RMM of stochastic nuclear properties is mainly due to the work by the Columbia group in the 1970's. Using time-of-flight neutron spectroscopy on a series of heavy nuclei, this group investigated a large number of s-wave neutron resonances (typically 150-200 per nucleus) and obtained resonance energies, total widths, and partial neutron widths. Similar evidence on proton resonances

in lighter nuclei was obtained by the TUNL group. The analysis of the entire data set by the Orsay group showed, within statistics, perfect agreement with a specific RMM, defined by the conditions of Hermitecity and time-reversal symmetry: The Gaussian orthogonal ensemble (GOE). This evidence is sufficient by itself to justify RMM of nuclear spectra and reactions in the domain of excitation energies corresponding to the energy of the neutron threshold and above.

A deepened understanding of, and a further justification for, the use of RMM came from the analysis of conservative systems with a few degrees of freedom which are fully chaotic in the classical limit. In the first half of the 1980's, numerical work by the Orsay group on the Sinai billiard showed that the eigenvalues of this system show GOE statistics. And Berry was able to show analytically that classical chaos is linked with GOE behaviour. These results and the empirical evidence on nuclei together suggest that the success of RMM in nuclei is a consequence of chaotic dynamics. The connection between time-reversal invariant dynamical systems which are classically chaotic, and GOE spectral fluctuation properties suggested by these results, holds only if (almost) all points of classical phase space are equally accessible. A counterexample is provided by microwave scattering on a set of irregularly shaped cavities linked sequentially by thin pipes. In a single such cavity, the condition just mentioned holds, but it fails for the set at large. Systems of this type are of interest in the physics of disordered systems where they pose the problem of localization. In the context of nuclear physics, the case of a single cavity corresponds to CN scattering. Here, the internal equilibration time of the CN is small in comparison with the decay time. The case of many cavities, on the other hand, corresponds to precompound reactions: The chain of cavities is mapped onto the chain of classes of states of fixed exciton number. Within each class, the equilibration time is short compared to the mixing time between neighbouring classes, and the decay time of the system is comparable to its total internal equilibration time.

2.2 CN scattering as the simplest example of RMM

The material covered in this part is to be found partly in ref. [1], partly in ref. [3]. CN scattering is modelled by assuming that all CN resonances of fixed spin and parity are eigenstates of a Hamiltonian which is a member of the GOE. Coupling this Hamiltonian to a set of open channels, one obtains a stochastic model for CN scattering. The parameters determine the probability distribution of the elements of the scattering matrix. This model is of interest not only for CN reactions (where it is very successful and plays a fundamental role also in nuclear physics applications to other sciences, and to technology). It is actually a stochastic model for chaotic scattering, applicable universally provided only that the phase space points in the interaction region are (almost) all equally accesible from any channel.

In view of its general applicability, I have displayed several consequences of this model: The two-point function, the Hauser-Feshbach limit, the elastic enhancement factor, the pole distribution of the sacttering matrix, and the connection with Dyson's circular orthogonal ensemble.

2.3 A RMM for spectral fluctuations in the presence of isospin mixing

This part is based on the work presented in reference [4].

In the absence of Coulomb forces, isospin would be a good quantum number for nuclear systems. States with fixed isospin at several MeV excitation energy would show GOE spectral fluctuations, while states differing in isospin would be independent of each other. In particular, such states would not repel each other and could come arbitrarily close. What happens to this situation, and how does the spectral statistics behave, under the influence of the (weak) isospin-breaking Coulomb force? This question, triggered by investigations of ^{26}Mg by the TUNL group, is tackled as follows. Typically we deal with two values of the isospin quantum number, each value defining a class of states. States within a given class are modelled by a GOE. The two GOE Hamiltonians are coupled by the Coulomb interaction. The matrix elements of this interaction are modelled as Gaussian distributed random variables with the same variance. The value of the variance corresponds to the strength of the Coulomb force.

Consequences of this model for spectral fluctuation properties were displayed; a comparison with ^{26}Mg data yidded a strength parameter which was consistent with other information on the violation of isospin symmetry in nuclei.

3. Second lecture

The results displayed in lecture one on subsections 2.2 and 2.3 are based on techniques which are of interest whenever stochastic modelling is used in physics or chemistry. This is why in subsection 3.1 of lecture two, I discussed briefly some technical aspects. In subsection 3.2, I turned to another application of RMM in nuclei: The analysis of test of fundamental symmetries in CN reactions.

3.1 Technicalities

Stochastic modelling is by its very nature not able to reproduce the fine details of a fluctuating observable. Rather, it aims at predicting the moments and correlation functions of the observable. The calculation of such quantities from a stochastic input requires the calculation of an average over the independent random variables defining the stochastic model. Performing these calculations for higher moments becomes increasingly complex, both numerically and analytically, and poses very difficult problems.

I have briefly discussed potentialities and limitations of the numerical approach (the Monte Carlo method) and of the analytical approach. The latter uses a generating functional and employs either replica trick or Grassmann integration. The last method is the one mainly used by the Heidelberg group. It is explained in reference [5].

3.2 Test of fundamental symmetries in CN reactions

The material in this part was based on an invited paper at the 1990 Santa Fé meeting [6].

Following the discovery of big parity-violation effects in the transmission of polarized neutrons on ^{139}La at Dubna, a systematic investigation of this phenomenon on several nuclei is under way at Los Alamos. The size (several per cent) of the effect indicates large enhancement factors. This in turn given rise to the hope that the CN may also be useful for testing time-reversal symmetry, and jointly parity plus time-reversal symmetry.

The determination from such data of the size of the symmetry-breaking interaction requires a stochastic model since we deal with the CN. This model is quite similar to that of isospin mixing mentioned above: For instance, states of fixed parity are modelled by a GOE Hamiltonian. The two Hamiltonian matrices modelling states of opposite parity are coupled by the weak symmetry breaking interaction. The matrix elements of this interaction are assumed to be uncorrelated random variables, all with the same variance. That variance measures the strength of the matrix elements of the symmetry-breaking interaction between complex CN states and is the object of the analysis of the data. The next step consists in relating the variance with the strength of the fundamental symmetry-breaking nucleon-nucleon interaction.

Recent data on Th at Los Alamos have put this stochastic approach into question. It was found that all cases where a big signal was found have the same sign for the parity-violating observable, while the stochastic model predicts randomness in sign. The observation is not undestood at present. The Th data will be published shortly [7].

4. Third lecture

I thought it is important for the students to realize that RMM are not restricted to nuclear physics, and are useful in other areas as well. To exhibit similarities and differences between the results, I focussed attention on conductance fluctuations in mesoscopic physics, and a comparison of their properties with Ericson fluctuations in CN reactions. This lecture was based on references [8-10].

Advances in microelectronics have made it possible to manufacture probes of wires of typical length 1 m, typical transverse dimensions 0.1 m. At temperatures below 1 K, the <u>inelastic</u> mean free path for the scattering of electrons by phonons becomes larger than the dimension of the sample. Electron transport through such probes is therefore determined by the <u>elastic</u> scattering of electrons by impurities and dislocations. As the precise positions, density and scattering properties of the latter are unknown, one mimics electron transport through such "mesoscopic" probes (two dimensional semiconductor devices also belong to this class) by a random potential. The resulting scattering problem is formally quite similar to that of CN or precompound nucleus scattering.

The interest in the properties of such devices, apart from their use in microelectronics, derives from the fact that on its way through the sample, the electron, although scattered elastically on a random set of impurities, retains its phase coherence. (This coherence is lost only upon inelastic scattering by phonons). Therefore mesoscopic devices display quantum-mechanical interference although they have length scales of about 10^4 Angström!.

A fascinating example is provided by the "Universal Conductance Fluctuations" (UCF) in mesoscopic wires and rings. Under the influence of an external magnetic field, the conductance of a mesoscopic device displays random fluctuations. The intuitive reason is this: As the external magnetic field increases, the trajectory of the electron through the sample will be bent more strongly. The electron will hit another set of impurities; this is kin to taking, at fixed external field, another sample: The fluctuations are a direct signal of the random distribution of impurities over the sample. They are quantal in origin – classically, probes of 1 micron size lead to "self-averaging" and to a smooth dependence of the conductance on the magnetic field. And the conductance fluctuations have much in common with Ericson fluctuations, i.e. with the random fluctuations of CN cross sections versus energy. Naturally, they are also related to the fluctuations versus frequency of the intensity of a microwave traversing a single, or a chain of irregularly shaped cavities.

Perhaps the most fascinating aspect of UCF is their independence of size, geometry or material of the sample. It is this feature which has given rise to their name. Another interesting aspect of quantum interference in mesoscopic systems arises when one measures the conductance of a ring pierced by a magnetic field. Then, UCF are superposed by Aharonov-Bohm type oscillations: On its way through either arm of the ring, the electron picks up a phase determined by the magnetic flux through the ring. Depending on the flux, the two amplitudes interfere constructively or destructively.

RMM for such devices can be designed in close analogy to the precompound nucleus scattering problems. This is of interest for several reasons: By looking at mesoscopic devices as scattering problems, one is in a position to study the influence of their coupling to external leads. The calculation of the average conductance and of its fluctuations can be carried out in analogy to the nuclear scattering problem. And by comparing the results, one can learn a lot about either problem.

Such a comparison shows that Ericson fluctuations and UCF, although similar in origin and character, display fundamental differences. One may say that Ericson fluctuations are the semi-classical limit of a fluctuation phenomenon the full quantum manifestation of which is seen in UCF

References

[1] O. Bohigas and H.A. Weidenmüller, Ann. Rev. Nucl. Part. Science 38 (1988) 421
[2] H.A. Weidenmüller, Nucl. Phys. A520 (1990) 509c
[3] C. Lewenkopf and H.A. Weidenmüller, Ann. Phys. (N.Y.) (1991) in press
[4] T. Guhr and H.A. Weidenmüller, Ann. Phys. (N.Y.) 199 (1990) 412
[5] J.J. Verbaarschot, M.R. Zirnbauer and H.A. Weidenmüller, Phys. Rep. 129 (1985) 367
[6] H.A. Weidenmüller, Nucl. Phys. A522 (1991) 293c
[7] C.M. Frankle et al., Phys. Rev. Lett. (1991) in press
[8] H.A. Weidenmüller, in "Recent Progress in Many-Body Theories", Plenum Publishing Corporation, New York (1990), p. 261
[9] H.A. Weidenmüller, Physica A167 (1990) 21
[10] H.A. Weidenmüller, Nucl. Phys. A518 (1990) 1

Subject Index

$3s$-shell 179

absolute occupation numbers 182
algebraic
 – approach 111
 – model 130
antiquarks 236

bag model 218, 222
bare optical potential 32

cavity approximation 218
 –, static spherical 218
central collisions 165
charge densities 174
 – in the lead region 174
charm production 167
chiral
 – bag 226
 – models 233
 – quark models 233
 – symmetry 222
collective order parameters 96
colour 197
coloured quarks 194
constant
 – angular momentum 104
 – energy 102
 – rotational frequency 104
   temperature 102
constituent quarks 205
 – – models 236
correlations in baryons 198
Coulomb energy 218
coupled-channel elastic scattering 40
coupling form factors 38

damping 57
 – of rotational motion 57
decay 26
 – of superdeformed nuclei 26
deconfined matter 157

deep inelastic structure functions 223
deformation-driven effects 86
dense matter 240, 242
density
 – distribution of $3s$-shell 179
 – of lead 178
deuteron charge form factor 185
dispersion relation 1, 8
drip-line nuclei 48

electromagnetic decay rate 234
electron scattering 172
energy deposition 161
exotic nuclear shapes 70

far from stability 45
fermion systems 16
finite
 – baryonic density 236
 – size effects 16, 19
 – – – in superconductors 19
flavour algebra 137
fluctuations 165
 – in central collisions 165
fundamental symmetries 258

gluons in hadrons and nuclei 194
grand canonical ensemble 107

hadronic structure 130
heavy-ion
 – collisions 155
 – –, ultrarelativistic 155
 – interactions 31
hedgehog chiral fields 243
high-spin 70
 – physics 51, 71
hot rotating nuclei 61

inelastic processes 38
intermediate energies 31
kaon interferometry 168

large N limit 214
lead region 174
lepton pairs 167
liquid-gas transitions 98
low energy QCD

magnetic moments 200
mass formulas 138
matrix elements of operators 140
maximal Pauli effects 217
meson structure 240
microcanonical ensemble 107
modification of nuclear medium 238
molecular spectra 111

negative energies 1
neutron form factors 185
non-relativistic quark model 215
nuclear
 – collective motion 15
 – mean field 1, 46
 – phase transitions 94
 – potential 1
 – shapes 52
nucleons and nuclei 194

observables 160
occupation
 – numbers 182
 – probabilities 13
octupole shapes 52
one-body observables 217
one-channel elastic scattering 32
operators, matrix elements of 140
optical-model potential 1, 6, 32
orbital alignment 86
order parameters 96
orientation fluctuations 104

pair transfer 24
pairing transitions 98
parity violation 188
Pauli principle 197
phase transitions 94
 – at finite temperature 94
 –, nuclear 94
pion interferometry 168
polarization model 40
polarized
 – gluons 209
 – strange quarks 207

potential energy curves 125
proton spin 204

quark matter 219
quark-diquark composites 245
quark-gluon plasma droplets 17
quarks
 – in hadrons and nuclei 194
 – in light nuclei 216
 – in nuclei 212

random-matrix modelling 255
rotational motion 57

scalar mesons 247
scale invariance 202, 248
shape
 – fluctuations 100
 – transitions 96
shell-model potential 1, 5
single-particle energies 3
Skyrmions 225
spectral
 – fluctuations 257
 – functions 12
spectroscopic factors 10
spectroscopy 70
spectrum generating algebra 130
 – – – of hadronic physics 130
spin polarization 205
spin-flavour correlations 197
spontaneous chiral symmetry breaking 235
statistical
 – fluctuations 98
 – mechanics 94
stochastic nuclear properties 255
strange quarks 207
strangeness production 166
strength functions 61
structure functions 223
superconductors 19
superdeformed
 – bands 81
 – nuclei 26
 – shapes 55, 75
symmetries in baryons 198
symmetry breaking interactions 142

t'Hooft self-consistency condition 214
transverse momentum spectra 169
two-nucleon Coulomb energy 218

$U(2)$ model 112

List of Participants

Dr. C.E. Alonso
Departamento de FAMN
Facultad de Física
Universidad de Sevilla
Apdo. 1065. 41080 Sevilla
SPAIN

Prof. G. Baym
Loomis Laboratory of Physics
University of Illinois
1110 W. Green Street
Urbana. IL 61801
USA

Dr. M.V Andrés
Departamento de FAMN
Facultad de Física
Universidad de Sevilla
Apdo. 1065. 41080 Sevilla
SPAIN

Dr. J. Bea Gilabert
Instituto de Física Corpuscular
Avda. Dr. Moliner, 50
46100 Burjassot
Valencia
SPAIN

Dr. N. Arbex
Univ. de Sao Paulo
Instituto de Fisica
Departamento de Fisica Matematica
Cidade Univarsitaria
CP 12516. Sao Paulo
BRAZIL

Prof. M. Bentley
SERC Daresbury Laboratory
Daresbury
Warrington WA4 4AD
UK

Dr. J.M. Arias
Departamento de FAMN
Facultad de Física
Universidad de Sevilla
Apdo. 1065. 41080 Sevilla
SPAIN

Dr. M. Bergmann
Ruhr Universität Bochum
Fakultät für Physik und Astronomie
ITP II, Universitätstrasse 150
Postfach 102148. 4630 Bochum 1
GERMANY

Dr. F. Barranco
Departamento de Física Aplicada
ETSII
Universidad de Sevilla
Sevilla
SPAIN

Prof. O. Bohigas
Division de Physique Théorique
Institute de Physique Nucléaire.
F-91406 Orsay
Cédex
FRANCE

Dr. I. Bombaci
Dipartimento di Fisica
Universita di Catania
Corso Italia 57
95129 Catania
ITALY

Dr. G. Colo
Universita degli Studi di Milano
Dipartimento di Fisica
Via Celoria 16
20133 Milano
ITALY

Dr. P. Bozek
Institute of Nuclear Physics
Department of Theoretical Physics
ul. Radzikowskiego 152
31-342 Kraków
POLAND

Dr. S. Cruz-Barrios
Universidade de Sao Paulo
Instituto de Fisica
Departamento de Fisica Matematica
Cidade Univarsitaria
CP 12516. Sao Paulo. BRAZIL

Prof. D.M. Brink
Department of Theoretical Physics
1 Keble Road
Oxford OX1 3NP
UK

Dr. F. de Blasio
Universita degli Studi di Milano
Dipartimento di Fisica
Via Celoria 16
20133 Milano
ITALY

Dr. G. Bunatian
Joint Institute for
Nuclear Research, LNPh,Dubna
Head Post Office, P.O. Box 79
101000 Moscow
USSR

Dr. E. Drukarev
Leningrad Nuclear Physics Institute
Gatchina
Leningrad 188350
USSR

Dr. W. Bürger
Institut für Kernphysik
Technische Universität Wien
Wiedner Hauptstrasse 8-10/142
1040 Wien
AUSTRIA

Dr. J. Dukelsky
Departamento de Física Teórica C-XI
Universidad Autónoma de Madrid
Cantoblanco
28049 Madrid
SPAIN

Dr. M. Centelles Aixalá
Departament d'Estructura i
Constituents de la Matèria
Fac. de Fisica
Univ. de Barcelona. Diagonal 647
08028 Barcelona. SPAIN

Prof. L. Egido
Departamento de Física Teórica C-XI
Universidad Autónoma de Madrid
Cantoblanco
28049 Madrid
SPAIN

Prof. F.E Close
Rutherford Laboratory
Didcot
Oxfordshire
UK

Dr. M. Elahrash
Tajura Research Center
P.O. Box 1863
Tripoli
LYBIA

Dr. C. Esebbag
Departamento de Física Teórica C-XI
Universidad Autónoma de Madrid
Cantoblanco
28049 Madrid
SPAIN

Dr. E. Garibay Ruíz
Egyetem tér 1/I
4010 Debrecen
HUNGARY

Dr. J.M. Espino
Departamento de FAMN
Facultad de Física
Universidad de Sevilla
Apdo. 1065. 41080 Sevilla
SPAIN

Dr. E. Garrido Bellido
Inst. de Estructura de la Materia
CSIC
Serrano 123
28006 Madrid
SPAIN

Dr. M. Fiolhais
Departamento de Fisica
Universidade de Coimbra
3000 Coimbra
PORTUGAL

Dr. N. Gelli
Dipartimento di Fisica
Largo Enrico Fermi 2
50125 Firenze
ITALY

Prof. A. Frank
Instituto de Ciencias Nucleares
Univ. Nacional Autónoma de México
Circuito Exterior C.U.
A. Postal 70-543
04510 Mexico D.F., MEXICO

Dr. J. Gómez-Camacho
Departamento de FAMN
Facultad de Física
Universidad de Sevilla
Apdo. 1065. 41080 Sevilla
SPAIN

Dr. M.I. Gallardo
Departamento de FAMN
Facultad de Física
Universidad de Sevilla
Apdo. 1065. 41080 Sevilla
SPAIN

Dr. J. González-Labajo
Departamento de Física Aplicada
E.U.P. de la Rábida
Ctra. Palos de la Frontera s/n
Huelva
SPAIN

Dr. L.M. García Raffi
Instituto de Física Corpuscular
Avda. Dr. Moliner, 50
46100 Burjassot
Valenia
SPAIN

Prof. A.L. Goodman
Physics Department
Tulane University
New Orleans, Luisiana 70118
USA

Dr. F. Garcías
Departamento de Física
Facultad de Ciencias
Universitat Illes Balears
07071 Palma de Mallorca
SPAIN

Dr. R. Heck
HMI-Berlin
Glienickerstrasse 100
Postfach 390128
1000 W-Berlin 39
GERMANY

Dr. I.M. Hibbert
Nuclear Group, Schuster Laboratory
University of Manchester
Brunswick Street
Manchester M13 9PL
UK

Prof. C. Mahaux
Institut de Physique, B5
Université de Liège
Sart Tilman
B-4000 Liège 1
BELGIQUE

Prof. F. Iachello
Yale University
Center for Theoretical Physics
Sloane Physics Laboratory
P.O. Box 6666, New Haven
Connecticut 06511. USA

Dr. I. Martel
Departamento de FAMN
Facultad de Física
Universidad de Sevilla
Apdo. 1065. 41080 Sevilla
SPAIN

Dr. A. Kabir
Physics Department
University of Surrey
Guildford GU2 5XH
UK

Dr. V. Martín Ayuso
Departamento de Física Teórica C-XI
Universidad Autónoma de Madrid
Cantoblanco
28049 Madrid
SPAIN

Dr. E.G. Lanza
INFN
Sezione di Catania
Corso Italia 57
95129 Catania
ITALY

Dr. C. Martínez de la Torre
Departamento de FAMN
Facultad de Física
Universidad de Sevilla
Apdo. 1065. 41080 Sevilla
SPAIN

Dr. M. Lassaut
Division de Physique Theorique
Institut de Physique Nucleaire Bp 1
91406 Orsay Cedex
FRANCE

Dr. M. Muñoz
Departamento de Física Aplicada
Escuela Arquitectura Técnica
Universidad de Sevilla
Sevilla
SPAIN

Dr. P. Lotti
INFN Sezione di Padova
Via F. Marzolo 8
35131 Padova
ITALY

Dr. Y. Nedjadi
Physics Department
University of Surrey
Guildford GU2 5XH
UK

Dr. M. Lozano
Departamento de FAMN
Facultad de Física
Universidad de Sevilla
Apdo. 1065. 41080 Sevilla
SPAIN

Dr. M.C. Nemes
Universidade de Sao Paulo
Instituto de Fisica
Departamento de Fisica Matematica
Cidade Univarsitaria
CP 12516. Sao Paulo. BRAZIL

Dr. R.M. Nikolaeva
Joint Inst. for Nuclear Research
LNPh, Dubna
Head Post Office, P.O. Box 79
101000 Moscow
USSR

Prof. G. Ripka
Service de Physique Théorique
Centre d'Etudes Nucleaires de Saclay
F-91191 Gif sur Yvette
Cédex
FRANCE

Dr. G. Pasquali
Dipartimento di Fisica
Largo Enrico Fermi 2
50125 Firenze
ITALY

Dr. L.M. Robledo
Departamento de Física Teórica C-XI
Universidad Autónoma de Madrid
Cantoblanco
28049 Madrid
SPAIN

Dr. A. Polls
Departament d'Estructura i
Constituents de la Matèria
Fac. de Fisica
Univ. de Barcelona. Diagonal 647
08028 Barcelona. SPAIN

Dr. W. Sakuler
Institut für Kernphysik
Technische Universität Wien
Wiedner Hauptstrasse 8-10/142
1040 Wien
AUSTRIA

Prof. A. Poves
Departamento de Física Teórica C-XI
Universidad Autónoma de Madrid
Cantoblanco
28049 Madrid
SPAIN

Dr. I. Saleh Juan
Tajura Research Center
P.O. Box 1863
Tripoli
LYBIA

Dr. C. Providencia
Departamento de Fisica
Universidade de Coimbra
3000 Coimbra
PORTUGAL

Prof. I. Sick
Institut für Kernphysik der
Universität
Klingelbergstrasse 82
4056 Basel
SWITZERLAND

Dr. L. Qin
Institut für Kernphysik der
Universität
Klingelbergstrasse 82
4056 Basel
SWITZERLAND

Dr. N. Tanimura
Institut für Theoretisch Physik
J-L Universität Giessen
Heinrich-Buff-Ring 16
6300 Giessen
GERMANY

Dr. A. Rapisarda
INFN
Sezione di Catania
Corso Italia 57
95129 Catania
ITALY

Dr. J.M. Udías Moinelo
Inst. de Estructura de la Materia
CSIC
Serrano 123
28006 Madrid
SPAIN

Prof. V. Vento
Departamento de Física Teórica
Universitat de València
Burjassot
Valencia
SPAIN

Prof. H.A. Weidenmüller
Max Planck Institut für Kernphysik
D-6900 Heidelberg
GERMANY

Dr. J.E. Villate
Centro De Física Nuclear
Av. Prof. Gama Pinto, 2
1699 Lisboa Codex
PORTUGAL

Dr. H. Wolters
Centro De Física Nuclear
Av. Prof. Gama Pinto, 2
1699 Lisboa Codex
PORTUGAL

Dr. F.J. Viñas
Departament d'Estructura i
Constituents de la Matèria
Fac. de Fisica
Univ. de Barcelona. Diagonal 647
08028 Barcelona. SPAIN

Dr. L. Zuffi
Universita degli Studi di Milano
Dipartimento di Fisica
Via Celoria 16
20133 Milano
ITALY

Prof. A. Vitturi
Universitá degli Studi di Trento
Dipartimento di Fisica
38050 Povo
Trento
ITALY

A. G. Sitenko, Academy of the Ukrainian SSR

Scattering Theory

1991. XI, 294 pp. 32 figs. (Springer Series in Nuclear and Particle Physics)
Hardcover ISBN 3-540-51953-X

This book is an introduction to nonrelativistic scattering theory. The presentation is mathematically rigorous, but is accessible to upper level undergraduates in physics. The relationship between the scattering matrix and physical observables, i.e. transition probabilities, is discussed in detail. Among the emphasized topics are the stationary formulation of the scattering problem, the inverse scattering problem, dispersion relations, three-particle bound states and their scattering, collisions of particles with spin and polarization phenomena. The analytical properties of the scattering matrix are discussed. Problems round off this volume.

P. Ring, P. Schuck, Technical University of Munich

The Nuclear Many-Body Problem

1980. XVII, 716 pp. 171 figs.
(Texts and Monographs in Physics)
Hardcover ISBN 3-540-09820-8

This book, while covering a fair amount of physical observations, stresses the methodology and technical aspects of the different theories presently used in the description of the nucleus. The authors present the more modern theories such as Boson expansions, generator coordinates, time-dependent Hartree-Fock method, and semiclassical models which so far have found only limited mention in textbooks.
The book also covers subjects like the liquid drop and the shell model, both presented in an updated version in, for example, rotations and random phase approximation. The full presentation of mathematical details, illustrated by observational data, will help the student fully understand the present views on the nuclear many-body problem.

K. L. G. Heyde, University of Gent, Belgium

The Nuclear Shell Model

1990. XII, 376 pp. 171 figs. (Springer Series in Nuclear and Particle Physics)
Hardcover ISBN 3-540-51581-X

This book evolved from a course in theoretical nuclear physics taught over seven years at the University of Gent and is thus well suited to and tested for lecture courses. The nuclear shell model is introduced from basic techniques such as angular momentum and tensor algebra. The material is developed from the beginning up to the present state-of-the-art calculations using self-consistent residual interactions. Problem sets and simple computer codes are included to facilitate a better acquaintance with the subject. The appendices constitute an integral part of the text going into depth on a number of technical derivations to provide the reader with a detailed background facilitating active research.
The book introduces the subject to advanced undergraduate and to graduate students providing them with knowledge and techniques for own research in this field. It is a highly useful prerequisite for lecturers teaching modern nuclear physics.

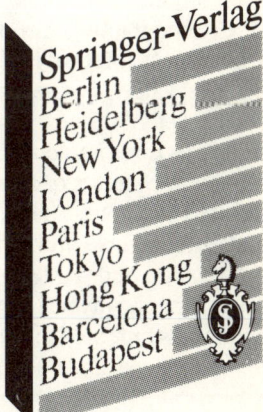

B. N. Zakhariev, A. A. Suzko

Direct and Inverse Problems

Potentials in Quantum Scattering

1990. XIII, 223 pp. 42 figs. Softcover ISBN 3-540-52484-3

This textbook can almost be viewed as a "how-to" manual for solving quantum inverse problems, that is, for deriving the potential from spectra or scattering data and also, as somewhat of a quantum "picture book" which should enhance the reader's quantum intuition. The formal exposition of inverse methods is paralleled by a discussion of the direct problem. Differential and finite-difference equations are presented side by side. The common features and (dis)advantages of a variety of solution methods are analyzed. To foster a better understanding, the physical meaning of the mathematical quantities are discussed explicitly. Wave confinement in continuum bound states, resonance and collective tunneling, energy shifts and the spectral and phase equivalence of various interactions are some of the physical problems covered.

P. C. Sabatier (Ed.)

Inverse Methods in Action

Proceedings of the Multicentennials Meeting on Inverse Problems, Montpellier, November 27th – December 1, 1989

1990. XIV, 636 pp. 125 figs. (Inverse Problems and Theoretical Imaging) Hardcover ISBN 3-540-51994-7

The basic idea of inverse methods is to extract from the evaluation of measured signals the details of the object emitting them. The applications range from physics and engineering to geology and medicine (tomography). Although most contributions are rather theoretical in nature, this volume is of practical value to experimentalists and engineers and as well of interest to mathematicians. The review lectures and contributed papers are grouped into eight chapters dedicated to tomography, distributed parameter inverse problems, spectral and scattering inverse problems (exact theory), wave propagation and scattering (approximations); miscellaneous inverse problems and applications and inverse methods in nonlinear mathematics.

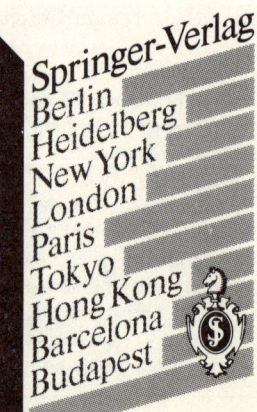

Research Reports in Physics

The categories of camera-ready manuscripts (e.g., written in T_EX; preferably both hard and soft copy) considered for publication in the **Research Reports** include:

1. Reports of meetings of particular interest that are devoted to a single topic (provided that the camera-ready manuscript is received within four weeks of the meeting's close!).
2. Preliminary drafts of original papers and monographs.
3. Seminar notes on topics of current interest.
4. Reviews of new fields.

Should a manuscript appear better suited to another series, consent will be sought from the author for its transfer to the other series.

Research Reports in Physics are divided into numerous subseries, e.g., nonlinear dynamics or nuclear and particle physics. Besides covering material of general interest, the series provides the possibility for topics that are too specialized or controversial to be published within the traditional avenues. The small print runs make a consistent price structure impossible and will sometimes have to presuppose a financial contribution from the author (or a sponsor). In particular, in the case of proceedings the organizers are expected to place a bulk order and/or provide some funding.

Within **Research Reports** the timeliness of a manuscript is more important than its form, which may be unfinished or tentative. Thus in some instances, proofs may be merely outlined and results presented that will be published in full elsewhere later. Since the manuscripts are directly reproduced, the responsibility for form and content is mainly the author's.

Springer-Verlag
Berlin
Heidelberg
New York
London
Paris
Tokyo
Hong Kong
Barcelona
Budapest

☐ Heidelberger Platz 3, W-1000 Berlin 33, F. R. Germany ☐ 175 Fifth Ave., New York, NY 10010, USA ☐ 8 Alexandra Rd., London SW19 7JZ, England ☐ 26, rue des Carmes, F-75005 Paris, France ☐ 37-3, Hongo 3-chome, Bunkyo-ku, Tokyo 113, Japan ☐ Room 701, Mirror Tower, 61 Mody Road, Tsimshatsui, Kowloon, Hong Kong ☐ Avinguda Diagonal, 468-4° C, E-08006 Barcelona, Spain

Research Reports in Physics

Manuscripts should be no less than 100 and no more than 400 pages in length. They are reproduced by a photographic process and must therefore be typed with extreme care. Corrections to the typescript should be made by pasting in the new text or painting out errors with white correction fluid. The typescript is reduced slightly in size during reproduction; the text on every page has to be kept within a frame of 16 × 25.4 cm (6⁵⁄₁₆ × 10 inches). On request, the publisher will supply special stationery with the typing area outlined.

Editors or authors (of complete volumes) receive 5 complimentary copies and are free to use parts of the material in later publications.

All manuscripts, including proceedings, must contain a subject index. In the case of multi-author books and proceedings an index of contributors is also required. Proceedings should also contain a list of participants, with complete addresses.

Our leaflet, *Instructions for the Preparation of Camera-Ready Manuscripts*, and further details are available on request.

Manuscripts (in English) or inquiries should be directed to

Dr. Ernst F. Hefter
Physics Editorial 4
Springer-Verlag, Tiergartenstrasse 17
W-6900 Heidelberg, Fed. Rep. of Germany
(Tel. [0] 6221-487495;
Telex 461 723;
Telefax [0] 6221-413982)

☐ Heidelberger Platz 3, W-1000 Berlin 33, F. R. Germany ☐ 175 Fifth Ave., New York, NY 10010, USA ☐ 8 Alexandra Rd., London SW19 7JZ, England ☐ 26, rue des Carmes, F-75005 Paris, France ☐ 37-3, Hongo 3-chome, Bunkyo-ku, Tokyo 113, Japan ☐ Room 701, Mirror Tower, 61 Mody Road, Tsimshatsui, Kowloon, Hong Kong ☐ Avinguda Diagonal, 468-4° C, E-08006 Barcelona, Spain